Microstrip Antenna Design for Wireless Applications

Microstrip Antenna Design for Wireless Applications

Edited by
Praveen Kumar Malik
Sanjeevikumar Padmanaban
Jens Bo Holm-Nielsen

CRC Press
Taylor & Francis Group
Boca Raton London New York

CRC Press is an imprint of the
Taylor & Francis Group, an **informa** business

First edition published 2022
by CRC Press
6000 Broken Sound Parkway NW, Suite 300, Boca Raton, FL 33487-2742

and by CRC Press
2 Park Square, Milton Park, Abingdon, Oxon OX14 4RN

CRC Press is an imprint of Taylor & Francis Group, LLC

ISBN: 978-0-367-55438-5 (hbk)
ISBN: 978-1-032-04788-1 (pbk)
ISBN: 978-1-003-09355-8 (ebk)

DOI: 10.1201/9781003093558

Typeset in Times
by Newgen Publishing UK

Dedication

This book is dedicated to my late father, who taught me to be an independent and determined person, without whom I would never be able to achieve my objectives and succeed in life.

Late (Sr.) Dharamveer Singh

Contents

Preface

This edited book aims to bring together leading academic scientists, researchers and research scholars to exchange and share their experiences and research results on all aspects of wireless communication and planer antenna design.

It also provides a premier interdisciplinary platform for researchers, practitioners and educators to present and discuss the most recent innovations, trends and concerns as well as practical challenges encountered and solutions adopted in the fields of wireless communication and planer antenna design.

Organization of the Book

The book is organized into eight parts. A brief description of each of the parts follows.

PART I: OVERVIEW AND INTRODUCTION

This part comprises introduction to microstrip antennas; analysis and analytical models of microstrip antennas; advances in the design of microstrip antennas and design challenges; reconfigurable wideband microstrip antennas and circularly polarized wideband microstrip antennas.

PART II: PERFORMANCE ANALYSIS OF MICROSTRIP ANTENNAS

This part provides insight into microstrip antenna design standards; microstrip antenna design parameters; microstrip antenna performance parameters; ultra-wideband antenna design parameters; ultra-wideband antenna design challenges and remedies; monopole-based ultra-wideband antennas and mathematical analysis of ultra-wideband antennas.

PART III: MULTIPLE-INPUT MULTIPLE-OUTPUT (MIMO) ANTENNA DESIGN AND ITS APPLICATIONS

This part includes design issues of MIMO antennas; MIMO antenna design for various applications; MIMO antenna design analysis and performance analysis of MIMO antennas.

PART IV: FRACTAL AND DEFECTED GROUND STRUCTURE MICROSTRIP ANTENNAS

This part includes defected ground structures; design of defected ground structures; fractal antenna design basics; design of fractal antennas and recent advances in design of fractal and defected ground structures.

PART V: MICROSTRIP ANTENNAS IN VEHICULAR COMMUNICATION

This part includes the design of antennas for vehicular communication; design aspects and challenges in vehicular antennas and recent antenna designs for vehicular communication.

PART VI: IMPORTANCE AND USE OF MICROSTRIP ANTENNAS IN IoT

This part focuses on the introduction to IoT; design of microstrip antennas for IoT applications; design challenges of antennas for IoT applications and current trends in the design of antennas for IoT applications.

PART VII: ULTRA-WIDEBAND ANTENNA DESIGN FOR WEARABLE APPLICATIONS

This part highlights the design of ultra-wideband antennas; design challenges of ultra-wideband antennas; antennas for biomedical applications and early diagnosis; wearable sensors for medical implants and advances in the field of ultra-wideband antenna design for biomedical applications.

PART VIII: MICROSTRIP ANTENNA DESIGN FOR MISCELLANEOUS APPLICATIONS

This part presents other antennas used in wireless communications.

Dr. Praveen Kumar Malik
Dr. Sanjeevikumar Padmanaban
Dr. Jens Bo Holm-Nielsen

Editors

Praveen Kumar Malik, PhD, is a Professor in the School of Electronics and Electrical Engineering, Lovely Professional University, Phagwara, Punjab, India. He received a B.Tech in 2000, an M.Tech (Honors) in 2010 and a PhD in 2015 with specialization in Wireless Communication and Antenna Design. He has authored or coauthored more than 40 technical research papers published in leading journals and conferences from the IEEE, Elsevier, Springer, Wiley and so on. Some of his research findings are published in top cited journals. He has also published three edited/authored books with international publishers. He has guided many students leading to M.E./M.Tech and has guided PhD students. He is Associate Editor of different journals. His current interests include microstrip antenna design, MIMO, vehicular communication and IoT. He has been invited as a guest editor and an editorial board member of many international journals, as a keynote speaker in many international conferences held in Asia and as a program chair, publications chair, publicity chair and session chair in many international conferences. He has been granted two design patents and several are in the pipeline.

Sanjeevikumar Padmanaban, PhD, (Member 12 – Senior Member 15, IEEE) received a bachelor's degree in electrical engineering from the University of Madras, Chennai, India, in 2002; a master's degree (Honors) in electrical engineering from Pondicherry University, Puducherry, India, in 2006; and a PhD degree in electrical engineering from the University of Bologna, Bologna, Italy, in 2012. He was an associate professor with VIT University from 2012 to 2013. In 2013, he joined the National Institute of Technology, India, as a Faculty Member. In 2014, he was invited as a visiting researcher at the Department of Electrical Engineering, Qatar University, Doha, Qatar, funded by the Qatar National Research Foundation (Government of Qatar). He continued his research activities with the Dublin Institute of Technology in Ireland, in 2014. Further, he served as an associate professor with the Department of Electrical and Electronics Engineering, University of Johannesburg, South Africa, from 2016 to 2018. Since 2018, he has been a faculty member with the Department of Energy Technology, Aalborg University, Esbjerg, Denmark. He has authored more than 300 scientific papers. He was the recipient of the Best Paper cum Most Excellence Research Paper Award from IET-SEISCON'13, IET-CEAT'16, IEEE-EECSI'19, IEEE-CENCON'19 and five best paper awards from ETAEERE'16 sponsored Lecture Notes in Electrical Engineering, a Springer book. He is a fellow of the Institution of Engineers, India; the Institution of Electronics and Telecommunication Engineers, India; and the Institution of Engineering and Technology, UK. He is an editor, associate editor and on the editorial board for refereed journals, in particular the *IEEE Systems Journal*, *IEEE Transaction on Industry Applications*, *IEEE Access*, *IET Power Electronics*, *IET Electronics Letters* and *Wiley-International Transactions on Electrical Energy Systems,* Subject Editorial Board Member – *Energy Sources – Energies Journal*, MDPI and the Subject Editor for the *IET Renewable Power Generation*, *IET Generation, Transmission and Distribution* and *FACTS* journal (Canada).

Jens Bo Holm-Nielsen, PhD, is Head of Research Group of Bioenergy and Green Engineering, Department of Energy Technology, Aalborg University, Denmark. He has 30 years of experience in the field of biomass feedstock production, biorefinery concepts and biogas production. He is a board member of R&D committees of the cross-governmental body of biogas developments 1993–2009, Denmark; secretary and/or chair of NGO biogas and bioenergy organizations, chair and presenter of Sustainable and 100 per cent Renewables and SDG goals. He has experience in variety of EU projects, has organized international conferences, workshops and training programs in the EU, United states, Canada, China, Brazil, India, Iran, Russia, Ukraine, among others. His research includes managing research, development and demonstration programs in integrated agriculture, environment and energy systems; fulfilled biomass and bioenergy R&D projects with a main focus on biofuels, biogas and biomass resources. He has supervised training programs and international courses for MSc and PhD students in these research fields.

Associate Editors

Dr. J. Vijayalakshmi, Associate Professor
Department of Electronics and Communication Engineering
Kongu Engineering College (Autonomous)
Perundurai, Erode, Tamil Nadu, India

Part I

Overview and Introduction

1 Microstrip Patch Antenna Techniques for Wireless Applications

Mahesh Kumar Aghwariya and Amit Kumar

1.1 INTRODUCTION OF ANTENNAS

An antenna is a device used for converting radio frequency (RF) signals propagating in a conductor into electromagnetic waves in free space or for converting guided electromagnetic waves into electrical signals. The antenna acts as a converter between freely propagating waves in space and electromagnetic waves. The name is taken from zoology, in which the Latin word *antennae* is used to describe the long, thin feelers owned by many insects. Antennas exhibit the property of reciprocity; this shows the identical property of the antenna while transmitting and receiving the signals [1]. Almost all widely used antennas are resonant devices, which are used at the narrowband of frequency [2]. All antennas perform two basic and complementary functions: firstly, they convert electromagnetic radiation into voltage and current; secondly, they convert the voltage and current into electromagnetic radiation that is transmitted into open space. The electromagnetic radiation that is transmitted into space consists of electric fields that are measured in V/M and magnetic fields that are measured in amp/m. According to the type of area to be detected, the antenna adopts a specific structure. The earliest existing antennas are not much different from RF generators in principle, such as the antenna used by Heinrich Hertz when he first proved the existence of electromagnetic waves in 1888. An antenna can be made from a parallel combination of inductors and capacitors. If the plates of the capacitor are opened and the inductor is reduced to the inductance of the coil, a dipole antenna will eventually be produced.

DOI: 10.1201/9781003093558-2

In fact, resonant circuits are still often used as a means to explain the various properties of antennas. It was not until about 1900, when a transmitting station and a receiving station were established, that a clear distinction was made and the antenna was classified as an independent component of the radio system. The microstrip antenna (MSA), also called the patch antenna, was a new design structure patented in 1955. It did not find many applications in the first couple of decades; however, for the last few years these have been widely used in wireless communication. In a microstrip antenna, dielectric material called substrate is sandwiched between two plates of the conducting materials [3]. The lower conducting surface and the upper conducting plate are called the ground plane and a patch, respectively. The patch and ground plane of the antenna are linked to the supply byline called the microstrip feed line. Many types of feeding methods are used to provide the supply to antenna. The microstrip antennas are also known as the printed antennas because of the similar fabrication process of printed circuit board (PCB). These are very small-sized planar antennas that are very useful in wireless communication. Installation of these antennas is very easy due to the fact that they are compact and lightweight. The size and shape of the patch are important aspects on which the performance of the microstrip antenna depends. Various shapes of the patch are being used to design the MSA. Some commonly used shapes are illustrated in Figure 1.1.

All these shapes can be used as patches in the MSA based on the operating frequency and required performance. The rectangular and circular patches are the extensively used shapes in the design of the microstrip patch antenna.

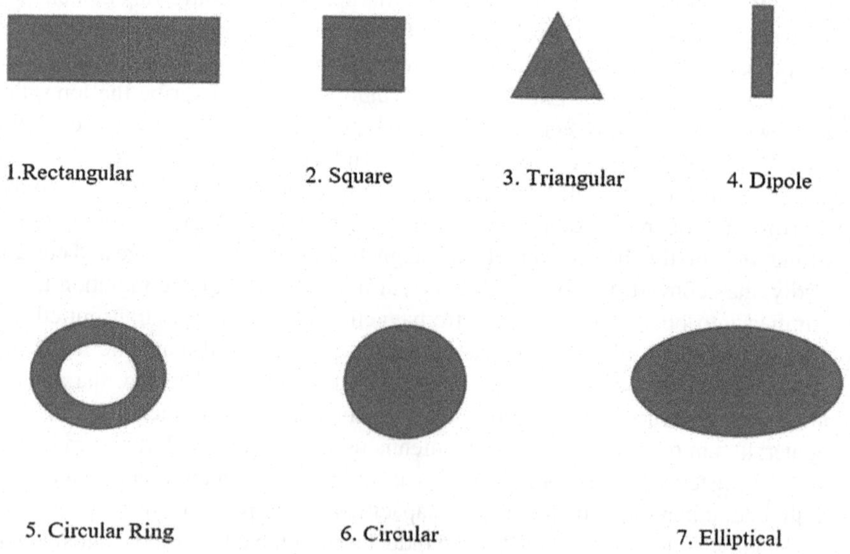

FIGURE 1.1 Various shapes of radiating patches.

1.2 RECTANGULAR MICROSTRIP PATCH ANTENNAS

The Rectangular Microstrip Patch Antenna (RMSPA) is the most commonly used microstrip antenna. In the RMSPA, a conducting rectangular-shaped structure is used as a patch to design the antenna. The basic structure of the antenna is given in Figures 1.2 and 1.3.

Figure 1.1 represents the complete structure of an RMSPA, Figure 1.2 the top view of the RMSPA and Figure 1.3 the side view of the RMSPA.

In the given figures:

L = length of the microstrip patch
W = width of the microstrip patch
H = height of the substrate

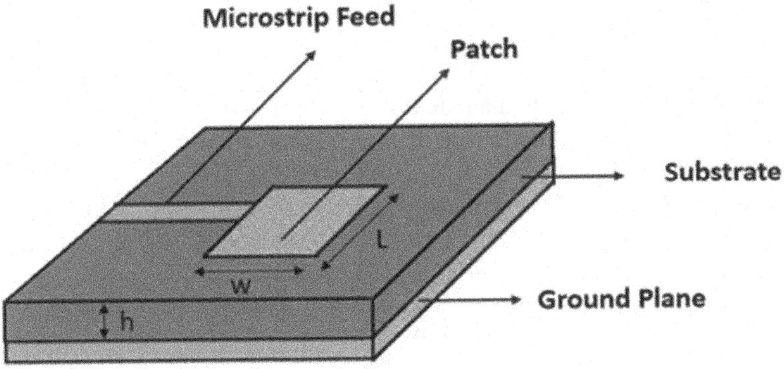

FIGURE 1.2 Microstrip patch antenna.

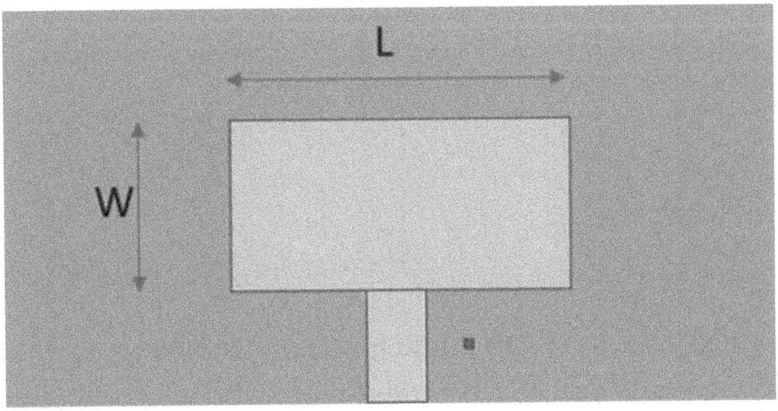

FIGURE 1.3 Top view of the microstrip patch antenna.

TABLE 1.1
Commonly used substrate materials

Substrate	Dielectric constant (ϵ_r)	Loss tangent ($\tan \delta$)
Glass Epoxy	4.4	0.02
96% Alumina	9.4 (5%)	0.0010
99.5% Alumina	9.8 (5%)	0.0001
Sapphire	9.4, 1.6	0.0001
Unreinforced PTFE, Cuflon	2.1	0.0004
Reinforced PTFE, RT Duroid 5880	2.2 (1.5%)	0.0009
Foam	1.05	0.0001
Air	1	0

The performance of the microstrip patch antenna depends on various parameters like length (L), width (W), substrate thickness (H) dielectric constant of the substrate material and also on the location of the feed line. In a properly designed RMSPA, the radiation intensity is normal to the radiating element [3, 4]. The antenna radiates from the width and not from the long side of the patch; hence, the radiation or radiation efficiency of the antenna is determined by the width of the patch (W). The smaller the value of W, the smaller the radiation, and the larger the value of W, the larger the radiation. The bandwidth of the antenna depends on W along with the substrate thickness H of the dielectric materials. The large value of W represents a larger bandwidth. The gain of the antenna is also determined by the W and it also depends on various other parameters. The large value of W represents the larger gain of the antenna. The resonating frequency of the antenna can be found by the length (L) of the patch. For the rectangular microstrip antenna, the typical value of L is one-third of one-half the wavelength depending upon the relative dielectric constant of the substrate material, which is commonly from 2.0 to 10.0. The lower values of the dielectric constants reflect higher efficiency. The operating frequency, also called the resonance frequency, of the antenna is based on the length (L) of the patch and effective dielectric constant of the substrate material as shown in Table 1.1.

The mathematical equation of the resonance frequency is given by the following equation:

$$f_0 = \frac{C}{2L\sqrt{\varepsilon_{\text{eff}}}} \tag{1.1}$$

1.3 FRINGING EFFECT IN MICROSTRIP ANTENNAS

When we feed the patch with respect to the ground, the electromagnetic waves start coupling from patch to the ground. In most parts of the patch, antenna radiation travels directly from patch to ground, but at the edge of a patch antenna the radiation

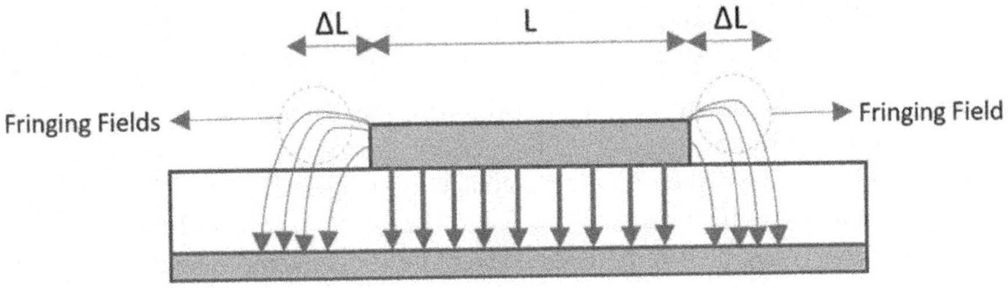

FIGURE 1.4 Fringing field lines of the microstrip antenna.

moves in the air and then comes to the ground. This fraction of the electromagnetic waves that traveled in the air is called the fringing of electromagnetic waves and this phenomenon is called the fringing effect. Based on this phenomena, the MSA radiates in the free space (Figure 1.4).

The antenna radiates the total physical length of the patch but because of the fringing effect, the electromagnetic waves travel in the air first before going back to the ground, and because of this the total effective length of the patch increases from both sides by the value ΔL [4]. So, the total effective length of the patch is increased.

$$\text{Effecting length of the patch } (L_{eff}) = \Delta L + L + \Delta L = L + 2\,\Delta L$$

where

L = total physical length of the patch

ΔL = incremental length of the patch because of the fringing

A large amount of fringing indicates large radiation from the antenna. Therefore, for better radiation efficiency, the fringing should be high.

Following are the three ways by which we can increase fringing in the microstrip antenna:

1. Selecting the lower value of the dielectric constant of the dielectric substrate material
2. Increasing the height of the substrate
3. Increasing the width of the patch

The electromagnetic wave that was initially traveling from the patch to the ground plane of the antenna was only traveling into the dielectric material but now because of the fringing effect the wave travels in the dielectric material as well as in the air. Because of the traversal of the wave in the air, the dielectric constant of the wave will also change, which will be the resultant of the dielectric constant of the substrate material and dielectric constant of the air. The operating frequency that was initially depending on the length of the patch and the relative dielectric constant of the material will now change because of the change in effective length of the patch and effective dielectric constant of the material.

1.4 MICROSTRIP ANTENNA FEEDING TECHNIQUES

After designing the microstrip antenna, we need to provide the supply to the antenna, which aids the antenna to start radiating [5, 6]. This is called feeding of the antenna. Basically, the feeding techniques are divided into contacting-type feeds and non-contacting-type feeds. Further, contacting-type feed is classified into two types: microstrip line feed and coaxial feed. The non-contacting-type feed is also divided into two types: proximity feed and aperture coupled feed. The whole classification of the microstrip feed is shown in Figures 1.5 and 1.6.

Microstrip line feed: The microstrip line feeding techniques are among the easiest methods to feed the antenna. In this feeding method, a strip of conducting transmission line is connected to the patch. It can also be considered as the extended patch of the antenna. It is a simple method in which impedance matching can be done in two ways: by varying the location of the microstrip feed line and by using inset feeding [5]. In inset feeding, the feed line is inserted inside the patch antenna. There are several advantages of using the microstrip line feed, such as it is easy to fabricate and impedance matching is simple. It also has some disadvantages, such as low bandwidth, cross-polarization increases substrate thickness, presence of surface wave and cross-polarization.

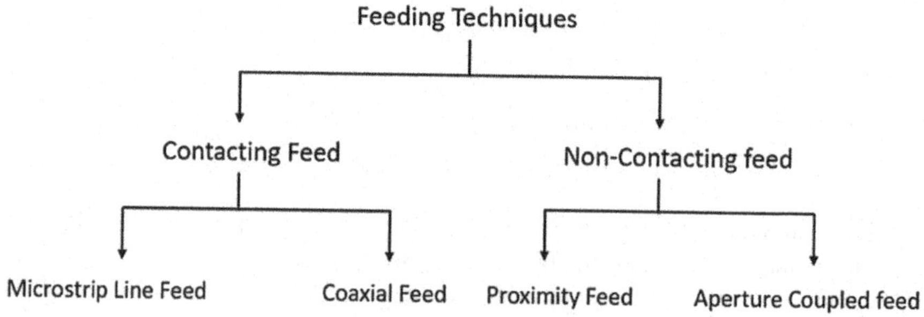

FIGURE 1.5 Classification of microstrip antenna feeding techniques.

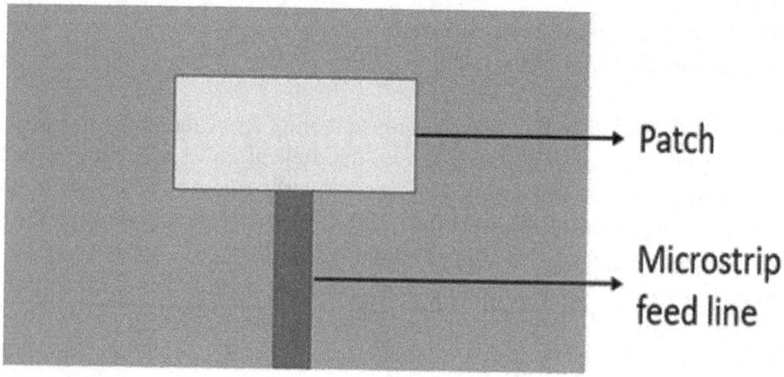

FIGURE 1.6 Microstrip line feed.

Coaxial feed: In the coaxial feed technique, the coaxial probe is used to feed the antenna. The outer conductor of the coaxial probe is connected to the ground plane of the antenna and the inner conductor is connected to the patch. By shifting the location of the inner conductor of the probe from the center of the patch, we can provide the impedance matching to the antenna. The location of the probe is adjusted to match the impedance and to control the input impedance. There are several advantages of coaxial probe: it is easy to fabricate, easy to model, easy impedance matching and there is little spurious radiation. The probe also generates inductance that is considered if the height of the substrate increases. There are some disadvantages to this feeding method such as low bandwidth and cross-polarization. In addition, the probe can also generate radiation in undesirable random directions (Figure 1.7).

Proximity feed: This is a non-contacting-type feeding mechanism. In this method, two dielectric materials are used. The patch is placed on upper dielectric material and the feed line is sandwiched between two dielectric materials, and below to the second dielectric material, the ground plane exists. In this feeding technique, the patch is not directly coupled to the patch. Thus, this is known as electromagnetically coupled feeding techniques, where we don't connect anything physically with the patch of the antenna. Using the thin lower dielectric material and placing the patch on top of the surface, the bandwidth increases and the spurious radiation decreases. The impedance matching is done by varying the length and width of the feed line. The design is always symmetrical about the axis of the patch. This feed method has low cross-polarization and high bandwidth. Implementation of this technique is difficult in comparison with the other feeding techniques because of the difficulties in alignment and finding the proper location of the patch as shown in Figure 1.8.

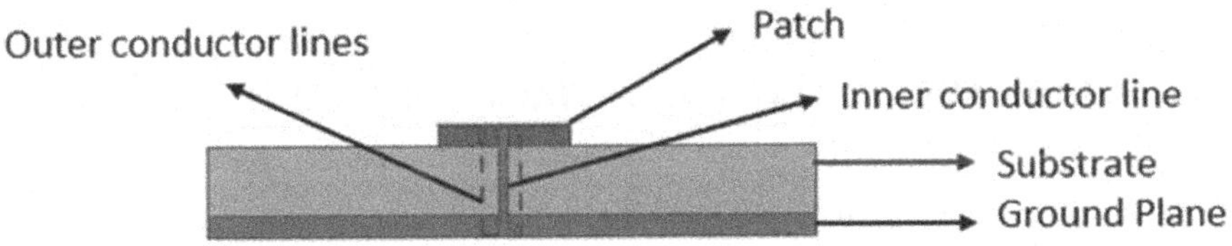

FIGURE 1.7 Side view of coaxial feed line.

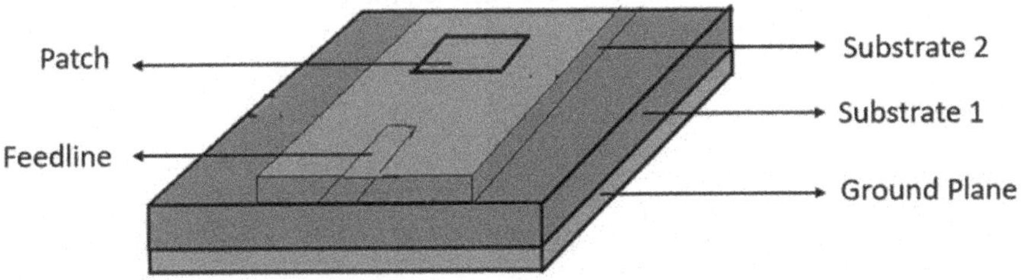

FIGURE 1.8 Proximity feed.

Aperture coupled feed: This is one of the non-contacting-type feed in which there is no direct physical contact between the patch and feed line. In this feeding technique, the ground plane is sandwiched between the two dielectric substrates. The patch is placed in the upper dielectric substrate, a slot is placed on the ground plane and the feed line is placed below the lower dielectric substrate. The patch is electromagnetically connected to the ground plane through the slot as shown in Table 1.2.

The structure is symmetric about the axis of the feed because of this low cross-polarization. This technique is easier to model and also it has moderate spurious radiation. In this method, fabrication is very difficult, and very low bandwidth is achieved.

1.5 COMPARISON OF MICROSTRIP ANTENNAS WITH CONVENTIONAL ANTENNAS

TABLE 1.2
Comparison of microstrip antennas with conventional antennas

Characteristics	Microstrip patch antenna	Microstrip slot antenna	Printed dipole antenna
Profile	Thin	Thin	Thin
Fabrication	Very easy	Easy	Easy
Polarization	Linear and circular	Linear and circular	Linear
Shape flexibility	Any shape	Rectangular and circular shape	Rectangular and triangular shape
Bandwidth	2–50% (resonant frequency)	5–30% (resonant frequency)	30% (resonant frequency)

1.6 ADVANTAGES AND DISADVANTAGES OF MICROSTRIP PATCH ANTENNAS

Microstrip patch antennas have attracted the attention of the research community due to their advantages as listed below. There are many advantages and applications of the microstrip antenna over the conventional antenna and these include the planar surface, integration with circuit elements, small size and the possibility of multiband operation [6]. Along with the advantages, there are also many disadvantages. Scientists and researchers are still working to overcome the problems related to the microstrip antennas.

Advantages:

- Small size
- Requires no cavity backing
- Low fabrication cost
- Capable of dual and triple frequency operations
- Linear and circular polarizations

Disadvantages:

- Low efficiency due to losses in the dielectric substrate
- Excitation of surface waves
- Low gain
- Requires complex feed structures
- Low power handling capacity
- Large ohmic loss

1.7 APPLICATION OF MICROSTRIP ANTENNAS

Microstrip patch antennas are well known for their robust design, manufacturing and range of applications. Due to substrate materials and low manufacturing costs, microstrip antennas are booming in business. Because of the large use of patch antenna, conventional antennas can be used for maximum applications [2, 6]. Some of the applications of microstrip antennas are as follows:

Global positioning system applications: Nowadays, high permittivity sintered materials with microstrip patch antenna substrates are being used in global positioning systems. It is expected that ordinary people will use millions of GPS receivers on land vehicles, airplanes and seaplanes so that their location can be accurately determined.

Radar applications: For radar applications or to detect moving targets, the microstrip antenna is a good choice. The photolithography-based manufacturing can mass-produce microstrip antennas with dual performance in a shorter time and lower cost than conventional antennas.

Rectenna applications: In this application, the antenna with high command performance always requires long-distance connectivity. Since the purpose is to use the rectenna to transmit DC power over a long distance through a wireless link, it can be achieved using microstrip antennas [7, 8].

Mobile and satellite communication applications: Microstrip antenna plays a key role in mobile and satellite communications because of small size, polarization (linear and circular) and so on. Nowadays, mobiles come in small size because of the microstrip antenna used in it. In satellite communications, a circularly polarized pattern is required; it can be achieved by a circular patch configuration.

Medical applications: It has been investigated by the researchers that microstrip antenna is an ideal choice for medical applications, such as tissue detection and treatment of tumors, because of its small size and most importantly of its modified electrical performance.

1.8 CONCLUSION

From the above discussions, it can be concluded that the microstrip patch antennas have optimal results in all the parameters when compared to the conventional antennas. These antennas have planar structure, low cost and are easy to design. To feed the antenna, many feeding techniques are available; impedance matching is the simplest feeding technique. These antennas can perform multiband operation; we can also use microstrip antenna in array form. The disadvantages of microstrip antennas include low bandwidth, low power handling capacity, low radiation efficiency and so on. These problems can be overcome by applying various techniques and methods. Thus, we can conclude that microstrip antennas are a better choice in comparison with conventional antennas.

REFERENCES

[1] R. Sharma, "Prototype of Slotted Microstrip Patch Antenna for Multiband Application," *International Journal of Engineering & Technology*, 2018, Vol. 7(3.12), pp. 1199–1201.

[2] R. Sharma and P. K. Singhal, "Miniaturized Patch Antenna Loaded with Metamaterial for Microwave Sensor Quantum Matter," *Quantum Matter*, 2013, Vol. 2(421).

[3] I. T. Agarwa and R. Sharma, "Frequency Scanning SIW Leaky Wave Horn Antenna for Wireless Application." In: G. Singh Tomar, N. Chaudhari, J. Barbosa, and M. Aghwariya (eds) *International Conference on Intelligent Computing and Smart Communication 2019*. Algorithms for Intelligent Systems. Springer, Singapore, 2020. https://doi.org/10.1007/978-981-15-0633-8_79.

[4] I. Kaur, A. K. Singh, J. Makhija, S. Gupta, and K. K. Singh, "Designing and Analysis of Microstrip Patch Antenna for UWB Applications," *2018 3rd International Conference on Internet of Things: Smart Innovation and Usages (IoT-SIU)*, Bhimtal, 2018, pp. 1–3, https://doi: 10.1109/IoT-SIU.2018.8519867.

[5] U. Raithatha and S. Sreenath Kashyap, "Microstrip Patch Antenna Parameters, Feeding Techniques and Shapes of the Patch – A Survey," *International Journal of Scientific & Engineering Research*, 2015, Vol. 6(4).

[6] J. Salai Thillai Thilagam and T. R. Ganesh Babu, "Rectangular Microstrip Patch Antenna at ISM Band," *2018 Second International Conference on Computing Methodologies and Communication (ICCMC),* Erode, 2018, pp. 91–95, https://doi:10.1109/ICCMC.2018.8487877.

[7] P. Tiwari and P. K. Malik, "Design of UWB Antenna for the 5G Mobile Communication Applications: A Review," *2020 International Conference on Computation, Automation and Knowledge Management (ICCAKM)*, 2020, pp. 24–30, https://doi: 10.1109/ICCAKM46823.2020.9051556.

[8] A. Kaur, P. K. Malik, and R. Singh, "Planar Rectangular Micro-strip Patch Antenna Design for 25 GHz." In: P. K. Singh, Y. Singh, M. H. Kolekar, A. K. Kar, J. K. Chhabra, and A. Sen (eds) *Recent Innovations in Computing*. ICRIC 2020. *Lecture Notes in Electrical Engineering*, vol. 701. Springer, Singapore, 2021. https://doi.org/10.1007/978-981-15-8297-4_18.

2 A Review
Microstrip Patch Antennas for Ultra-Wideband

Ashish Singh, Krishnananda Shet, Durga Prasad, Satheesh Rao, Anil Kumar Bhat and Ramya Shetty

2.1 INTRODUCTION

Modernization has led to technological advancement, apart from the motor bike and motor car and high-speed internet. All these were only possible due to miniaturization of electronic components that are now easily mounted on a printed circuit board. Today the world is dependent on the miniaturized communication devices and vastly growing use of modern communication devices in everyday routines. Presently, the growth in wireless communication has increased in cellular phones, wireless local area networks (WLANs), Global Positioning Systems (GPSs) and satellite telephones [1–7]. This advancement in wireless communication is due to the rapid development of microwave integrated circuits that can embed all the components of the circuits on the same printed boards. In all these communication devices, one major device is always present, which is the antenna. Nowadays, microstrip patch antennas have been used in almost all wireless communication devices such as mobile phones, laptops, data cards and so on. These devices can function at more than one frequency range for various applications. Because band performance of the antenna degrades when functioning at third and higher order frequencies, further bandwidth is limited for the applications. Thus, future demand is for higher bandwidth that can be utilized for expected bands applications. In previous years, the development of microstrip patch for various wireless communications [8] has increased.

In this view, a literature survey has been presented for ultra-wideband (UWB) patch antennas as per Federal Communications Commission (FCC) standard. The overview of UWB patch antennas, such as notch characteristics, bandwidth above 500 MHz and 3.1–10.6 GHz along with the conclusion and future scope, is discussed in the following section.

DOI: 10.1201/9781003093558-3

2.2 BAND-NOTCH CHARACTERISTICS

The band-notch characteristics significant features by dual and multiband [9–11] behaviour can be observed on UWB band range 3.1–10.6 GHz. Further, band-notched characteristic is similar to dual and multiband characteristics. It has one major advantage, that is, it provides less interference between two highly used bands. For example, WLAN operates at two close frequency bands 5.15–5.35 GHz and 5.725–5.825 GHz. This interesting shape was used by various scientists and researchers for designing patch antennas as shown in Figure 2.1. The band-notch characteristics were observed by Kim et al. and Choi et al. in 2005 [12–13], as shown in Figure 2.1(a) and (b). Later, various researchers proposed research on this characteristics using the patch antenna, which are reviewed below.

Zhao [14] proposed a dipole antenna similar to microstrip structure, which is fed in middle of both patch [Figure 2.1(c)]. He found that antennas radiate equally in both the directions whose bandwidth lies between 3 and 10 GHz and notch characteristics are observed at 8 GHz. Further, he found that by varying the patch length, notch characteristics are achieved. Input impedance found to be 100 Ω for real and imaginary parts was approximately zero.

Moghadasi et al. [15] designed a rhombus-shaped antenna feed via a microstrip line feed at the top surface and at the bottom surface slot etched just below the patch and designed on an RT Duroid substrate [Figure 2.1(d)]. They found that the variation of diagonals of the rhombus and slot on the bottom side of the substrate plays a vital role in exploring UWB. They also found E and H plane plots for the designed antenna at 3, 5, 7, 9 and 11 GHz. The gain of the antenna is constant throughout the impedance bandwidth.

Liu et al. [16] designed UWB antenna fed via a microstrip line. The antenna size was $15 \times 15 \times 1.6$ mm^3 and has bandwidth from 3.05 to 11.15 GHz with band-notched characteristics [Figure 2.1(e)]. They found omnidirectional radiation pattern at 4 and 8 GHz. The width and gap between the slots were varied to understand the behaviour of antennas and it was observed that an antenna characteristic depends on the width and gap of the slot. An antenna gain almost constantly varies from 3.02 to 3.92 dBi.

Movahedinia and Azarmanesh [17] proposed that planar monopole antenna has bandwidth that lies between 3.1 and 12.6 GHz and band-notch characteristic was observed at 5.6 GHz with effective ground plane. The current distribution was shown

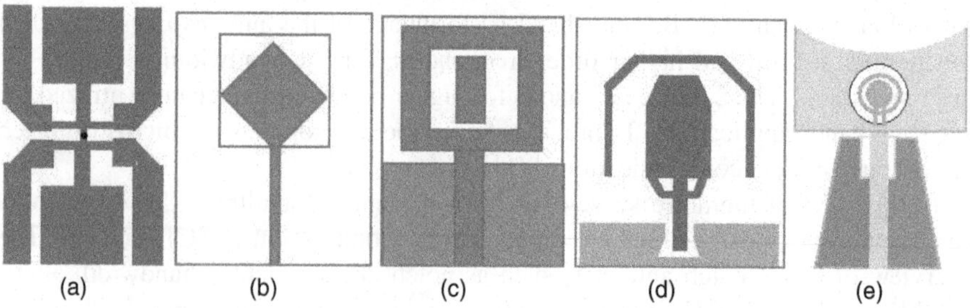

FIGURE 2.1 Antenna designs for band-notch characteristics [12–16].

at 3, 5.6 and 11 GHz frequencies and noticed that current was uniform along the patch length, which is a good sign for antenna radiation; whereas, weak current flows along the strip surrounding the patch. The radiation pattern observed was nearly omnidirectional for the operating frequencies. The gain of the antenna was maximum at 3.1 and 10.6 GHz and minimum at 5.4 GHz.

Li et al. [18] developed and designed an antenna with dimensions of 24 × 25 × 1.6 mm^3 with a curved Y-shaped patch in which circular split ring resonators were etched on the top side of the substrate and a trapezium patch with rectangular notch was designed on the bottom of the substrate. The author achieved band-notch characteristics at 5.2 and 5.8 GHz and between 3.1 and 14 GHz frequencies. A variation of the split ring resonator circular curve from 0.7 to 0.9 mm was presented and triple band-notch characteristics were achieved.

Mehranpour et al. [19] designed a band-notched characteristic monopole antenna with dimensions of 10 × 16 × 1.6 mm^3 on an FR4 substrate $\varepsilon_r = 4.4$ and fed through a microstrip line. They achieved 140% bandwidth at 2.89–17.83 GHz on an etching a two L-shaped strip on top of the radiating substrate where the V-shaped structure was implemented. Two notch bands were observed between 5.2/5.8 GHz WLAN and 3.5–5.5 GHz WiMax and 4 GHz C band. They have shown current distribution on the patch antenna and studied the effect of the ground plane V-shaped structure and the radiation pattern was nearly omnidirectional.

Gao et al. [20] designed an antenna with a size of 20 × 27 × 1 mm^3 having a U-shaped patch in which a U-shaped slot was etched and the ground plane also had a U-shaped slot. They achieved UWB from 2.89 to 11.52 GHz with dual-band-notched characteristics at 3.4–3.69 GHz and 5.15–5.825 GHz. The antenna was excited via a 50 Ω microstrip line and reached a maximum peak gain of about 5 dBi with a radiation pattern similar to an omnidirectional pattern.

Ojaroudi et al. [21] designed an antenna for microwave imaging. The antenna size was 12 × 18 × 1.6 mm^3 on an FR4 substrate. To achieve the UWB on the ground plane, the authors etched a U-slot, a T-slot and an E-slot. Current distribution was plotted for the antenna having a T-slot at 10.4 GHz and E-, T-, U-slots at 12.3 GHz frequencies. They found that the current distribution was more at 12.3 GHz frequencies for the antenna having E-, T- and U-slots on the ground plane. They have discussed the input and output signals from antenna simulation at $\Theta = 0$, 45 and 90, the out pulse was 3.9 nsec. The radiation pattern was also plotted for the resonating frequencies at 4.15, 10.4 and 12.3 GHz.

Zhu et al. [22] designed and simulated a dipole antenna. It has a similar structure at the top and bottom of the substrate. The antenna was designed on Roger 6010 of thickness 0.635 and dielectric constant $\varepsilon_r = 10.2$ with antenna dimensions of 28 × 28.5 mm^2. They plotted the current distribution of the antenna at resonating 3.5, 5.2 and 5.8 GHz and found that the current was evenly distributed at lower resonant frequencies. They also discussed the variation of the reflection coefficient for thickness of substrate and dielectric constant.

Ojaroudi and Ojaroudi et al. [23] designed a monopole-type antenna with T-shaped ground plane etched on the rectangular patch and C-shape is also etched surrounding it. Thereafter, the T-shape is placed over the etched T-shape in such a way that the etched T-patch surrounds the shape. The antenna is fed via a microstrip line and the size of the antenna is 12 × 18 × 0.8 mm^3 which was designed on an FR4 substrate. They achieved dual-band-notch characteristics on placing the T-shaped patch in the T-slot. The gain of antenna is constant over the entire frequency range and cross- and co-polarization was found at 4.8, 7.3 and 10 GHz.

Ojaroudi et al. [24] achieved a UWB from the unique geometry of an antenna in which a pair of U-slots were on the top of the radiating surface and a T-shaped patch was etched at the bottom of the ground. A notch-band characteristic was found at 3.3 and 5.2 GHz while the antenna band extended from 2.81 to 15.12 GHz; the current distribution was also shown at 3.9 and 5.5 GHz. The antenna was nearly omnidirectional for both the x–z and y–z planes at 3, 9 and 15 GHz for the H-plane and E-plane.

Azim et al. [25] achieved dual-band-notch behaviour on etching an F-shaped slot on the ground plane that was just beneath the radiating patch feed through a 50 Ω SMA connector. An antenna was designed on an FR4 substrate of a rectangular patch with dimensions of 14.5 × 14.75 × 1.6 mm^3. The simulation and experimental results were verified for a UWB frequency band of 2.97–10.67 GHz. They also found that an F-shaped structure at the backside controls the UWB frequency band. Gain of the antenna was nearly constant at lower resonance, that is, 2.5 dBi at 6 GHz

S. R. Emadian and J. A. Shokouh [26] designed an antenna with dimensions of 15 × 15 mm^2 and achieved 2.6 to 23 GHz impedance bandwidth. The design of the antenna was fed by a coplanar waveguide (CPW) with semi-circular and rectangular slots etched on the ground plane and an angular ring slot was cut on the radiating patch. The length of elliptical ring slots (ERS) was determined for particular notch frequency as follows:

$$L \approx \frac{c}{2 f_n \sqrt{\varepsilon_e}} = \frac{\lambda_g}{2}$$

where f_n is notch frequency, ε_e is effective dielectric constant, c is the velocity of light and $\lambda_g/2$ is the wavelength. The antenna characteristics are maximum gain 5 dBi, maximum radiation efficiency 80% and nearly omnidirectional radiation pattern at radiation frequencies.

Oraizi et al. [27] proposed a compacted log periodic antenna on RT/Duroid 5880 and achieved a gain of 7 dBi over operating frequencies 3.1 to 10.6 GHz. The fractal techniques like triangular Koch, square Koch and tree Koch were applied to the branches of a log periodic antenna and studied the characteristics of antenna on these fractal techniques. Gain and efficiency of these antennas were compared on applying the fractal techniques and found that there were no huge variations in gain and efficiencies of the antennas on resonating bandwidth 3.1–10.6 GHz. Input and output of the radiation signal for these fractal techniques were plotted and found that input and output response time of the antenna is good. Their design was focused to get band-notch characteristics, so another fractal technique was applied on log periodic antenna, that is, Giuseppe Peano and UWB frequency bands were achieved with notch-band characteristics at 5.8 GHz and 8.3 GHz. Gain of antenna for Giuseppe Peano fractal design is maximum at 5.2 dBi but sharp dip at band-notch characteristics and the output response of the antenna are good except at 5.8 GHz. The overall dimension of antenna is 110 × 70 × 1.6 mm^3. The designed antenna can be effectively used for wireless applications such as biomedical imaging and impulse radios.

2.3 10.6 GHz BANDS

UWB patch antenna became very famous because of its interesting feature of very large bandwidth. This characteristic of UWB patch antenna was exploited by various researchers and scientists; some related figures are shown in Figure 2.2. The first UWB patch antenna was proposed by Kiminami et al. [28], the bow-tie-shaped antenna with effective ground plane. The antenna was designed on dielectric substrate $\varepsilon_r = 6.15$ of thickness 1.27 and is fed via a microstrip line [Figure 2.2(a)]. They simulated antenna on IE3D and HFSS simulation software and compared the derived results with measured results. They observed broadside radiation pattern for E- and H-plane and average gain of 3.4 dBi.

Ghaderi and Mohajeri [29] developed an antenna with wide slots on hexagonal patch fed via a microstrip line [Figure 2.2(b)] and achieved UWB from 2.1 GHz to 18 GHz with reasonable good gain, varying between 4 and 4.9 dBi over entire bandwidth. They plotted 2D radiation pattern for E-plane along y–z and H-plane along x–z for co- and cross-polarization. The radiation pattern is bidirectional for E-plane and omnidirectional for H-plane for 4.5 and 8.5 GHz frequencies. Further, an antenna at higher resonance frequency do not have good radiation pattern because of its radiating structure. These antenna design covers WLAN and WiMax bands.

Wu et al. [30] proposed the Vivaldi antenna with stepped connection structure as shown in Figure 2.2(c) and achieved the bandwidth from 3 to 15.1 GHz with average gain of 6.6 dBi. The radiation pattern was broadside with a radiation efficiency of 70%. The antenna size was 41 × 48 × 0.8 mm^3 fed via a microstrip line excited by an SMA connector. Further, group time delay was calculated for studying the transmitting and receiving signals from the designed antenna.

Gong et al. [31] designed compact slot antenna of dimension 23.7 × 23.7 × 0.8 mm^3 [Figure 2.2(d)] and had UWB ranging from 3.05 to 11.9 GHz. They achieved band-notched characteristics while varying length of strips and omnidirectional radiation pattern are achieved at 3.5, 6 and 9 GHz. Further, maximum gain of 5 dBi was obtained at 8 GHz while 2.1 dBi was obtained at 3.1 GHz.

Karimain et al. [32] presented a compact UWB monopole antenna with notch-band characteristics. The antenna was designed on an FR4 substrate and has tree-shaped radiating structure with total size of antenna as 14 × 16 × 1 mm^3 [Figure 2.2(e)].

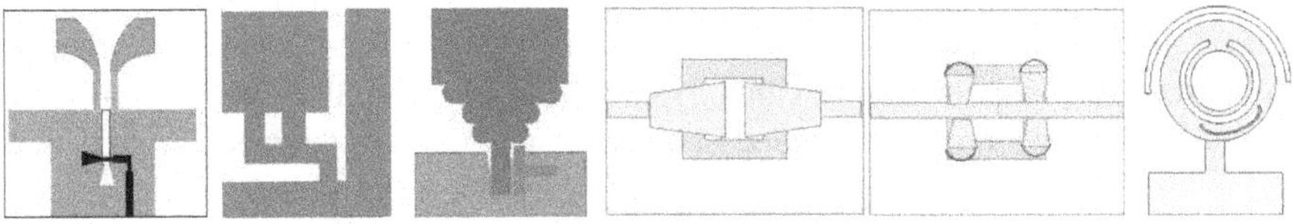

(a) (b) (c) (d) (e)

FIGURE 2.2 Antenna designs for 3.1–10.6 GHz band characteristics [28–32].

They achieved 170% (2.8–19.2) GHz bandwidth with rejection characteristics at 5.1–5.9 GHz. They control the band-notched characteristics by adjusting the length of slots and the performance characteristics of the antenna was studied for dimensional variation.

Haraz et al. [33] designed an antenna with UWB characteristics that has multilayer slots. This is one of the special types of antennas designed with a common ground plane, two radiating structures fed by a different microstrip line and the ground plane has a rectangular slot. They proposed two radiating structures with trapezoidal and butterfly shapes. Parameter analysis has been done to exploit the behaviour of the antenna at different parameters. The simulation of the antenna was done on CST and HFSS; in addition, these results were compared by experimentally testing the results of antenna network analyser. Both antennas provide UWB frequency band between 3.1 and 10.6 GHz with reasonably good gain.

Mohammed et al. [34] proposed a uniplaner printed triple-band rejection characteristic UWB antenna. The antenna was designed on FR4 substrate that has angular ring patch feed with microstrip line. They achieved notch-band characteristics on etching C-slot on the angular ring. The dimension of antenna is $30 \times 30 \times 1.6$ mm^3 in which 3×12 mm^2 is the structure of microstrip line. They used particle swarm optimization and firefly techniques for the analysis of antenna design. Multiband characteristics were achieved from 2.1 to 10.6 GHz with variable gain from 2 to 6 dBi over the UWB frequency range.

Yadav et al. [35] designed compact band-notch characteristics with a planar antenna by integrating a Bluetooth band. The antenna was designed using an L-shaped defected ground structure (DGS) and triangle-shaped radiating patch. The overall size of the antenna was $13 \times 24 \times 1.6$ mm^3 and excited by an L-shaped strip to achieve a Bluetooth band of 2.40–2.480 GHz and circular polarization was observed at the same frequency. Measured and simulated return loss was compared and bandwidth covers from 3.1 to 24.5 GHz. The gain and efficiency of the antenna over the entire band varied whereas an average gain of 3 dBi and 80% was found. The antenna had UWB of 21.4 GHz from 3.1 to 24.5 GHz with integrated Bluetooth band rejection at 6.3 and 7.2 GHz.

2.4 UNDER FCC STANDARD BAND GREATER THAN 500 MHz

The band greater than 20% or 500 MHz bandwidth can be said to be a UWB band as provided by the FCC in 2002. Previously, lots of research papers were published under broadband and wideband characteristics. These types of antennas have existing bandwidth enhancement techniques; review papers and books are available too. In this view, a diminutive paper has been reviewed under this characteristic [36–50] and few related figures are shown in Figure 2.3. Matin et al. [36] designed a U-slot antenna on the bottom patch and stacked it with the same substrate material [Figure 2.3(a)]. They concluded that wideband impedance characteristics can be obtained through their proposed design. They compared the theoretical results with simulated and measured results. The obtained gain is about 9.31 dBi at 4.4 GHz and the broadside radiation pattern is found at both resonating frequencies.

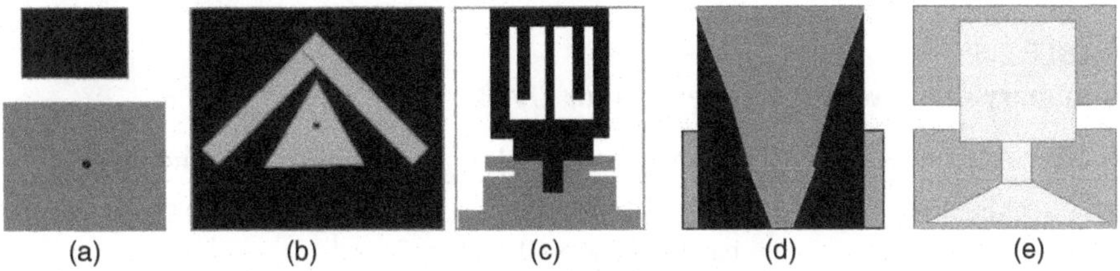

FIGURE 2.3 Antenna designs for UWB characteristics with bandwidths greater than 500 MHz [36–40].

Elsadek and Nashaat [37] proposed a V-shaped antenna with a coaxial feed on the isosceles triangle patch [Figure 2.3(b)]. They found multiband from 2.9 to 5.5 GHz, with resonating frequencies of 2.88, 3.64, 4.38, 4.81 and 5.6 GHz. Further, on varying the position of the feeding point, the multiband was changed to dual band at first and higher resonant 3.5% and 53%, respectively, and found an impedance bandwidth at 4.65 GHz.

Ojaroudi et al. [38] developed a UWB antenna with band rejection characteristics. They designed a T-shaped patch in which two U-shaped slots were etched on the top layer and had a rectangular shape at the bottom layer [Figure 2.3(c)] and designed near feed to provide proper ground shielding and maximum gain was achieved at 3.1 GHz of 4.2 dBi whereas 2 dBi gain was constant between 5 and 12 GHz.

Kamakshi et al. [39–40] designed two antennas; one antenna was fabricated on bakelite with step-shaped horn structure and the other desktop-shaped antenna; both antennas showed broadside radiation pattern [Figures 2.3(d) and (e)].

Ansari et al. [41–42] demonstrated antenna designs with a theoretical concept based on the circuit theory concept. Parasitic patch antennas with square and circular shapes were reported with a bandwidth above 500 MHz. Further, an angular ring antenna was also proposed by the same group of researchers with simulation and theoretical concepts [43].

In 2018, Guo et al. [44] reported the patch which electromagnetically excited and had notches and a microstrip placed on ground plane. They found a UWB bandwidth of 147% but not within the FCC standard range of 3.1–10.6 GHz. The antennas showed a good radiation pattern at the entire bandwidth for the antenna size of 37.4×64 mm^2.

Khan et al. [45] designed an angular ring patch antenna with a huge frequency bandwidth from 3.1 to 16 GHz. They also found that the antenna characteristics, that is, VSWR, in an acceptable range of less than 1.9 was the desired characteristic of the antenna for efficient transmission.

Gautam et al. [46] demonstrated the design of a J-shaped antenna with multiple-input and multiple-output characteristics that have shown notch bands from 5.1 to 5.8 GHz and 6.7 to 7.1 GHz for various wireless applications.

Chen et al. [47] proposed a fish-shaped patch antenna for wireless communication and UWB features. They achieved a very high bandwidth of 132% that covers a frequency range from 4.28 to 21 GHz for a defined antenna size of $0.199\ \lambda_L \times 0.256\ \lambda_L \times 0.092\ \lambda_L$ (where λ_L is the wavelength at 4.28 GHz).

TABLE 2.1
Summary of a few antenna designs for UWB

S. No.	Antenna shape	Frequency band (GHz)	Gain (dBi)	Ground plane
1	Microstrip [16]	3.5–11.15 GHz, Notch band at 4 and 8 GHz	3.02–3.92	$15 \times 15 \times 1.6$ mm^3
2	Y-shaped [18]	3.1–14 GHz, Notch band at 5.2–5.8 GHz	–	$24 \times 25 \times 1.6$ mm^3
3	V-shaped [19]	2.89–17.83 GHz, Notch band at 2.2/5.8/3.5–3.8 GHz	–	$10 \times 16 \times 1.6$ mm^3
4	U-shaped [20]	2.89–11.5 GHz, Notch band at 3.4–3.69 GHz and 5.15/5.825 GHz	5 dBi	$20 \times 27 \times 1$ mm^3
5	Vivaldi [30]	3.1–10.6 GHz	6.6 dBi	$41 \times 48 \times 0.8$ mm^3
6	Slot [31]	3–15.1 GHz	5 dBi	$23.7 \times 23 \times 0.8$ mm^3
7	Compact [32]	2.8–19.2 GHz	–	$14 \times 10 \times 1$ mm^3
8	Angular ring [34]	2.1–10.6 GHz	2–6 dBi	$30 \times 30 \times 1.6$ mm^3
9	Ellipse [44]	1.5–10.4 GHz	–	$37.4 \times 64 \times 1.6$ mm^3

Kissi et al. [48] designed an antenna device for UWB application that has reflector cavity. Antenna has 3.15 dBi gain at 4 GHz showing omnidirectional pattern and achieved antenna characteristics over the frequency band of interest. Further, reflector cavity-backed antenna has a gain of 6.4 dBi, which also proves good antenna features at entire frequency band.

Singh et al. [49–50] proposed angular ring antenna with defected ground plane. They found frequency bands between 2.9 and 13.1 GHz fed via a microstrip line and fabricated on FR4 substrate. Gain of 4.1 dBi was observed with radiation efficiency of 70.99%. They performed theoretical analysis using circuit theory. Further, Singh et al. presented the review paper for circuit theory analysis on a patch antenna with UWB band, dual-band and multiband. They presented basic circuit diagrams of patch antenna designs their research [51–52]. Table 2.1 summarizes a few antenna designs for UWB antennas falling under 3.1–10.6 GHz, notch-band characteristics and frequency bands greater than 500 MHz.

2.5 CONCLUSION

A review of UWB patch antennas has been presented, which includes special characteristics when designing antenna such as application, size and shape, compactness, radiation pattern and bandwidth. Emphasis has been given to the FCC standard on UWB and its notch-band characteristics for antenna designs. Further, lots of research papers have been published for UWB antennas; therefore, it is difficult to include all papers in this review; only recent manuscripts are included. It has been observed that greater than 20% bandwidth, 100.6% bandwidth and notch-band characteristics are achieved by DGS, microstrip line feed, co-planar wave guide, stacking and reactive loading. It has been noticed that Fr4 as dielectric constant has been used in obtaining higher bandwidth. Moreover, UWB antennas reported having a maximum gain of nearly 6 dBi. Further, notch-band characteristics for antenna design was focused on three bands in particular, that is, 3.1–3.5, 5.2–5.4 and 5.8 GHz and was achieved through DGS. The bandwidth above 500 MHz was achieved using different bandwidth enhancement techniques as well as with DGS. Mostly all the reported papers have been designed for wireless communications for various applications such as Bluetooth, Wi-Fi, WiMax, DCS and bio-medical wireless applications. The current trend UWB antennas are focussing on is the band rejection characteristics for dual-band and multiband frequencies operation. Further, UWB antennas are required for millimetre wave application in advance wireless communication for high-speed vehicular communication system.

REFERENCES

[1] B. K. Pandey, A. K. Pandey, S. B. Chakrabarty, S. Kulshrestha, A. L. Solanki, and S. B. Sharma, "Dual frequency single aperture microstrip antenna element for SAR applications," *IETE Technical Review*, Vol. 23, No. 6, pp. 357–366, 2006.

[2] C. Reig and E. A. Navarro, "Printed antennas for sensor applications: A Review," *IEEE Sensors Journal*, Vol. 14, pp. 2406–2218, 2014.

[3] L. H. Abderrahmane, M. Benyettou, and M. N. Sweeting, "An S band antenna system used for communication on earth observation microsatellite," *Aerospace Conference 2006 IEEE*, pp.1–6, 2006.

[4] N. Fayyaz, E. Shin, and S. S. Naeini, "A novel dualband patch antenna for GSM band," *Antenna and Propagation for Wireless Communications, IEEE-APS Conference,* pp. 156–159, Nov. 1998.

[5] J.-Y. Jan, "Low profile dual frequency circular microstrip antenna for dual ISM bands," *Electronics Letter*, Vol. 37, pp. 999–1000, Aug. 2001.

[6] H. T. Chen, K. L Wong, and T. W. Chiou, "PIFA with a meandered and folded patch for the dual band mobile phone application," *IEEE Transactions on Antennas and Propagation*, Vol. 51, pp. 2468–2471, Sept. 2003.

[7] S. Y. Lin and K. C. Huang, "A compact microstrip antenna for GPS and DCS application," *IEEE Transactions on Antennas and Propagation,* Vol. 53, pp. 1227–1229, Mar. 2005.

[8] J. W. Wu, Y. D. Wang, and J. H. Lu, "Novel dual broadband rectangular slot antenna for 2.4/5- GHz wireless communication," *Microwave Optical Technology. Lett*ers, Vol. 46, pp. 197–201, Aug. 2005.

[9] Ashish Singh, Mohammad Aneesh, Kamakshi, and J. A. Ansari, "Circuit theory analysis of aperture coupled patch antenna for wireless communication," *Radioelectronics and Communications Systems,* Vol. 61, No. 4, pp. 168–179, 2018.

[10] Ashish Singh, Mohammad Aneesh, Kamakshi, and J. A. Ansari, "Analysis of microstip line fed patch antenna for wireless communications," *Open Engineering Journal*, Vol. 7, pp. 279–286, 2017.

[11] S. Verma, J. A. Ansari, and Ashish Singh, "Truncated equilateral triangle microstrip patch with and without substrate," *An International Journal of Wireless Personal Communication*, Vol. 95, pp. 873–889, July 2017.

[12] K.-H. Kim, Y.-J Cho, S.-H. Hwang, and S.-O. Park, "Band-notched UWB planar monopole antenna with two parasitic patches," *Electronic Letters*, Vol. 41, pp. 783–785, 2005.

[13] W. Choi, K. Chung, J. Jung, and J. Choi, "Compact ultra-wideband printed antenna with band-rejection characteristics," *Electronics Letters*, Vol. 41, pp. 990–991, 2005.

[14] Chuan-Dong Zhao, "Analysis on the properties of a coupled planar dipole UWB antenna," *Antennas and Wireless Propagation Letters, IEEE*, Vol. 3, pp. 317–320, 2004.

[15] M. N. Moghadasi, M. Koohestani, M. Golpour, and B. S. Virdee, "Ultra-wideband square slot antenna with a novel diamond open-ended microstrip feed," *Microwave and Optical Technology Letters*, Vol. 51, pp. 1075–1080, 2009.

[16] Hsien-Wen Liu, Chia-Hao Ku, Te-Shun Wang, and Chang-Fa Yang, "Compact monopole antenna with band-notched characteristic for UWB applications," *Antennas and Wireless Propagation Letters, IEEE*, Vol. 9, pp. 397–400, 2010.

[17] R. Movahedinia and M. N Azarmanesh, "Ultra-wideband band-notched printed monopole antenna," *Microwaves, Antennas & Propagation, IET*, Vol. 4, pp. 2179–2186, 2010.

[18] L. Li, Z.-L. Zhou, J.-S. Hong, and B.-Z. Wang, "Compact dual-band-notched UWB planar monopole antenna with modified SRR," *Electronics Letters*, Vol. 47, pp. 950–951, 2011.

[19] M. Mehranpour, J. Nourinia, C. Ghobadi, and M. Ojaroudi, "Dual band-notched square monopole antenna for ultrawideband applications," *Antennas and Wireless Propagation Letters, IEEE*, Vol. 11, pp. 172–175, 2012.

[20] P. Gao, L. Xiong, J. Dai, S. He, and Y. Zheng, "Compact printed wide-slot UWB antenna with 3.5/5.5-GHz dual band notched characteristics," *Antennas and Wireless Propagation Letters, IEEE*, Vol. 12, pp. 983–986, 2013.

[21] M. Ojaroudi, N. Ojaroudi, and N. Ghadimi, "Dual band-notched small monopole antenna with novel coupled inverted U-ring strip and novel fork-shaped slit for UWB applications," *Antennas and Wireless Propagation Letters, IEEE*, Vol. 12, pp. 182–185, 2013.

[22] Fuguo Zhu, Steven Gao, Anthony, T. S. Ho, Raed A. Abd-Alhameed, Chan H. See, Tim W. C. Brown, Jianzhou Li, Gao Wei, and Jiadong Xu, "Multiple band-notched UWB antenna with band-rejected elements integrated in the feed line," *Antennas and Propagation, IEEE Transactions*, Vol. 61, pp. 3952–3960, 2013.

[23] N. Ojaroudi and M. Ojaroudi, "Novel design of dual band-notched monopole antenna with bandwidth enhancement for UWB applications," *Antennas and Wireless Propagation Letters, IEEE*, Vol. 12, pp. 698–701, 2013.

[24] M. Ojaroudi, N. Ojaroudi, and N. Ghadimi, "Dual band-notched small monopole antenna with novel coupled inverted U-ring strip and novel fork-shaped slit for UWB applications," *Antennas and Wireless Propagation Letters, IEEE*, Vol. 12, pp. 182–185, 2013.

[25] R. Azim, M. T. Islam, and A. T. Mobashsher, "Dual band-notch UWB antenna with single tri-arm resonator," *Antennas and Wireless Propagation Letters, IEEE*, Vol. 13, pp. 670–673, 2014.

[26] S. R. Emadian and J. Ahmadi-Shokouh, "Very small dual band-notched rectangular slot antenna with enhanced impedance bandwidth," *Antennas and Propagation, IEEE Transactions*, Vol. 63, pp. 4529–4534, 2015.

[27] H. Orazi and H. Soleiman, "Miniaturization of UWB triangular slot antenna by the use of DRAF," *IET Microwaves, Antennas & Propagation*, Vol. 11, pp. 193–202, 2017.

[28] K. Kiminami, A. Hirata, and T. Shiozawa, "Double-sided printed bow-tie antenna for UWB communications," *Antennas and Wireless Propagation Letters, IEEE*, Vol. 3, pp. 152–153, 2004.

[29] M. R. Ghaderi and F. Mohajeri, "A compact hexagonal wide-slot antenna with microstrip-fed monopole for UWB application," *Antennas and Wireless Propagation Letters, IEEE*, Vol. 10, pp. 682–685, 2011.
[30] J. Wu, Z. Zhao, Z. Nie, and Q.-H. Liu, "A printed UWB Vivaldi antenna using stepped connection structure between slotline and tapered patches," *Antennas and Wireless Propagation Letters, IEEE*, Vol. 13, pp. 698–701, 2014.
[31] B. Gong, X. S. Ren, Y. Y. Zeng, L. H. Su, and Q. R. Zheng, "Compact slot antenna for ultra-wide band applications," *Microwaves, Antennas & Propagation, IET*, Vol. 8, pp. 200–205, 2014.
[32] R. Karimian, H. Oraizi, and S. Fakhte, "Design of a compact ultra-wide-band monopole antenna with band rejection characteristics," *Microwaves, Antennas & Propagation, IET*, Vol. 8, pp. 604–610, 2014.
[33] O. M. Haraz, A.-R. Sebak, and S. A. Alshebeili, "Design of ultra wideband multilayer slot coupled vertical microstrip transitions employing novel patch shapes," *Microwave and Optical Technology Letters*, Vol. 57, No. 3, pp. 747–756, 2015.
[34] H. J. Mohammed, A. S. Abdullah, R. S. Ali, R. A. Abd-Alhameed, Y. I. Abdulraheem, and J. M. Noras, "Design of a uni-planar printed triple band-rejected ultra-wideband antenna using particle swarm optimization and the firefly algorithm," *IET Microwaves, Antennas & Propagation*, Vol. 10, pp. 31–37, 2016.
[35] S. Yadav, A. K. Gautam, B. K. Kanaujia, and K. Rambabu, "Design of band-rejected UWB planar antenna with integrated Bluetooth band," *IET Microwaves, Antennas & Propagation*, Vol. 10, pp. 1528–1533, 2016.
[36] M. A. Matin, B. S. Sharif, and C. C. Tsimenidis, "Dual layer stacked rectangular microstrip patch antenna for ultra wideband applications," *Microwaves, Antennas & Propagation, IET*, Vol. 1, pp. 1192–1196, 2007.
[37] H. Elsadek and Dalia M. Nashaat, "Multiband and UWB V-shaped antenna configuration for wireless communications applications," *Antennas and Wireless Propagation Letters, IEEE*, Vol. 7, pp. 89–91, 2008.
[38] M. Ojaroudi, Iran Urmia, G. Ghanbari, N. Ojaroudi, and C. Ghobadi, "Small square monopole antenna for UWB applications with variable frequency band-notch function," *Antennas and Wireless Propagation Letters, IEEE*, Vol. 8, pp. 1061–1064, 2009.
[39] K. Kamakshi, J. A. Ansari, A. Singh, M. Aneesh, and A. K. Jaiswal, "A novel ultrawide band toppled trapezium-shaped patch antenna with partial ground plane," *Microwave Optical Technology Letters*, Vol. 57, pp. 1983–1986, 2015.
[40] Kamakshi, J. A. Ansari, A. Singh, and M. Aneesh, "Desktop shaped broadband microstrip patch antennas for wireless communications," *Progress in Electromagnetics Research Letters*, Vol. 50, 13–18, 2014.
[41] J. A. Ansari, Kamakshi, A. Singh, and A. Mishra, "Ultra wideband co-planer microstrip patch antenna for wireless applications," *An International Journal of Wireless Personal Communication*, Vol. 69, No. 4, pp. 1365–1378, 2013.
[42] J. A. Ansari, Kamakshi, A. Singh, and M. Aneesh, "Analysis of L-probe proximity fed annular ring patch antenna for wireless applications," *An International Journal of Wireless Personal Communication*, Vol. 77, No. 2, pp. 1449–1464, 2014.
[43] G. Bozdag and M. Seemen, "Compact wideband tapered-fed printed bow-tie antenna with rectangular edge extension," *Microwave and Optical Technology Letters*, pp. 1–6, 2019. DOI: 10.1002/mop.31733
[44] L. Guo, M. Min, W. Che, W. Yang "A novel miniaturized planar ultra-wideband antenna," *IEEE Access*, Vol. 7, pp. 2769–2773, 2018.

[45] Z. Khan, A. Razzaq, J. Iqbal, A. Qamar, and M. Zubair, "Double circular ring compact antenna for ultra-wideband applications," *IET Microwaves, Antennas & Propagation*, Vol. 12, pp. 2094–2097, 2018.

[46] A. K. Gautam, S. Yadav, and K. Rambabu, "Design of ultra-compact UWB antenna with band-notched characteristics for MIMO applications," *IET Microwaves, Antennas & Propagation*, Vol. 12, pp. 1895–1900, 2018.

[47] B. Chen, X. Yang, and D. Hu, "Fish-shaped ultra-wideband patch antenna with tilt-folded patch feed," *IET Microwaves, Antennas & Propagation*, Vol. 13, pp. 1312–1317, 2019.

[48] C. Kissi, M. Särestöniemi, T. Kumpuniemi, S. Myllymäki, M. Sonkki, M. N. Srifi, H. Jantunen, and C. Pomalaza-Raez, "Reflector-backed antenna for UWB medical applications with on-body investigations," *International Journal of Antennas and Propagation*, Vol. 2019, 17 pages.

[49] A. Singh, K. Shet, and D. Prasad, "Ultra wide band antenna with defected ground plane and microstrip line fed for Wi-Fi/Wi-Max/DCS/5G/satellite communications," *IntechOpen*. DOI: 10.5772/intechopen.91428.

[50] A. Singh, K. Shet, D. Prasad, A. K. Pandey, and M. Aneesh. "A review: circuit theory of microstrip antennas for dual-, multi-, and ultra-widebands," *IntechOpen*. DOI: 10.5772/intechopen.91365.

[51] P. K. Malik, M. Singh. "Multiple bandwidth design of micro strip antenna for future wireless communication," *International Journal of Recent Technology and Engineering*, Vol. 8, No. 2, pp. 5135–5138, July 2019. DOI: 10.35940/ijrte.B2871.078219.

[52] A. Kaur and P. K. Malik, "Tri state, T shaped circular cut ground antenna for higher 'x' band frequencies," *2020 International Conference on Computation, Automation and Knowledge Management (ICCAKM)*, pp. 90–94, Dubai, 2020. DOI: 10.1109/ICCAKM46823.2020.9051501.

Part II

Performance Analysis of Microstrip Antennas

Part II

3 Design and Performance Analysis of Microstrip Antennas Using Different Ground Plane Techniques for Wireless Applications

J. Silamboli and N. Thangadurai

DOI: 10.1201/9781003093558-5

3.1 INTRODUCTION

In telecommunications, the microstrip antenna would be fabricated on a printed circuit board (PCB) using the microstrip techniques. A microstrip antenna would be printed directly as an internal antenna. It is also called a printed antenna. Supported by a ground plane, a metallic patch would be mounted on a substrate to form a microstrip antenna [1]. This antenna is widely used because the microstrip antenna would be fabricated easily in a PCB. Nowadays, in many mobile applications, the microstrip antenna is being installed. This is very much possible because of its small size, less weight and inexpensive. The microstrip antennas have become very popular in recent years because of their thin planar, which can be integrated into consumers' products [1].

The most common microstrip antenna is the patch antenna. This antenna uses continuous patches in an array. Square, dipole, rectangular, circular, circular ring and elliptical are the common shapes of the microstrip antenna; however, it is also possible to make any continuous shapes. The most commonly used shape is the rectangle, which looks like a truncated microstrip transmission line. Low profiles, conformal, lightweight, inexpensive and robust are the advantages of these types of antennas [1].

3.2 PARAMETERS OF MICROSTRIP ANTENNAS

3.2.1 Gain

Gain is one of the parameter in the theory of antenna. Usually, directivity is lesser when compared with each other. Gain can be given as the fraction of the intensity of radiation in a given direction from the antenna to the total input power accepted by the antenna divided by 4π.

$$G = 4\pi U/Pin$$

3.2.2 VSWR

VSWR stands for Voltage Standing Wave Ratio. It can be defined as the ratio of the maximum voltage to the minimum voltage of the given antenna.

$$VSWR = Vmax/Vmin$$

3.2.3 Bandwidth

Bandwidth can be defined as the range of frequencies in which the operation of the antenna is proper. Bandwidth can be given by the formula:

$$BW = 100 \times (fh - fl)/fc$$

where, fh is highest frequency; fl is lowest frequency and fc is centre frequency. Every antenna has its own bandwidth as per its design.

3.2.4 Return Loss

In a transmission line, the reflection of the power of a signal is called the return loss.

Return loss can be written as:

$$\text{Return loss} = 10\ \log 10\ Pi/Pr$$

The reflection coefficient can be given as the fraction of amplitude of the incident wave to amplitude of the reflection wave.

3.2.5 Shapes

Different output parameters are obtained with different shapes of the microstrip patch. In order to get a better output efficiency, the shape and dimensions should be defined clearly. The directivity of a microstrip patch can be realised from the cavity models of patches. It measures the degree to which the radiation is concentrated in a single direction.

3.3 DESIGN CONSIDERATIONS

A microstrip patch antenna consists of a ground plane, a substrate (dielectric material) and a metal plate (microstrip patch). It is also a semiconductor material. A metal microstrip radiation patch mounts on a substrate that is dielectric material with a ground plate in the other side [1]. A dielectric material with a mounted patch antenna is provided with a feed through the microstrip transmission line. When using lumped components, a wideband response is received; whereas, in a microstrip patch we get a narrowband response. For RF and microwave, we only need a narrowband response.

Figure 3.1 represents the cross-sectional view of a microstrip antenna with ground, dielectric and patch [2].

Following are the two basic important factors of any characteristic antenna: one is impedance and the other is radiation. In impedance, reflection coefficient, return loss and VSWR are very important. Gain, directivity, efficiency and polarization are included in radiation. Impedance matching is the most important factor. The device communication works in some value and antenna has another value. For matching

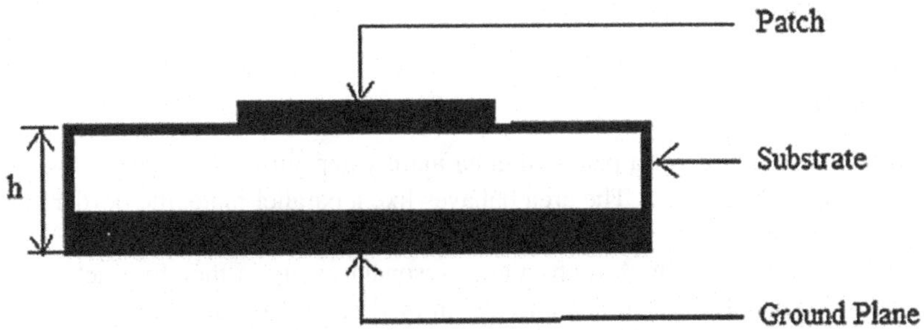

FIGURE 3.1 Cross-sectional view of MSA.

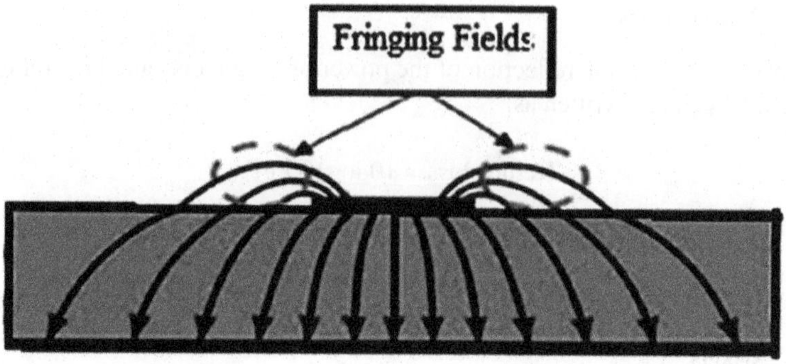

FIGURE 3.2 Fringing effect.

these two, we need impedance matching. It requires good impedance matching between the source and the antenna [2].

3.4 FRINGING EFFECT

The electromagnetic waves couple with the patch to ground when a feed is provided to the patch antenna with respect to the ground. At the patch, it is directly connected to the ground plane. But when it comes to the edges of the patch, electromagnetic waves travel to the space and then enters the ground. The electromagnetic waves create a field that travels into the space and this is called the fringing effect [2].

From Figure 3.2, we are able to observe that at the centre the electromagnetic waves directly radiate into the ground plane; whereas, at the edges the electromagnetic waves first travel through the space and then enters the ground plane. Because of fringing the antenna radiates into the space [3]. Fringing can be increased by the following three methods:

1. By reducing the dielectric constant
2. By increasing the value of height of the substrate
3. By having higher width of the substrate

The effective length of the antenna increases because of the fringing effect on the length of the patch. (Effective length = L + 2L.)

3.5 PRINCIPLES

The operating principle of a patch antenna mainly depends on the area between the ground plane and the patch. The area behaves like a parallel plate, the border acts as an open circuit of the patch by neglecting radiation loss; so the power gets reflected and stays below the patch. A high-quality resonant cavity is therefore achieved by the patch antenna. The patch antenna has limited bandwidth due to high Q system (narrow bandwidth). This makes the antenna's input impedance stay imminent to the desired value over a low range of centre frequencies [1].

3.5.1 Properties

The microstrip patch antenna is a low-profile antenna. Lightweight, low cost, easy to integrate are some of the advantages of this antenna compared to other antennas. The patch antenna could also be fabricated in square, dipole, rectangular, circular, circular ring and elliptical shapes. These antennas are conformal. The fabrication process of this antenna is very much easy on a PCB [2].

3.5.2 Feeding Methods

A feeding line would be used to excite or radiate by direct or indirect contact. The microstrip feeding line employs to feed the patch antenna; other feeding lines like coaxial line would also be used. These methods fall into two major groups: contacting and non-contacting types [3].

3.5.3 Contacting Type

With the help of a connecting element such as a microstrip line, the radio frequency energy feeds directly to the radiating patch [3]. This type of feeding is known as the contacting type. Line feed microstrip and probe feed microstrip are the different kinds of contacting types.

3.5.4 Line Feed Microstrip

The microstrip antenna mounted on an insulating material is being fed by the microstrip transmission line. The dielectric material is held up by the ground plane. The impedance matching would be achieved in two ways. One is by shifting the position of the microstrip line [4]. By moving the transmission line, impedance matching is done. The second is by having inset feeding. In insert feeding, we insert lines inside the patch antenna. By insertion we can have impedance matching between the patch antenna and the microstrip transmission line (the feed line). Since we are inserting, the substrate raises the surface and thus the spurious feed emission.

The advantages of the line feeding methods are simple to manufacture, easy to match (impedance matching), easy to create a replica. It has a disadvantage of shallow bandwidth (less than 2%). Since we are shifting the feed line for impedance matching, the axis point changes and it becomes asymmetric; because of this cross-polarization occurs [4].

3.5.5 Probe Feed Microstrip

Feeding microstrip patch antenna using a probe feed or a coaxial feed is a very common technique. By connecting the inner conductor of the coaxial cable to the radiation patch and by connecting the outer conductor to the ground plane, this type of probe is achieved. Impedance discrepancy occurs when we feed this microstrip patch at the middle with the innermost conductor at a particular time. To give impedance matching, we should change the innermost conductor from the middle towards the perimeter. Hence, the impedance matching is provided by shifting the innermost conductor from the middle.

In this microstrip antenna, the coaxial cable is connected to the patch from the ground. The benefits of this feed method are simple to make, simple to match, less spurious emission. The drawbacks are narrow bandwidth (2%) and cross-polarization. The reason for cross-polarization is that we are changing the innermost conductor from the middle, so the patch design will be asymmetric. Cross-polarization is produced because of this asymmetric design. A hole should be made into the substrate so the connector projects out of the ground plane; this is very hard to model [4].

Advantages

- Ease of fabrication
- Easy to match
- Low spurious radiation

Disadvantages

- Narrow bandwidth
- Difficult to model specially for thick substrate
- Possess inherent asymmetries that generate higher order modes which produce cross-polarization radiation. To avoid these drawbacks, we go for non-contacting type.

3.5.6 Non-Contacting Type

In the non-contacting type, the energy transmission is between the radiating patch and the microstrip line by coupling of electromagnetic fields. There are two kinds of non-contacting types: proximity coupled feed microstrip and aperture coupled feed [3].

3.5.7 Proximity Coupled Feed

There are two substrates (two insulating substances) that form the proximity coupled feed. Here the patch antenna is mounted on the first insulating substance that is placed on the second insulating material and the ground plane on another, one substrate for the patch and another substrate for the microstrip transmission line. The feed line is inserted between the first stand and second insulating substances. Since these are not connected physically, the feed line and the patch antenna have no physical contact. By changing the length and breadth of the feed line, we incur impedance matching. Here the design always stays symmetrical with view to the axis of the patch. Here there is no direct contact. The antenna radiates by coupling mechanism.

Figure 3.3 represents the proximity coupled microstrip antenna. The advantages of this proximity coupled feed method are it removes spurious feed emission, low cross-polarization and greater bandwidth (13%), it is simple to create a replica. The drawback is that it is complex to manufacture due to the presence of two insulation materials. The cost is also high due to the use of two substrates [4].

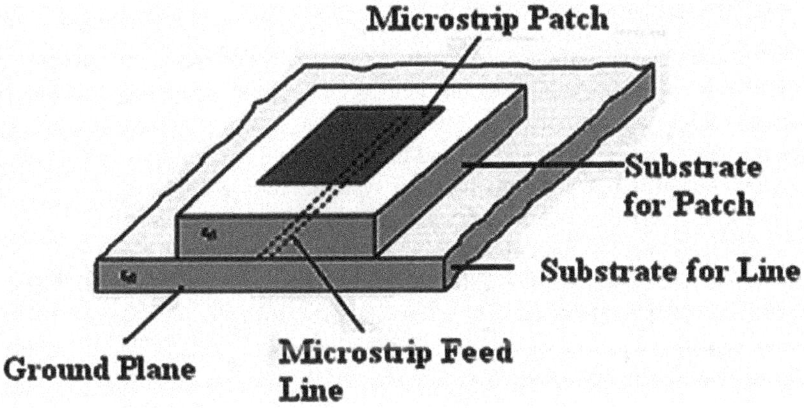

FIGURE 3.3 Microstrip antenna.

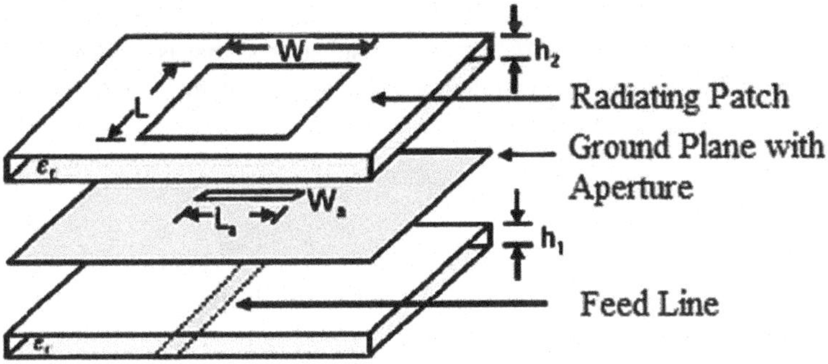

FIGURE 3.4 Antenna aperture.

3.5.8 Aperture Coupled Feed

In case of aperture coupled feed, we use two dielectric materials. Two dielectric materials are inserted above and below the ground plane.

Above the first insulating substance the patch antenna is placed, and below the second material the microstrip feed line is placed. There is a slot on the bottom because the line couples with the patch antenna and radiates. It is symmetric with the view to the axis of the patch.

Figure 3.4 represents the aperture coupled microstrip antenna. Here the ground plane is placed between the two insulating substances and the feed line is kept below the second insulating substance. The advantages of this feed method are low cross-polarization, moderate spurious radiation, easy to model. The drawback is that it has limited bandwidth and is more complex to make because a ground plane is inserted between the two substrates [4].

The advantages of feeding is that it allows independent optimization of feed mechanism element.

3.6 PERFORMANCE ANALYSIS

Microstrip antenna is widely used in telecommunications because of its low profile, lightweight, conformable and inexpensive. These antennas play a vital role in aerospace, radar, wireless and satellite communications, and so on. The WLAN is popularly apprehended as expensive and the data speed connectivity is also high, in the user vigour of wireless communication system. Here, we are increasing the performance characteristics like VSWR, directivity, efficiency, bandwidth, return loss, gain and reflection coefficient of the patch antenna for the WLAN using the different ground plane techniques. In this, FR4 is used as a substrate where the patch antenna is mounted and the feed line is done by feeding methods [5].

At the resonant frequency of 2.4 GHz with 50 Ω input impedance and by using the FR4 substrate, the rectangular patch antenna is designed with εr = 4.4, lose tangent δ = 0.02 and thickness h = 1.6 mm [5].

First, we have to design the specifications of an antenna such as frequency operation, input impedance, VSWR, return loss, reflection coefficient, directivity and gain.

The substrate in the patch antenna also acts as the mechanical support of the antenna. A substrate with a dielectric material is needed to increase the antenna's electrical performance. The characteristics of substrates, that is, dielectric constant, loss tangent and their variation with temperature and frequency, are to be considered for the selection of a substrate.

- Calculation of width (W):

$$W = \frac{c}{2f\sqrt{s_r} + \frac{1}{2}} \text{ where c} = 3 * 10^8 \text{ m/s}$$

- Calculation of actual length (L):

$$L = L_{\text{eff}} - 2\Delta L$$

$$L_{\text{eff}} = \frac{c}{2fo\sqrt{\varepsilon_{reff}}}$$

$$\frac{L}{h} = \Delta 0.412 \frac{(\varepsilon_{reff} + 0.3\left(\frac{W}{h} + 0.264\right)}{(\varepsilon_{reff} - 0.258)\left(\frac{W}{h} + 0.8\right)}$$

where

$$\varepsilon_{reff} = \frac{\varepsilon_r + 1}{2} + \frac{\varepsilon_r - 1}{2}\left[1 + 12\frac{h}{W}\right]^{-1/2}$$

- Calculation of inset feed depth (y_0):

$$y_o = 10^{-4}\begin{Bmatrix} 0.016922\varepsilon_r^{\,7} + 0.3761\varepsilon_r^{\,6} - 6.1783\varepsilon_r^{\,5} + 93.187\varepsilon_r^{\,4} \\ -6823.69\varepsilon_r^{\,3} + 2.5619\varepsilon_r^{\,2} - 4043\varepsilon_r + 6697 \end{Bmatrix}\frac{L}{2}$$

- Calculation of feed width (W_f):

$$\frac{8e^A}{e^{2A}-2}\frac{W_o}{h} \le 2$$

$$\frac{W_f}{h} = \left\{ \frac{2}{\pi}\left\{ B-1-\ln(2B-1)+\frac{\varepsilon_r-1}{2\varepsilon_r}\left[\ln(B-1)+0.39-\frac{0.61}{\varepsilon_r}\right]\right\}\frac{W_o}{h}\ge 2 \right\}$$

$$A = \frac{Z_o}{60}\sqrt{\frac{\varepsilon_r+1}{2}} + \frac{\varepsilon_r+1}{\varepsilon_r-1}\left(0.23+\frac{0.11}{\varepsilon_r}\right)$$

$$B = \frac{377\pi}{2Z_o\sqrt{\varepsilon_r}}$$

- Calculation of notch gap (g):

$$g = \frac{V_o}{\sqrt{2*\varepsilon_{reff}}}\frac{4.65*10^{-12}}{f_o(inGH_Z)}$$

Along with the dimensions and considering the ground plane dimensions, that is, length and width of the substrate as $6h + L$ and $6h + W$ (short edge), the antenna can be drawn. The antenna is stimulated by using HFSS and photolithography fabrication [5].

3.7 SIMULATION AND RESULTS

The simulated and fabricated frequency responses were presented with respect to the performance characteristics like VSWR, directivity, efficiency, bandwidth, return loss, gain and reflection coefficient of the patch antenna for the WLAN using different ground plane techniques. Since the antenna is operated at 2.4 GHz, the simulation is performed on fixed inset length or place. The antenna can also be tested using VNA E5071C [5].

3.8 APPLICATIONS OF MICROSTRIP PATCH ANTENNAS

Due to less weight, less profile, less volume, small dimension, low cost, easy fabrication and conformity, the microstrip patch antennas are employed in many fields. The applications of microstrip antennas are as follows [6]:

- Space vehicles such as rockets, satellite, space craft, space shuttle, high-performance aircrafts and missiles
- Satellite navigation receivers and in direct broadcast services
- Mobile radios, phones and pager, feed elements in complex antenna and integrated antenna and in rectifying antenna
- Radar altimeters, command and control systems, remote sensing and popper and other radars
- Analysis of various aspects of the environmental instrumentation, integrated antenna, feed elements in complex antenna, intruder alarms and biomedical radiators
- Base stations for personal communications
- Circularly polarized microstrip antennae are used in geopositioning satellite (GPS) systems
- Used in RFID field (WiMax)
- Telemedicine, in medical applications it is used in the treatment of dangerous tumours [6]

3.9 MULTIPLE-INPUT MULTIPLE-OUTPUT

The multiple-input multiple-output (MIMO) characteristic means using multiple transmission and multiple receiving antennas to make capital out of multipath propagation for multiplying the radio link capacity. In the wireless communication standards, including Wi-Fi (IEE 802.11), 3G, 4G LTE and WIMAX, MIMO becomes a crucial element. Recently in power line communications, parts of ITU standard and home plug specification for three-wire installations, MIMO has been applied. MIMO precisely calls attention to a practical technique for one data signal that has been sent and received simultaneously over the same radio channel by making capital out of multipath propagation [7]. It is different from smart antenna technique. The functions of the MIMO has been divided into three main categories: precoding, spatial multiplexing (SM) and diversity coding.

Precoding: A multistream beam formation is the narrow definition of precoding. In general, it is thought to be spatial processing that will take place at the transmitter.

Spatial multiplexing: In this, a high-rate signal is split into multiple lower rate streams and each stream is transmitted in the similar frequency channel from the different transmitting antennas.

Diversity coding: When there is no channel knowledge at the transmitter, the diversity coding technique is used [7].

In MIMO, for implementing space–time transmit, 3G allows it in combination with transmit beam to form at the base station. It is used in 4G and in the upcoming 5G technologies. MIMO is also worked out to be used in the mobile radio (3GPP and 3GPP2).

In non-wireless communication systems such as home networking standard ITU-TG.9963, MIMO technology can be used [7].

The MIMO is a wireless technology that contains multiple transmitters and receivers. Because of this more data can be sent at a time. MIMO is implemented in phone handsets (Figure 3.5).

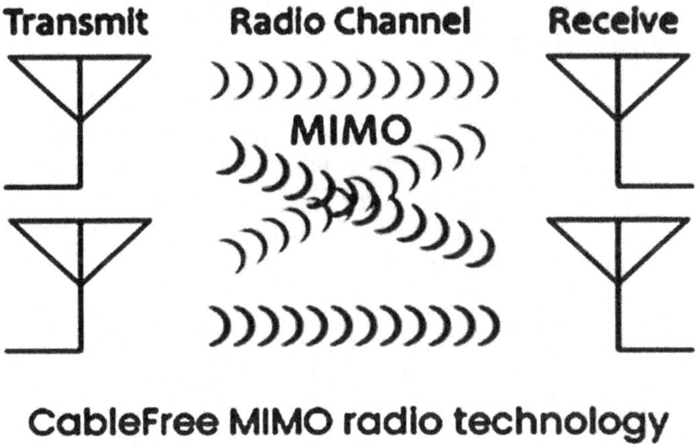

FIGURE 3.5 Multiple-input multiple-output.

3.10 FRACTAL AND DEFECTED GROUND STRUCTURE MICROSTRIP ANTENNAS

In the microstrip antenna, the broadband radiating patch is mounted on an insulating material and is supported by a ground plane, which forms the basic structure. For any possible shapes of conducting material such as gold or copper, a patch is constructed. In this, the main radiator is a rectangular patch. The benefits of these types of broadband antennas are easy to make, simple in structure, small size, less weight and inexpensive [8]. Hence this type of broadband rectangular microstrip patch antenna is fabricated for wireless communications.

An etched slot or fault is merged on the ground plane of the planar circuit which is called a defected ground structure (DGS). The DGS has become popular among all techniques because of its simple structural design. In ractal DGS, single or multiple slots are present in the ground plane [9]. This method is implemented to do the band stop characteristics. DGS has been used in the microstrip field for the enhancement of bandwidth, gain, radiation characteristics of the microstrip antenna and to cut cross-polarization [9].

Defects on the ground plane disturbs the distribution of current, and the transmission line characteristics also change by changing some characters such as resistance, capacitance and inductance. This is the working principle. The most defect shapes which have been merged on the ground plane are dumbbell, circular dumbbell, spiral, "U", "V", "H", cross, concentric rings, split rings resonators and fractals. Fractal DGS has bandgap characteristics [8]. To increase the gain, the area of the patch shapes is employed with a copper layer which is being carved [9]. It suppresses the current distribution on the patch antenna. The main disadvantage is that it affects the radiation pattern and reduces the front to back ratio. DGS also finds applications in microwave filter, coupler, power divider and so on.

3.11 MICROSTRIP ANTENNAS FOR VEHICULAR COMMUNICATION

Vehicle to vehicle communication is a technology that can be used for vehicle's interaction. The proposed antenna is an omnidirectional microstrip antenna designed to avoid road problems, especially in the blind spot zone, by tracking the nearby vehicles' band range used in the proposed antenna for vehicle to vehicle communication. This range is from 5.85 to 5.925 GHz, which includes Dedicated Short-Range Communication (DSRC) and Wireless Access in Vehicular Environment (WAVE). The size of the proposed antenna is $20.2 \times 24.1 \times 16$ mm^3. As the designed antenna is omnidirectional, a better impedance matching is achieved. The FR4 and PVC materials are used for designing the proposed antenna [9]. The gain obtained is 4.78 dBi in case of FR4 materials and 5.9 dBi in case of PVC materials. This covers the complete DSRC band. In this antenna design, the genetic algorithm is used. This algorithm increases the gain of the antenna to about 5.82 dBi and provides omnidirectional characteristics. It also reduces the size of the antenna to 82%. The antenna can be designed in multiple ways to increase its functionality [8]. The antenna return loss and mutual coupling can be reduced by using the multilayered patches, and hexagonal patches can be used to increase the antenna's directivity. In patch antenna, DGS is used to reduce mutual coupling. Offset feeding method is also employed in this antenna design. Stimulation is done by the ANSYS Electromagnetic Suite [9]. The proposed antenna is simple in design and cheap ompared to other works. VSWR is calculated to be nearly 1. For good performance, the antenna should be fixed at the centre of the vehicle. When compared to earlier works, this design provides a high gain of 5.6 dBi after optimization. The designed antenna using the genetic algorithm results in providing high gain, reduced size, good characteristics, good impedance matching and VSWR in vehicle to vehicle communication by covering the entire DSRC band [8].

The DSRC system can be explained as follows: With the specified frequency range, the system allows the vehicles to communicate with other vehicles and with the environment wirelessly. It also indicates the roadside hazards. The antenna implemented provides the essential performance and works at all times, even during difficult RF environment. The antenna design stipulates good connectivity and effective transmission and reception. The DSRC system involved here provides two-way communications – short range to medium range [5, 10].

3.12 IMPORTANCE AND USE OF MICROSTRIP ANTENNAS IN IoT APPLICATIONS

Development of technology day-by-day has led to the emerging application of IoT. The IoT devices are designed with small microstrip antennas and the demand for it has increased. There are lots of IoT applications such as real-time weather monitoring system, in aircraft for noticing turbulence, irregular atmospheric motion, pressure differences, weather variations and to take remedial action within short period. A GPS antenna uses a frequency range from 1.2 to 1.6 GHz in which 1.5 GHz is the most widely used frequency. The microstrip antennas are widely used because they have low profile, less weight and they are inexpensive. To get an improved bandwidth two methods are used: patch meandering and shorting pin. These two methods are also used to miniaturize antenna. The designing of antennas for IoT applications is always dependent on the frequency range, transmission power or energy and space availability. The IoT devices use various frequency ranges for different applications. The frequency for a given IoT device should fall into the antenna's bandwidth. The microstrip antennas are usually employed in IoT. Applications in IoT devices with GPS potentialities as signals transferred by satellites are usually RHCP (right-handed circular polarization) and LHCP (left-handed circular polarization). So, choosing of a population match is very critical. A microstrip antenna that has a multiband response at 2.39, 4.3, 5.8 and 6.56 GHz with bandwidths of 230, 380, 390 and 350 MHz respectively is used for IoT applications. The FR4 material is chosen to be the substrate of the antenna. This antenna is mainly employed in Bluetooth and Wi-Fi RF devices and radio-site applications.

3.13 ULTRA-WIDEBAND ANTENNA DESIGN FOR WEARABLE APPLICATIONS

The microstrip antennas have a great demand to be designed as ultra-wideband antennas that can be employed for wearable applications. Personal communication technology is one of the major reasons for this demand of antenna for wearable applications. Wearable devices are mainly used by patients for healthcare, battle field communications, environment monitoring, radio operations, help and so on. These kinds of devices should be easy to wear and lightweight. In order not to disturb the wearing comfort, ellipse, rectangle, triangle, circle and other geometric shapes are being used. They can be divided as electrically small ($c \ll 0.1\ \lambda$) and electrically large ($c \gg \lambda$) where c is the circumference and $\lambda$is the wavelength. For most of the pagers, RF tags, controllers, the loop antennas are used because the less conductivity of our body will not lessen the quality of the antenna.

Dual band microstrip antennas are also used in wearable devices. These types of antennas are small in their size and weight. Selection of a substrate is very important while designing these kinds of antennas. The dual band antenna has a copper foil patch that is *h* type and the unit of this is millimetre. The ground is made of copper and the substrate size and the ground size are same. For the fabrication to be easy and for the simple matching purposes, the antenna has a side-fed configuration. The foil used is for the ground and the substrate is manually cut to the expected dimensions. This dual band antenna obtains impedance matching bandwidth from 33.395 to 2.505 GHz and 5.52 to 6.15 GHz. The antenna is designed to be acceptable by users in the consumer market and is very small in size and weight. The substrate is a step forward to wearable antennas.

The antenna in wearable devices should be placed at a maximum distance from the human body to avoid hazardous effects. These antennas should be designed in such a way that it increases the battery life of the device. Wireless modules are suitable for better connectivity and is also concerned with the antenna performance depending on its position.

REFERENCES

[1] Sunil Singh, Neelesh Agarwal, Navendu Nitin, A. K. Jaiswal, "Design consideration of microstrip patchantenna", *International Journal of Electronics and Computer Science Engineering*, Vol. 2, Iss. 1, pp. 306–316, 2013.
[2] Microstrip antennas: The patch antenna. Antenna Theory. www.Antenna-theory.com.
[3] Feeding methods microstripantennas.blogspot.com.
[4] Navpreet Kaur, et al., "Study the various feeding techniques of microstrip antenna using design and simulation using CST Microwave Studio", *International Journal of Microwaves Application*, Vol. 6, Iss.1, pp. 5–9, 2017.
[5] V. Arora and P. K. Malik, "Analysis and synthesis of performance parameter of rectangular patch antenna". In: Singh P., Pawłowski W., Tanwar S., Kumar N., Rodrigues J., Obaidat M. (eds) *Proceedings of First International Conference on Computing, Communications, and Cyber-Security (IC4S 2019). Lecture Notes in Networks and Systems*, vol. 121. Springer, Singapore, 2020. https://doi.org/10.1007/978-981-15-3369-3_12.
[6] Indrasen Singh and V. S. Tripathi, "Microstrip patch antenna and its applications: A survey", *International Journal of Computer Applications*, Vol. 2, Iss. 5, pp. 1595–1599, 2011.
[7] A. Rachakonda, P. Bang, and J. Mudiganti, "A compact dual band MIMO PIFA for 5G applications", *IOP Conference Series: Materials Science and Engineering*, Vol. 263, Iss. 5, pp. 1–10, 2017.
[8] R. Udit and S. Sreenath Kashyap, "Microstrip patch antenna parameters, feeding techniques and shapes of the patch – A survey", *International Journal of Scientific and Engineering Research*, Vol. 6, Iss. 4, 2015.
[9] Mukesh Kumar Khandelwal et al., "Defected ground structure: Fundamentals, analysis, and applications in modern wireless trends", *International Journal of Antennas and Propagation*, Vol. 2021, Iss. 06/10, pp. 1–22, 2016.
[10] P. Tiwari and P. K. Malik, "Design of UWB antenna for the 5G mobile communication applications: a review", *2020 International Conference on Computation, Automation and Knowledge Management (ICCAKM)*, 2020, pp. 24–30, Dubai. doi: 10.1109/ICCAKM46823.2020.9051556.

FIGURE REFERENCES

1. Manickam Karthigai Pandian and Thangam Chinnadurai, "Design and optimization of rectangular patch antenna based on FR4, teflon and ceramic substrates", *Advances in Electrical and Electronic Engineering*, Vol. 11, pp. 1–6, 1969.
2. *Journal of Microwaves, Optoelectronics and Electromagnetic Applications.* On-line version Vol. 18, Iss. 2, São Caetano do Sul June 2019. Epub June 19, 2019. https://doi.org/10.1590/2179-10742019v18i21554
3. Associate Professor: WANG Junjun School of Electronic and Information Engineering, Beihang University F1025, New Main Building. wangjunjun@buaa.edu.cn13426405497
4. Image rectanna: Insert-Fed and Edge-Fed patch antenna. https://images.app.goo.gl/ttCZCmUmv9w7EX926
5. "Bandwidth enhancement of modified square fractal microstrip patch antenna using gap-coupling", February 2015 *Engineering Science and Technology: An International Journal*, 2012(2). doi:10.1016/j.jestch.2014.12.001

6. Jibendu Sekhar Roy, "Investigations on a new proximity coupled dual-frequency microstrip antenna for wireless communication". June 2007. Project: Future 5G., 25.92 KIIT University; Dr. Milind Thomas Themalil, 2.96JK Lakshmipat University.
7. Hayat Errifi Baghdad Abdennaceur, "Design and optimization of aperture coupled microstrip patch antenna using genetic algorithm", May 2014. 14.66 University of Hassan II of Casablanca; Abdelmajid Badri, 19.95 University of Hassan II of Casablanca.
8. "5G cellular user equipment: From theory to practical hardware design."

4 Design and Simulation of High Gain Microstrip Patch Antennas Using Fabricated Perovskite Manganite Added Polymer Nanocomposite for Multiple-Input and Multiple-Output Systems

R. Karpagam, V. Megala, G. Rajkumar, K. Sakthipandi and G. K. Sathishkumar

4.1 INTRODUCTION TO MPA

The microstrip patch antenna (MPA) can be directly printed on a circuit board and offers advantages such as low profile, less weight and conformability to a shaped surface. MPAs are used extensively in mobile, missile, aerospace, radar, medical device and multiple-input multiple-output (MIMO) communication systems [1].

DOI: 10.1201/9781003093558-6

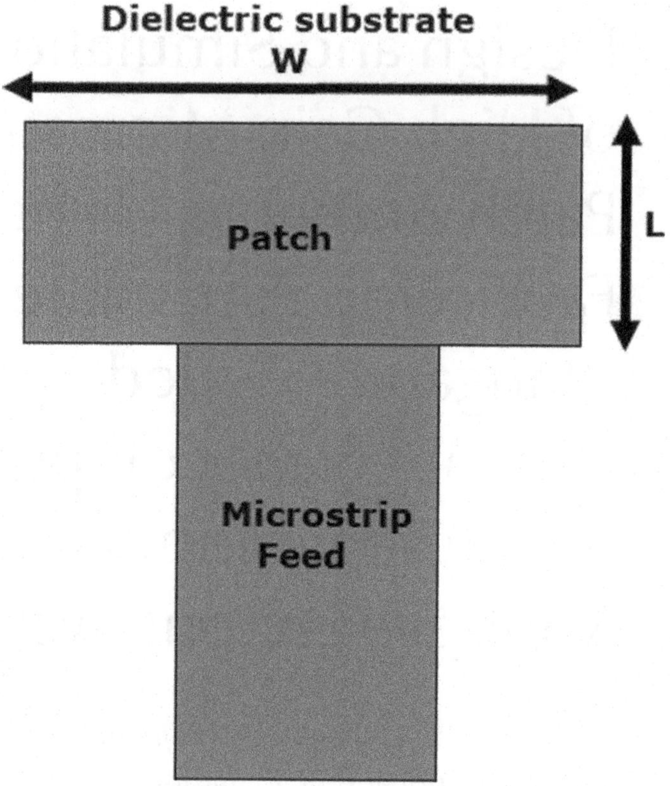

FIGURE 4.1 Geometrical top view of microstrip patch antenna.

Figure 4.1 shows the top view of a normal MPA. The patch, microstrip feed and the ground plane are made up of conducting material (metal). The substrate is made of dielectric material of permittivity ε_r. The patch of dimension L × W × H (length × width × height) is placed on the top of the dielectric substrate. Input impedance depends on 'W' of the antenna. Bandwidth increases with increase in 'W'. The 'H' is determined using the wavelength of operation.

The condition for selecting height is 1/40 of wavelength < H < wavelength of operation. If the condition is not satisfied, then it results in degradation of output. The frequency of operation 'f_c' depends on L and permittivity of the dielectric material ε_r as given by Equation (4.1):

$$f_c \approx \frac{c}{2L\sqrt{\varepsilon_r}} = \frac{1}{2L\sqrt{\varepsilon_r \varepsilon_o \mu_o}} \quad (4.1)$$

Although the specific design of antenna geometries differs, a general relationship for many antenna designs states that a dimension of the antenna, l m, is inversely proportional to the frequency and the square root of the properties of the dielectric medium through which the wave propagates in the antenna.

4.2 MINIATURIZATION OF MPA

Antennas designed to radiate very high frequency (VHF) and ultra high frequency (UHF) bands can be impractically large for mobile applications and compact packages [2]. The size miniaturization of MPA greatly affects the following of MPA:

- The gain
- The bandwidth

4.2.1 Dielectric Constant for Miniaturization of MPA

Designing an MPA by minimizing its size without compromising gain and bandwidth is achieved by controlling the dimensions of toroid structure on insulating materials. The search for suitable RF dielectric and insulating materials for the fabrication of miniaturized MPAs possessing high gain with good impedance matching has been a challenge to the research community in past decades [24, 25]. The major risk of low gain antennas is high SAR which affects the ecosystem [3].

$$l_a \alpha \lambda_a = \frac{1}{f_a\sqrt{\varepsilon_0 \varepsilon_r \mu_0 \mu_r}} \tag{4.2}$$

Since the speed of light in vacuum is constant, the size of an antenna as given in Equation (4.2) can be manipulated through the following three factors [8, 9]:

- The frequency of radiation (f_a)
- The relative permittivity (ε_r) of the medium through which the wave travels in the antenna
- The relative permeability (μ_r) of the medium through which the wave travels in the antenna

Compact antennas can be easily made for antennas operating at several gigahertz and higher. However, for applications requiring operation in the VHF and UHF bands from 30 MHz to 3 GHz, increasing the frequency to reduce the size of the antenna is not an option.

Besides flexibility in the selection of compact antenna geometries, the size of the antenna can be reduced by increasing the relative permittivity and/or the relative permeability of the material. Since the material must be electrically insulating, it is most often impractical to incorporate magnetic materials, and the antenna size can only be reduced by increasing the material's permittivity [10, 11].

The potential of antenna size reduction as the dimension decreases with increasing values of the relative permittivity with constant permeability is demonstrated in Figure 4.2. A size reduction factor is plotted for vacuum/air and polymer. The size of a dielectric-loaded antenna falls dramatically at relatively low values of the dielectric constant. The size reduction factor is 5 for a dielectric constant of 25 in comparison to an antenna incorporating air. At a relative permittivity of 100, the antenna size has

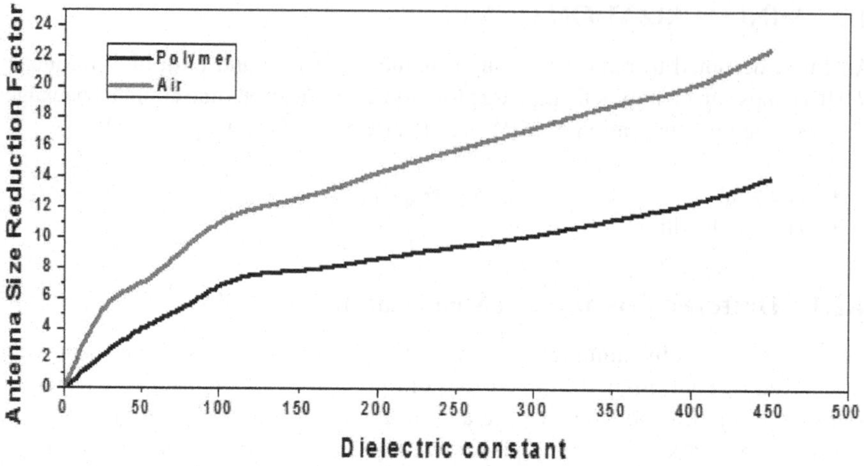

FIGURE 4.2 Antenna size reduction with increasing dielectric constant.

been reduced to 10% of its air-equivalent size. The antenna size can theoretically be further reduced with ever-increasing values of relative permittivity. However, a situation of diminished returns is evident due to the dependence on the square root of the relative permittivity [12].

Therefore, a point is reached at which further increases in the relative permittivity produce negligible benefits in size reduction. In addition to the need for a high dielectric constant to reduce the size of the antenna, the material must have adequate mechanical properties to be effectively used in high power antennas. For antennas operating in very high power systems, the material must be capable of handling voltages of the order of hundreds of kilovolts; therefore, dielectric strengths of the order from several tens of kilovolts per millimeter to hundreds of kilovolts per millimeter are required [27, 28]. Mechanical requirements include the capability of being formed or machined into the complex shapes required for some antenna geometries and the strength and resistance to fracture to withstand vibration and impact. Ceramics with high dielectric constants do not meet these dielectric strength and mechanical requirements, so a composite material is necessary [31].

4.2.2 PerPolyN for MPA Design

Perovskite-impregnated polymer nanocomposite (PerPolyN) is a novel RF antenna substrate consisting of epoxy polymer impregnated with perovskite-based fillers to form a nanocomposite toroid [4]. The characteristic improvements of the proposed nanocomposite depend on how the nanofiller is distributed and dispersed in the polymer matrix. The perovskite-based nanofiller used in the composite has exotic properties such as

- Colossal Magneto Resistance (CMR)
- High anisotropic field

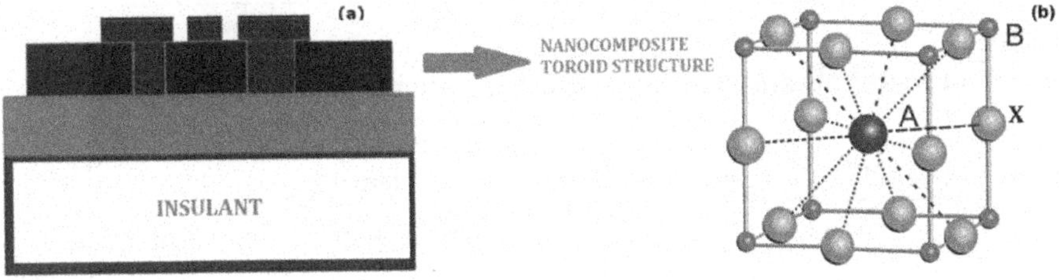

FIGURE 4.3 (a) Nanocomposite toroid structure of MPA; (b) structure of perovskite [4].

- High permeability
- Suppressed eddy current
- Low hysteresis loss
- Significant texability
- Low loss tangent

But it has poor mechanical strength and low breakdown strength that are compensated by epoxy perovskite encapsulated material as shown in Figure 4.3(a).

According to quarter wave principle, modifying the thickness of the PerPolyN between the two metal plates changes the switching frequency of the microstrip antenna.

The different composition of perovskite-based magnetic material added with polymer results in enhancing the electrical, magnetic and mechanical properties of the proposed antenna. Thus, antenna dielectric is able to tune microstrip antenna for lower permittivity (ε_r) to permeability(μ_r) ratio to achieve good impedance matching with that of free space and gives minimum absorption in X-band and minimum insertion loss in Ku-band.

4.3 INTRODUCTION TO PEROVSKITE – A DIELECTRIC MATERIAL

Perovskite was named after the Russian mineralogist L.A. Perovski (1792–1856) [29]. ABX_3 is the general formula for perovskite. A is larger cation, B is also a cation smaller than A and X is an anion binding A and B as shown in Figure 4.3(b) [7].

4.4 PerPolyN AS A DIELECTRIC MATERIAL OF MPA

Presently, MPA is subjected to narrow bandwidth, low gain, low efficiency, high return loss, decreased data rate and high specific absorption rate as shown in Table 4.1.

Drawbacks of a Antennas with high permittivity substrates include strong field confinement and poor impedance matching. High permittivity substrates result in narrowband characteristics and excitement of surface waves leads to lower antenna efficiency. Further, high permittivity substrates shorten the wavelength and it is difficult to achieve impedance matching on the substrates due to larger return loss problems [9].

TABLE 4.1
Different materials and their performance parameters

S. No.	Material	ε_r	μ_r	Frequency range	Remarks
1.	$Sr_{1-x}Ca_xMnO_3$ ($0.0 \le x \le 0.4$)	12.6	12.6	8.2–18 GHz	Electromagnetic characterization of microwave sintered strontium calcium manganite can be used as an impedance matched material
2.	$La_{1-x}Sr_xMnO3$	−8000	−2	0.5 MHz to 1 GHz	The tunability of the negative permittivity and negative susceptibility at high-frequency makes $La_{1-x}Sr_xMnO_3$ a promising candidate for double negative materials
3.	$La_{1-x}Sr_xMn_{1-y}B_yO3$ (B = Fe, Co, Ni)	20	0.8	2–18 GHz	The unique properties of doped rare-earth manganese oxides provide an important foundation for developing the materials with strong absorption and wideband for microwave and the materials have wide application prospect
4.	$Nd (Mg_{0.5-x} Ca_x Sn_{0.5}) O_3$	19.5		100–400 GHz	$Nd(Mg_{0.5-x}Ca_xSn_{0.5})O_3$ complex perovskite microwave ceramic used in hybrid dielectric resonator antenna
7.	$(Ba_{0.6-x}Sr_{0.4}Ca_x) TiO_3$ x = 0.10, 0.15, 0.20	2500		1 kHz	Barium strontium calcium Titanate ceramics dielectric properties made it suitable for microwave phase shifter
8.	$Ba_xSr_{1-x}TiO_3$	15		2.5 GHz	Barium strontium titanate dielectric resonator ceramic suits to load on array antennas [30]
9.	$Bi_{1-x-y} Gd_x La_y Fe_{1-y} Ti_yO_3$ (x¼ 0.1 and y¼ 0.0, 0.1, 0.2, 0.3 and 0.5)	2400			A minor loop traced for all these samples indicates an antiferromagnetic nature with weak ferromagnetism
10.	$BaTiO_3$ $SrTiO_3$	1,000		10 GHz	Ca + 2 is substituted at the Sr + 2 site, the value of '*x*' has a dramatic effect not only on the dielectric constant but also on the phases of the material. It ranges from orthorhombic to tetragonal
12.	$Pr_{0.7} Ca_{0.3} Mn O_3$			Tera Hertz	Terahertz radiation from strong electric fields

Disadvantages of high permittivity were circumvented using conventional ferrite materials used in the electronic devices in the market. But, ferrite is an inadequate material in the gigahertz band because of its small saturation magnetization which leads to low ferromagnetic resonance frequency. Compounds of ferrites and other oxides have suppressed the eddy current loss and operate in high frequency range. Traditional metal-based magnetic materials at microscale do not show adequate frequency stability and low loss [10]. In recent years, microwave communication and polymer microwave dielectric material made a huge breakthrough among all microwave dielectrics to improve the performance of MPA in the field of data rate, size, gain, bandwidth, dielectric strength and electrical degradation [26].

In the proposed novel polymer microwave, dielectric materials are used to surround the patch for the current simulated design. The proposed novel dielectric material acts as a waveguide that modifies the frequency and spatial structure of the electromagnetic fields to obtain the desired output of MPA.

A miniaturized MPA can be fabricated using optimized PerPolyN as microwave dielectrics. PerPolyN is a new class of nanocomposite that contains perovskite-structured nanofiller (ABO_3) impregnated in polymer epoxy resin. Systematic doping of A site and B site of nanofiller gives ferroelectric displacement requirement of d_0 state and magnetism requirement of partially filled d state.

The polymer epoxy and perovskite nanofiller have been preferred because of the following properties:

- It creates the electromagnetic properties in epoxy polymer matrix.
- The hydrophobic nature of polymer epoxy resin improves breakdown strength of nanofiller.
- PerPolyN is viable for lightweight MPA structures.
- PerPolyN shows its exotic properties such as high permittivity and high permeability.
- PerPolyN suppresses eddy current, hysteresis loss and loss tangent which can be achieved by altering the composition of perovskite-based nanofiller.

The size miniaturization of MPA greatly affects its gain, bandwidth and efficiency. It greatly relies on tuning of PerPolyN properties due to its flexible composition and structure of perovskite-based nanofiller. The major risk of low gain antenna is high specific absorption rate (SAR) that affects the ecosystem and hence a trade-off exists between the antenna dimensions and antenna performance. In order to solve this problem, the novel microstrip antenna is designed by controlling the properties and dimensions of PerPolyN.

According to quarter wave principle, modifying the thickness of the PerPolyN changes the switching frequency of the MPA. The essential parameter to reduce the size and bandwidth enhancement of MPA is obtained by inherent tailor-made permittivity and permeability of PerPolyN with minimum loss characteristics at higher frequencies. This controls the performance quality factors and ensures frequency tunability of MPA. PerPolyN helps to avoid high risk of emitting electromagnetic discharge by mitigating local electric field enhancement due to surface degradation.

The PerPolyN-based novel fabricated antenna provides tailor-made electromagnetic properties with less loss characteristics in the gigahertz band to meet the desired data rate for MIMO requirements.

4.5 GLOBAL IMPACT OF MPA

Nowadays, the source of generating electromagnetic waves is increasing globally from outdoor environmental RF sources, mobile phone base stations, radio and television transmitters [5]. These electromagnetic radiations (EMR) affect children and adolescents. The current requirement in mobile communication is higher operating frequency range mobile phones and wireless network. The increase in demand causes cross-talk and noise problems called electromagnetic interference (EMI). The microwave absorber in MPA is called a microwave dielectric and is the best solution to minimize EMI in various communication devices [6]. To avoid the effects due to EMI, attempts have been made to design and fabricate an MPA with a new class of nanocomposite materials. The importance of PerPolyN-based novel fabricated antenna is to avoid the potential health hazards due to EMR coming from MPA.

The PerPolyN microwave nanocomposite controls the excessive self-emission of electromagnetic waves due to lowering of magnetic and dielectric losses. This improves the life of children, especially at the risk of leukemia from a long time exposure to electromagnetic waves. The PerPolyN polymer composite material has enhanced electromagnetic properties due to lowering of non-ionizing radiation.

The results of PerPolyN MPA contribute an important role in the development of MPA for mobile and satellite communications. E-smog causes microwave syndrome such as sleep disturbances, headaches, exhaustion, impaired concentration, tinnitus, heat sensation on the ears and so on. The children are especially at risk of leukemia for a long time exposure of E-Smog. The proposed antenna avoids the risk in EMR in day-to-day use of mobile phones and communication devices. Thus, the development of novel perovskite MPA also protects the earth from EMI pollution.

The reinforcement of clay in nanocomposite as an insulant has eco-friendly characteristics. The clay-loaded composite has less exposure to gamma irradiation. The corona effect, gas discharge are due to ionization of gases occurring at low pressure levels in RF components. The lowered value of corona inception of clay-loaded composite prevents electromagnetic degradation. The perovskite structure has the advantages of global warming due to lower emission of CO_2 and isopropanol degradation [23, 24]. The proposed composite avoids spurious scattering and does not alter radiation patterns of antenna. The high gain antenna leads to achieving higher data rates which ensure coverage in all positions. The novel polymer composite material not only protects the patch antenna to the environment hazards, it also greatly improves the antenna performance.

4.6 METHODOLOGY OF PEROVSKITE SYNTHESIS

Step 1: *Preparation of nano perovskite manganite for microwave absorbing fillers*
Different compositions of perovskite manganite ($La_{1-x}Sr_xMnO_3$; $La_{1-x}Ca_xMnO_3$; $Pr_{(1-x)}Ca_xMnO_3$, where $x = 0.12, 0.165, 0.3$) are prepared in mol% using sonication reaction as mentioned in Figure 4.4 [13–23].

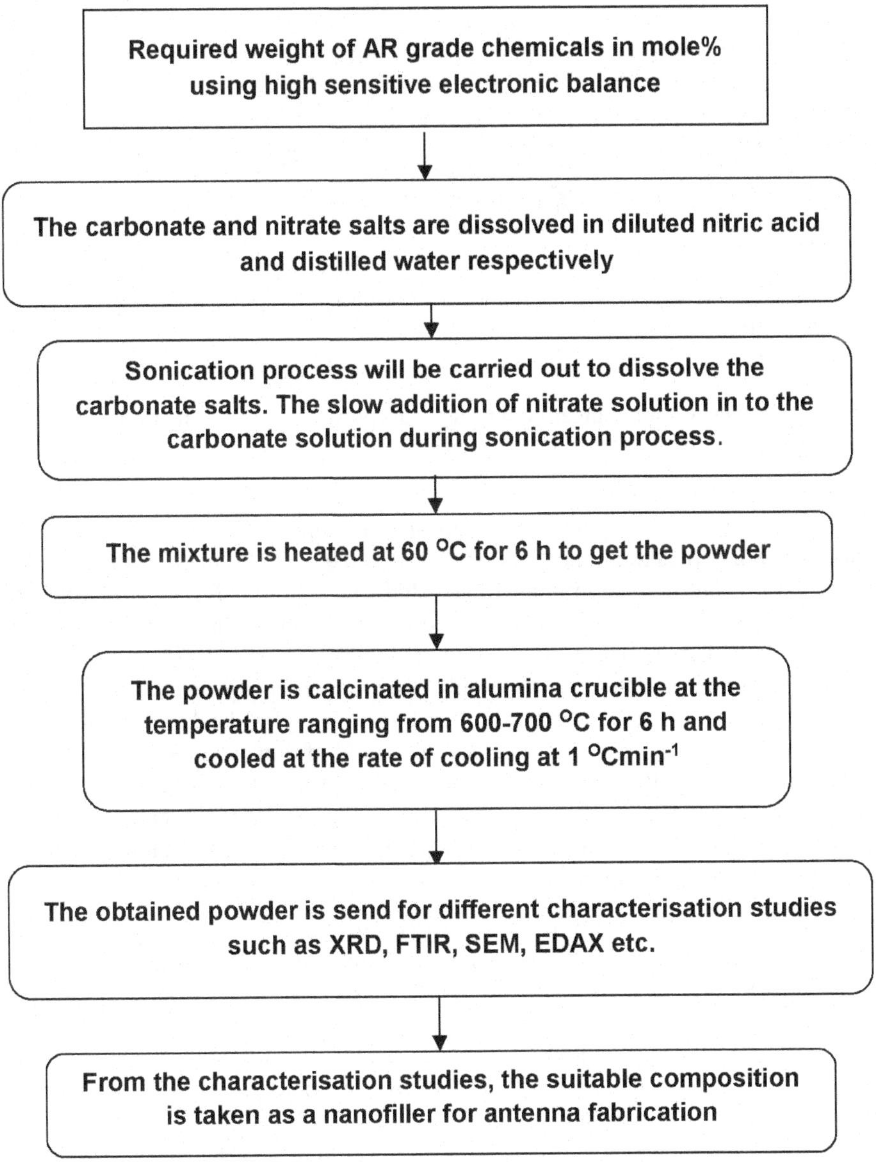

FIGURE 4.4 Flow chart of composite preparation and MPA fabrication.

Analytical grade (AR) chemicals of $Sr(NO_3)_2$, $La(NO_3)_2$, $Ca(NO_3)_2$, $Pr(NO_3)_3$ and $MnCO_3$ are weighted in mol% using a digital balance. The weighted carbonate powders and nitrate powders are separately dissolved in diluted nitric acid and distilled water, respectively.

Initially, sonication is carried out to dissolve $MnCO_3$ and the other nitrate salt solutions at 30 °C. The obtained solution is placed in hot air oven at 60 °C for 24 hours and dried to powder, which is calcined in alumina crucible at 400 °C for 2 hours

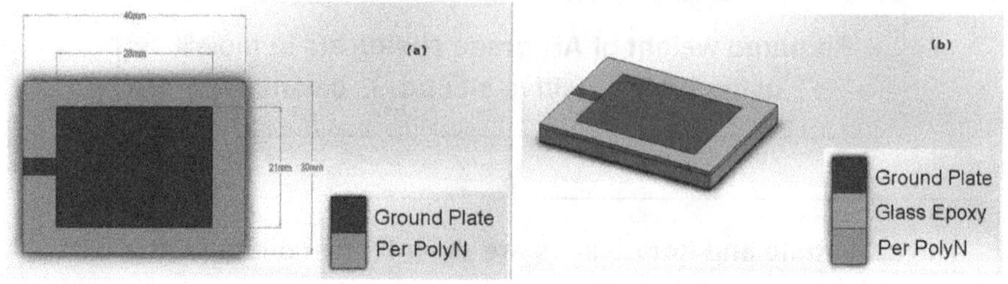

FIGURE 4.5 (a) Top view of PerPolyN MPA with specification; (b) isometric view of PerPolyN MPA.

to get a nanoperovskite manganite powder. This powder is used as nanofiller for MPA fabrication.

Step 2: *Fabrication of PerPolyN*
The perovskite-based polymer nanocomposite of compositions (80 – X) Epoxy-20, hardener-(X), perovskite, where x = 0, 0.3, 0.6, 0.9 and 1.0, in wt.% are prepared through mechanical shear mixing route. High pure di-glycidyl ether of biphenol epoxy resin (CY205, Ciba Geigy Inc.), triethylene-tetraamine hardener (HY 951) and prepared nanofiller are used to synthesize PerPolyN.

The perovskite nanofiller is uniformly dispersed in epoxy resin matrix using a mechanical shear mixer with 2000 rpm for 3 hours in a glass beaker of capacity 500 ml. This mixture is poured in a stainless steel mold of dimension 30 cm × 30 cm × 1 mm. The cast is taken out from the mold and kept in hot air oven at 75 °C for 3 hours and then cooled slowly at a rate of 1 °C per minute at room temperature. The prepared PerPolyN is cut into pieces of required dimension using stainless blade for the different characterization studies.

Step 3: *Simulation of MPA using fabricated PerPolyN*
The double-side copper-coated glass epoxy (3-mm thick FR4) in rectangular size is selected as the substrate materials. The bottom side of the clad is selected as ground plate.

Polyethylene ceramic composite (er ¼ 12:1 and tan d ¼ 0:004) and epoxy ceramic composite (er ¼ 14:1 and tan d ¼ 0:022) exhibit low dielectric losses with comparatively high dielectric constant.

The PerPolyN is used as a microwave absorber around the patch due its low dielectric loss and high permittivity ($\epsilon_r = 10.5$, $\tan\delta = 0.003$).

The proposed prototype model used for the design of MPA using the novel PerPolyN is shown in Figure 4.5.

4.7 RESULTS AND CONCLUSION

Designing an MPA by minimizing its size without compromising gain and bandwidth is achieved by controlling the dimensions of toroid structure on insulating materials. Figure 4.6 shows the design of MPA.

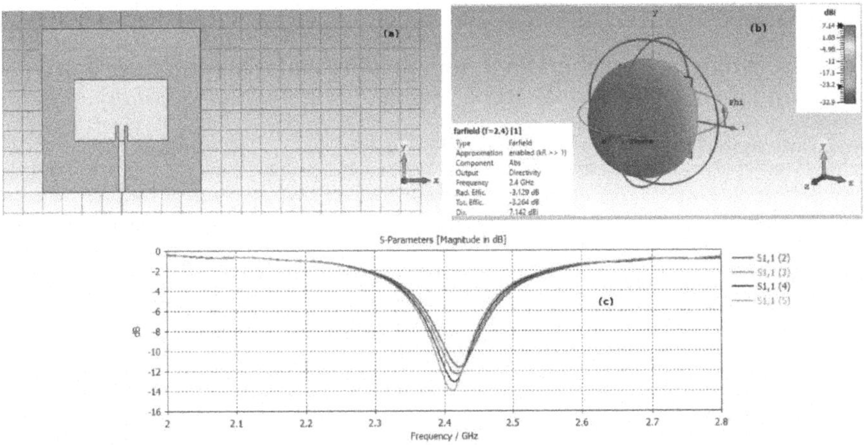

FIGURE 4.6 (a) Design of MPA; (b) radiation pattern of designed MPA; (c) comparison of return loss.

4.7.1 Antenna Design Parameters

Dielectric length = 80 mm
Dielectric width = 80 mm
Dielectric height = 1.5 mm
Patch length = 45 mm
Patch width = 30.2 mm
Transmission line width = 2.98 mm
Feeder width = 1.5 mm
Feeder transmission line length = 7.16

4.7.2 Validation of Antenna Parameters

Farfield (f = 2.4) (30 dBmW at Port1, Mode1)
Type: Farfield
Approximation: enabled (kR >> 1)
Component: Theta
Output: Directivity
Frequency: 2.4 GHz
Radiation efficiency: –3.194 dB
Total efficiency: –3.261 dB
Total radiation power [dBmW]: 26.74
Directivity (Abs): 7.113 dBi
Directivity (Theta): 7.113 dBi

A gain of 7.14 dB is obtained and is shown in Figure 4.6(b). The parameter 'gain' of the antenna shows the estimation of intensity in the direction where the peak value of radiation is obtained to that transmitted in an isotropic way.

TABLE 4.2
Comparing different combinations of perovskite materials and their impact on return loss of MPAs at 2.4 GHz frequency

Perovskite material	Dielectric constant	Return loss (dB)
PerPolyN	10.6	–14
$Sr_{1-x}Ca_xMnO_3$	12.6	–13
$Ba_xSr_{1-x}TiO_3$	15	–12.5
$Nd\ (Mg_{0.5-x}\ Ca_x\ Sn_{0.5})\ O_3$	19.5	–11.5

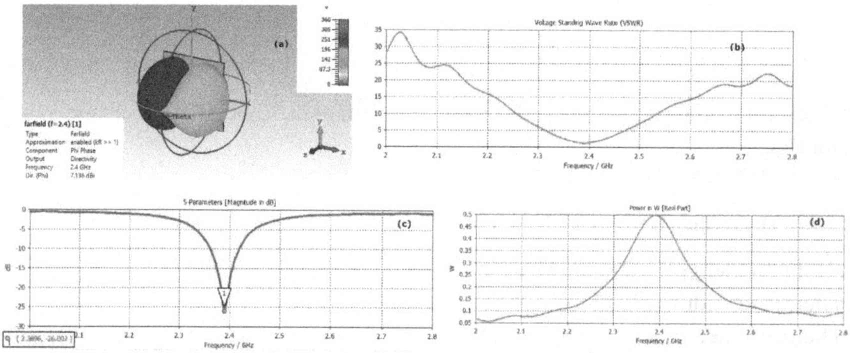

FIGURE 4.7 PerPolyN with dielectric constant 10.6; (a) radiation pattern; (b) voltage standing wave ratio; (c) return loss and (d) power radiated.

The S11 parameter is shown in Figure 4.6(c). Return loss is the amount of power which is lost due to the reflected signal and is known as the reflection coefficient. The value of S11 is the return loss obtained as –14 to –11.8 dB for PerPolyN, $Ba_xSr_{1-x}TiO_3$, $Nd(Mg_{0.5-x}Ca_xSn_{0.5})O_3$ and $La_{1-x}Sr_xMn_{1-y}B_yO_3$ at an operating frequency of 2.4 GHz.

The search for suitable RF dielectric and insulating materials for the fabrication of miniaturized MPAs with high gain with good impedance matching has been a challenge to the research community in past decades. It has been analyzed with various perovskite dielectric materials with a constant height of 0.787 mm and a frequency of 2 GHz, which is tabulated in Table 4.2.

As the dielectric constant of perovskite material increases, the L × W size and input impedance are decreased and the effective dielectric constant is increased. Perovskite material can be synthesized with different combinations to achieve good dielectric constant and their application in MPA minimizes the size further.

PerPolyN with dielectric constant 10.6 is considered as the best material for MPA because of reduced return loss. PerPolyN radiation pattern and voltage standing wave ratio are shown in Figure 4.7(a) and (b). The return loss in Figure 4.7(c) for PerPolyN with dielectric constant 10.6 is –26 dB. The power radiated is shown in Figure 4.7(d).

An overview of different materials used in MPAs and their applications has been presented in this chapter (see Table 4.2). This study can assist material scientists and MPA researchers in the selection of appropriate materials depending on the diverse applications.

REFERENCES

[1] E. Hammerstad and O. Jensen, "Accurate models for microstrip computer aided design", *IEEE MTT-S International Microwave Symposium Digest*, 1980, pp. 407–409, doi: 10.1109/MWSYM.1980.1124303.

[2] R. K. Hoffmann, *Handbook of Microwave* Integrated *Circuits*, 1987, Artech House Microwave Library, United States.

[3] K. C. Gupta, *Computer-Aided Design of Microwave Circuits*, 1981, Artech House Publishers, United States.

[4] Y. Chen, L. Zhang, Y. Zhang, H. Gao, and H. Yan, "Large-area perovskite solar cells – a review of recent progress and issues", *RSC Advances*, 2018, 8, 10489–10508, doi: 10.1039/C8RA00384J.

[5] M. Gallastegi, M. Guxens, and A. Jiménez-Zabala, "Characterisation of exposure to non-ionising electromagnetic fields in the Spanish INMA birth cohort: study protocol", *BMC Public Health*, 2016, 16, 167–177.

[6] L. N. Medeiros and T. G. Sanchez, "Tinnitus and cell phones: the role of electromagnetic radiofrequency radiation", *Brazilian Journal of Otorhinolaryngology*, 2016, 82, 97–104.

[7] K. McBride, J. Cook, S. Gray, S. Felton, L. Stellaab, and D. Poulidi, "Evaluation of $La_{1-x}Sr_xMnO_3$ ($0 \leq x < 0.4$) synthesised via a modified sol–gel method as mediators for magnetic fluid hyperthermia", *CrystEngComm*, 2016, 18, 407–416.

[8] F. Namin, T. G. Spence, D. H. Werner, and E. Semouchkina, "Broadband, miniaturized stacked-patch antennas for L-band operation based on magneto-dielectric substrates", *IEEE Transactions on Antennas and Propagation*, 2010, 58(9), 2817–2822.

[9] R. E. Collin, *Foundations for Microwave Engineering*, 2nd ed., New York: McGraw-Hill, 1992, p. 450.

[10] F. Mazaleyrat and L. K. Varga, "Ferromagnetic Nanocomposites", Journal of *Magn*etism and *Magn*etic *Mater*ials, 2000, 253, 215–216.

[11] H. Ohsato, "Research and development of microwave dielectric ceramics for wireless communications", *Journal of the Ceramic Society of Japan*, 2005, 113(1323), 703–711.

[12] R. C. Buchanan, *Ceramic Materials for Electronics*, Marcel Dekker: New York and Basel, 1986, pp. 1–8.

[13] M. T. Sebastian, "Dielectric materials for wireless communication", *Science Direct*, Elsevier: UK, 2008.

[14] G. Tong, W. Wu, Q. Hua, Y. Miao, J. Guan, and H. Qian. "Enhanced electromagnetic characteristics of carbon nanotubes/carbonyl iron powders complex absorbers in 2–18 GHz ranges", *Journal of Alloys and Compounds*, 2011, 509, 451–456.

[15] S. I. Kim, M. R. Kim, E. K. Cho, K. Y. Sohn, and W. W. Park, "Electromagnetic wave absorption properties of nanocrystalline Fe-based P/M sheets incorporating multi-walled carbon nanotubes", *Metals and Materials International*, 2020, 16, 121–123.

[16] M. Terada, M. Itoh, J. R. Liu, and K. Machida, "Electromagnetic wave absorption properties of Fe3C/carbon nanocomposites prepared by a CVD method", *Journal of Magnetism and Magnetic Materials*, 2009, 321, 1209–1214.

[17] A. S. Bhalla, R. Guo, and R. Roy, "The perovskite structure–a review of its role in ceramic science and technology", *Materials Research Innovations*, 2000, 4, 3–26.

[18] A. N. Yusoff, M. H. Abdullah, S. H. Ahmad, S. F. Jusoh, A. A. Mansor, and S. A. A. Hamid, "Electromagnetic and absorption properties of some microwave absorbers", *Journal of Applied Physics*, 2002, 92, 876–877.

[19] A. M. Nicolson and G. F. Ross, "Measurement of the intrinsic properties of materials by time-domain techniques", *IEEE Transactions on Instrumentation and Measurement*, 1970, 19, 377–383.

[20] P. Kanhere and Z. Chen, "A review on visible light active perovskite-based photocatalysts molecules", *Molecules*, 2014, 19, 19995–20022.

[21] S. Kimura, T. Kato, T. Hyodo, Y. Shimizu, and M. Egashira, "Electromagnetic wave absorption properties of carbonyl iron-ferrite/PMMA composites fabricated by hybridization method", Journal of Magnetism and Magnetic Materials, 312, 181–186, 2007.

[22] D. Wang, T. Kako, and J. Ye, "New series of solid-solution semiconductors (AgNbO3)1−x (SrTiO3)x with modulated band structure and enhanced visible-light photocatalytic activity", *The Journal of Physical Chemistry C*, 2009, 113, 3785–3792.

[23] H. Z. Friedlander, "Organized polymerization III. Monomers intercalated in Montmorillonite", *ACS, Division of Polymer Chemistry. Reprints*, 1963, 4, 300–306.

[24] G. Subodh, V. Deepu, P. Mohanan, and M. Sebastian, "Dielectric response of high permittivity polymer ceramic composite with low loss tangent", *Applied Physics Letters*, 2009, 95, 062903.

[25] G. Subodh, V. Deepu, P. Mohanan, and M. Sebastian, "Polystyrene/Sr2Ce2Ti5O15 composites with low dielectric loss for microwave substrate applications", *Polymer Engineering& Science*, 2009, 49, 1218.

[26] J. Varghese, N. Joseph, H. Jantunen, S. K. Behera, H. T. Kim, and M. T. Sebastian, *Advanced Ceramics and Composites for Defense, Security, Aerospace and Energy Applications*, Berlin: Springer, 2018.

[27] V. Megala and C. Pugazhendhi Sugumaran, "Enhancement of corona onset voltage using PI/MWCNT nanocomposite on HV conductor", *IEEE Transactions on Plasma Science*, 2020, 48(4), 1122–1129, doi: 10.1109/TPS.2020.2979843.

[28] V. Megala and C. Pugazhendhi Sugumaran, "Application of PI/MWCNT nanocomposite for AC corona discharge reduction", *IEEE Transactions on Plasma Science*, 2019, 47(1), 680–687, doi: 10.1109/TPS.2018.2877581.

[29] M. De Graef and M. E. McHenry. *Structure of* Materials*:* An Introduction *to* Crystallography, Diffraction *and* Symmetry. Cambridge University Press, United Kingdom, 2007, p. 671. ISBN 978-0-521-65151-6. 2010.

[30] S.-E. Lee, S. P. Choi, K.-S. Oh, J. Kim, S.M. Lee, and K. R. Cho, "Flexible magnetic polymer composite substrate with Ba1.5Sr1.5Z Hex ferrite particles of VHF/ low UHF patch antennas for UAVs and medical implant devices", *Materials*, 2020, 13, 1021.

[31] A. Raveendran, M. Sebastian, and S. Raman, "Applications of microwave materials: A review", *Journal of Electronic Materials*, 2019, doi:10.1007/s11664-019-07049-1.

5 Compact Microstrip Patch Antenna Design with Three I-, Two L-, One E- and One F-Shaped Patch for Wireless Applications

Mehaboob Mujawar

5.1 INTRODUCTION

With the steady change of innovation, frequency reconfigurable antennas have achieved significant advancement in the RF world. The new gadgets restrain the physical space utilized by wiping out the requirement for various antennas. This is particularly indispensable in cell phones, for example, mobile phones that get numerous bands of frequencies like cell tower gathering, Wi-Fi and GPS. The other option in contrast to recurrence reconfigurable reception apparatuses is a wideband antenna; notwithstanding, wideband antennas get huge ranges of frequencies acquainting

DOI: 10.1201/9781003093558-7

commotion with the framework. Frequency reconfigurable radio antennas slender the transfer speed to explicit frequencies, ordinarily reducing the measure of noise for the signal. The fundamental point of this chapter is to plan a minimized microstrip patch antenna for remote applications like wireless fidelity, worldwide interoperability for microwave access, wireless local area network and satellite communication. A sample of the antenna was created using IE3D Zeland software. The MSA was planned on a FR4 substrate with a dielectric constant of 4.4 and a depth of 0.162 mm. It resounds at 3.1, 7.8, 6.3 and 4.7 GHz with a data transfer capacity of 1.41 GHz at 2.59–3.99 GHz, 3.03 GHz at 3.99–7.02 GHz and 1.6 GHz at 6.99–8.59 GHz. The proposed plan has I, L, E and F patches on the ground plane. It works on three groups of recurrences: S, C and X bands. It has a solitary port with the input impedance of 50 Ω matched to it. The different parameters of antenna, such as voltage standing wave ratio, directivity of the antenna, gain, return loss and pattern of radiation, were examined. In the course of the last few decades, there is a quick development of remote correspondence applications. The presentation of all such remote frameworks relies upon the plan of the receiving antenna. With the advances in media transmission, the prerequisite for minimized reception apparatus has expanded fundamentally. In portable correspondence, the prerequisite for littler reception apparatuses is very enormous, so noteworthy advancements are done to configure smaller, insignificant weight, low profile receiving antennas for both scholastic and mechanical networks of media transmission. The technologist centered into the plan of microstrip patch radio antennas. Numerous assortments in structuring are conceivable with microstrip antennas.

5.2 IMPORTANCE OF A GOOD ANTENNA DESIGN

The antenna is considered to be one of the major hardware components of a mobile phone. It is not expensive for its production but it is considered to be important for the design and performance of a mobile phone. A mobile phone reception is solely dependent on how good the antenna functions. That is, an improperly designed antenna will not be able to receive a clear signal from the base station and therefore will disable the use of a mobile phone. Furthermore, while talking over mobile phones, the screen is usually turned off. The part draining most of the energy from the battery is the mobile phone transmitter. Therefore, the energy consumption is heavily dependent on the mobile phone antenna's efficiency. When mobile phones were invented in the late 1990s, they were totally dependent on the antenna as they were mounted external to the design. Nowadays with antennas being incorporated inside the mobile phones, their visual design dependence is less obvious but still important. Finally, there is yet another important aspect that is radiation safety, where construction of antenna has a very important role. Mobile phones dissipate microwaves in frequencies similar to those of microwave ovens and likewise result in radiations that are harmful for the user while using the mobile phones. Design of the mobile phone antenna is the most important factor to consider when trying to keep inside radiations to the standard limitations that are set up by the regulatory bodies such as the IEEE and Federal Communications Commission (FCC). Single band antennas uphold just one or two frequencies and nowadays an ever increasing number of guidelines are being upheld by the gadgets. They utilize few antennas for every application. This prompts huge space prerequisite. One predicted issue related to antenna for such gadgets is to cover 4G LTE bands while also covering WiMAX and WLAN/Bluetooth bands. Because of space imperatives in mobile phones, covering various bands with a solitary antenna structure is a need of great importance. The main objectives of this chapter are: (a) to analyze the characteristics of a reconfigured microstrip patch antenna for various shapes of the patches; and (b) to target the wireless applications such as Wi-Fi, Wimax, WLAN and satellite communications in the frequency range of 2–8 GHz. In this chapter we will design a microstrip patch antenna with defects in the ground plane to improve the functioning of antenna.

5.3 REVIEW OF THE EXISTING TECHNIQUES

With more development in the field of radio frequency technology, innovation in antenna design leads to work on numerous groups of frequencies with more extensive ranges as seen in Ref. [1]. A reduced multiband patch antenna has a wide range of uses in the field of wireless communications. The actual design of the antenna comprises of I, L and F shapes. It supports three frequency bands: the S, C and X bands. A quad band antenna introduced in Ref. [2] is used in numerous applications as it is intended to have a solitary radiator with a capacity to communicate and get different frequencies. The quad band antenna can be utilized for GSM, Wi-Fi, WiMax and WLAN applications. We can observe from Ref. [3] that the multiband LPA nearly covers the wireless LAN band. The issue of frequency shifting has happened which made the ideal working frequency at 2.4 GHz move to 5.2 GHz. Utilizing LPA as components to make a multiband antenna is a smart thought so that chosen frequency can work utilizing a solitary antenna. It is unique in relation to fractal antenna, where the working frequency scales up and works at undesired frequency. A small-size monopole antenna with three bands is appropriate for WiMax and WLAN applications [4] utilizing a Y-formed strip implanted in the round ring and a defected ground plane; the antenna parameters, for example, size, radiation patterns and higher gains, are useful. A triple band antenna has a rectangular patch with identical L and U slots. The proposed antenna can be a good decision for WLAN/WiMAX applications because of its size, structure, great multiband qualities and radiation pattern that radiates equally in all directions over various bands of frequencies [5]. A double band monopole antenna [6, 7] is such that it covers the lower WiMax frequency band but has a frequency tunable circuit to operate for the higher WiMax frequency band that comprises a biasing circuit for the varactor with an L-shaped stub along with resistance R1 and connected to the second branch using the resistor. The next node is connected to the resistor R2 and finally to the ground. An antenna operating for three bands 10.8, 2.8 and 5.8 GHz, which is typically in X band and it practically fits for applications like satellite, spacecraft and WLAN applications. It has an omnidirectional radiation pattern that improves the efficiency. Slots of H shape are created on the antenna patch that basically acts as radiators and is combined with the microstrip triple band antenna, where the slot dimensions are varied to get the best results [8]. The antenna design [9] has two back-to-back F-shaped slots within a rectangular-shaped patch. It mainly operates in the C band for satellite communication applications. This antenna has a C-shaped slot within a rectangular patch microstrip antenna that operates only in the S band of frequency with a higher return loss, bandwidth and gain [10]. A new fork-type low-profile non-uniform grounded antenna (NM-FGA) has been designed in Ref. [10]. For three bands, wireless local area network applications the impedance bandwidth and omnidirectional radiation characteristics is of crucial importance. A compact triple band printed antenna was designed in Ref. [10]. Because of its compact size and electromagnetic property, it operates for multimode terminal design in integrated wireless communication systems. To obtain three bands effectively, we need to tune the three strips that are circular in shape. A microstrip-fed wideband antenna is designed by a ring strip which is annular and encompassed by a rhombus-shaped strip. It has points of interest, such as small size, more gain and good radiation patterns, which meet the necessities of wireless local area networks and worldwide interoperability for microwave access applications. The antenna is a single-feed compact seen in Ref. [10] as a rectangular very small strip antenna for three band application; the C-shaped slot is etched within the patch of the antenna and the ground of the antenna has T- and I-shaped patches within the ground structure.

5.4 ANTENNA DESIGN CONSIDERATIONS OF PROPOSED WORK

5.4.1 Antenna Structure

The antenna has a rectangular-shaped ground plane of dimensions $46 \times 48 \times 0.162$ mm^3. The design comprises of three I-, two L-, one E- and one F-shaped patches. The width of feed line is 4 mm to match with an antenna that has an input impedance of a 50 Ω excitation port. FR4 substrate is used with a dielectric constant 4.4, thickness 0.162 mm and loss tangent 0.02. The values of the optimized dimensions are listed in Table 5.1.

5.4.2 Antenna Design in IE3D

The antenna was designed and simulated using IE3D software (Figure 5.1). Shape of the antenna comprises a minimized ground plane with three I, two L, one E and one F molded patches of metal framed on it. It has a size of the ground plane as $50 \times 48 \times 0.162$mm^3 which forms a rectangular patch. It is manufactured on FR4 substrate by considering 4.4 as an absolute permittivity because it has a decent cost and accessibility. A 50 Ω excitation port is coordinated to it. Different formed patches help to achieve multiband of frequencies. Each patch has a width of 1 mm (Figure 5.2).

5.5 RESULTS AND SIMULATION

5.5.1 Voltage Standing Wave Ratio

Voltage standing wave ratio (VSWR) is the measure of difference between a reception apparatus and the feed line associated with it. This is otherwise called the standing

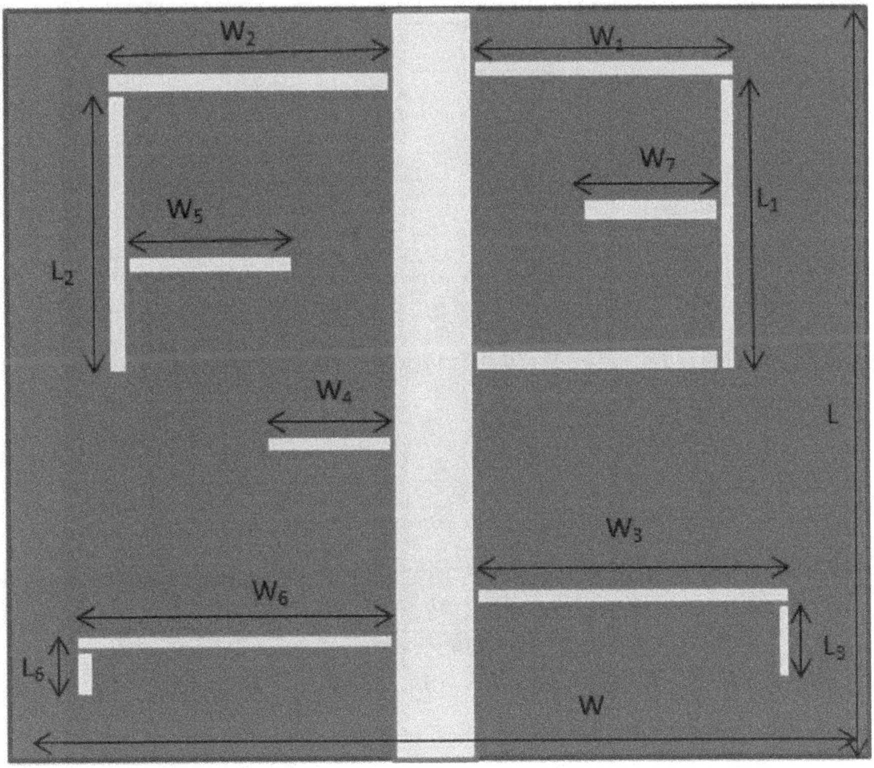

FIGURE 5.1 Antenna design.

TABLE 5.1
Patch sizes

Parameter	Dimension (mm)	Parameter	Dimension (mm)
L_1	8.5	W_5	9
W_1	12	L_6	5
L_2	18	W_6	15
W_2	16	W_7	8
L_3	6	L	46
W_3	17	W	48
W_4	8	t	0.162

wave ratio (SWR). A VSWR under 2 is viewed as reasonable for most antenna applications. The reception apparatus can be depicted as having a "Great Match". Thus, when it is said that the antenna is inadequately matched, frequently it implies that the VSWR is more than 2 for a frequency of interest.

As illustrated in Figure 5.3, when the antenna is resonating at a frequency of 3.1 GHz, the VSWR value obtained is 1.52; when the resonating frequency is 4.7 GHz,

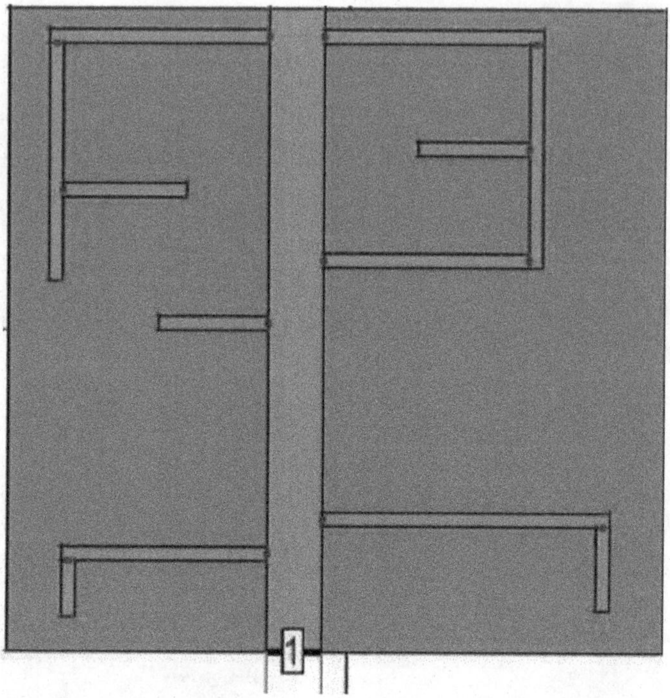

FIGURE 5.2 Antenna design in IE3D.

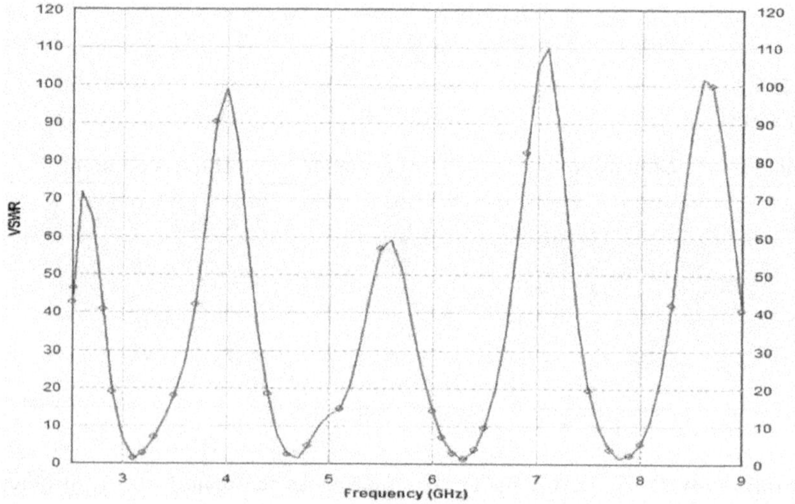

FIGURE 5.3 Voltage standing wave ratio (VSWR) curve.

we obtain 1.7 as VSWR value; at a frequency of 6.3 GHz, we get VSWR value as 1.4; and finally at a frequency of 7.8 GHz, we obtain VSWR value as 1.67.

5.5.2 S-Parameter/Return Loss

Return loss is a proportion of the viability of power transmitted from a transmission line to a load, for example, antenna. In the event that the power occurrence on the antenna under test is P_{in} and the power reflected back to the source is P_{ref}, the level of difference between the incident and reflected power in the traveling waves is given by the equation.

$$RL = 10\log_{10}\frac{P_{in}}{P_{ref}}$$

The return loss in dB is given in terms of reflection coefficient Γ

$$RL = 20\log|\Gamma|$$

As can be observed from Figure 5.4, at a frequency of 3.1 GHz we obtain return loss of −14 dB; when the frequency is 4.7 GHz, the return loss is −12.25 dB; at a frequency of 6.3 GHz, return loss value is −16.5 dB; and when the frequency is 7.8 GHz, it results in a return loss of −12.1 dB.

After going through the above observations, we can see that −16.5 dB is the minimum return loss at a frequency of 6.3 GHz.

5.5.3 Gain

The gain of an antenna is defined as the ratio of gain in the desired direction to the gain of the reference antenna (lossless isotropic source). When the direction of the

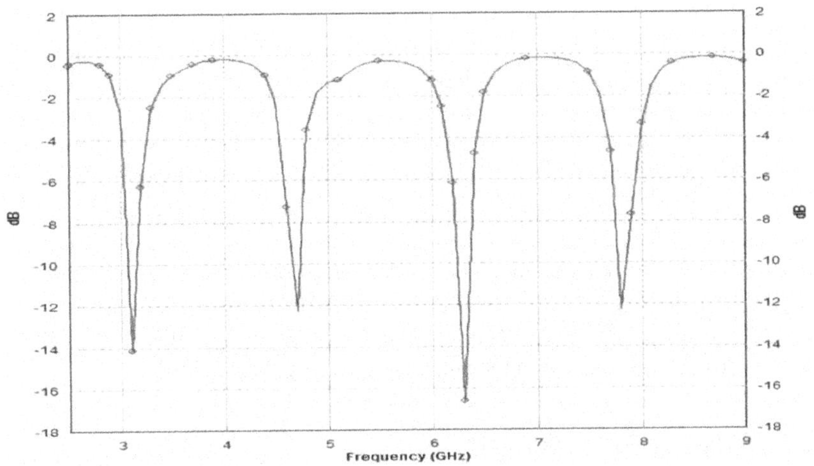

FIGURE 5.4 Return loss curve.

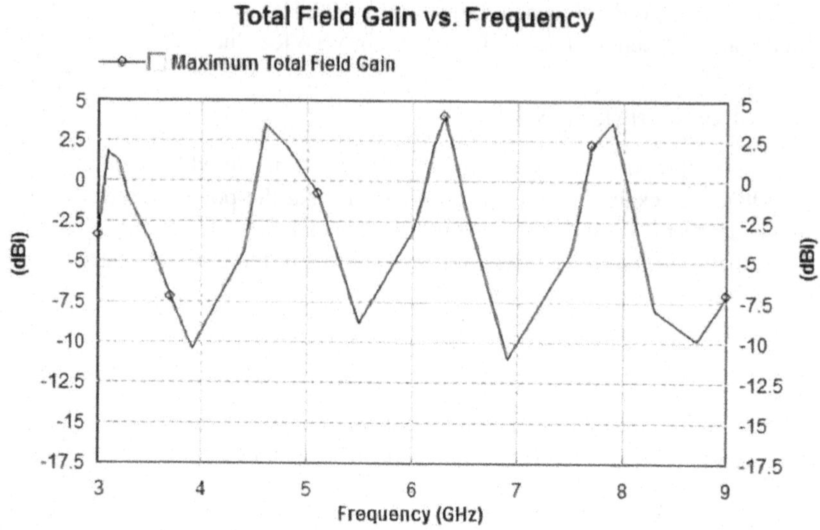

FIGURE 5.5 Gain curve.

antenna is not given, then we consider the direction in which maximum radiation occurs.

In Figure 5.5, we can see that the antenna has a gain which is more than 2 db. When the frequency is 3.1 GHz, antenna has a gain of 2.2 dBi. As we increase the frequency to 6.3 GHz, we obtain antenna gain as 3.7 dBi and finally for a frequency of 7.8 GHz, we obtain the antenna gain to be 3.4 dBi. After observing all the gain values of the antenna, we can conclude that the antenna has higher gain of 3.89 dBi at a frequency of 4.7 GHz.

5.5.4 Directivity

Directivity is the proportion of the convergence of antenna radiation pattern in a specific direction. The beam of an antenna will be more focused depending on the directivity of the antenna. A higher directivity additionally implies that the beam will travel further. A high directivity is not in every case better; for instance, numerous applications like cell phones require omnidirectional antennas and along these lines require antennas with low or no directivity. High directivity antennas are utilized in lasting establishments, for example, satellite TV, as they have to send and receive data over longer separations in a specific direction. The designed antenna at a frequency of 7.8 GHz has more directivity. When the frequency is 3.1 GHz, antenna has a directivity of 4.6 dBi. As we increase the frequency to 6.3 GHz, we obtain antenna directivity as 5.42 dBi and finally for a frequency of 7.8 GHz, we obtain the antenna directivity to be 5.7 dBi.

5.5.5 Radiation Efficiency

It is a parameter of antenna that describes accurately the performance of the antenna depending on how efficiently the transmission and reception of RF signals are done. Basically stated as the factor of radiated power by an antenna to the input power received, the designed antenna has more radiation efficiency at a frequency of 7.8 GHz. Radiation efficiency is varying in the range of 60–80%. When the frequency is 3.1 GHz, the antenna has a radiation efficiency of 68%. As we increase the frequency to 6.3 GHz, we obtain antenna radiation efficiency as 72% and finally for a frequency of 7.8 GHz, we obtain the antenna radiation efficiency to be 75%.

5.5.6 Antenna Efficiency

This gives the actual performance of the antenna. For an antenna to be ideal, it needs to have 100% efficiency. Theoretically, an antenna can have an efficiency of 100% but practically it is not possible; most antennas practically radiate only 50–60% of the total power provided to a particular antenna. The designed antenna has more efficiency at a frequency of 6.3 GHz. Efficiency is varying in the range of 45–70%. When the frequency is 3.1 GHz, antenna has an efficiency of 53%. As we increase the frequency to 6.3 GHz, we obtain antenna efficiency as 70% and finally for a frequency of 7.8 GHz, we obtain the antenna efficiency to be 62%.

5.5.7 Radiation Pattern

From Figures 5.6 and 5.7, we can analyze two-dimensional radiation patterns for resonating frequencies; major lobes will have a directional current which is between 60° to 30° and 120° to 150°.

Figure 5.8 shows surface current distribution that indicates the actual maximum current distribution in the antenna patch, which can be seen in the F patch of the antenna design with a max E-current of 15.018 A/m. The surface current distribution is seen at all the resonating frequencies. The current is confined to the corners of the patch. At higher frequencies, the current flows more in the curve paths than the straight ones; this is known as skin effect. Due to this, the current flows through the surface of the conductor instead of its central region. This is the reason why patch antennas are extremely useful in terms of size and performance.

5.6 CONCLUSION

A compact microstrip patch antenna was designed using different shapes of patches mainly three I-, two L-, one E- and one F-shaped patches. This antenna was implemented on a FR4 substrate. The different antenna parameters were analyzed, which mainly includes antenna gain, directivity, VSWR, reflection coefficient. This antenna had operated in three frequency bands which had resonated at frequencies 3.1, 7.8, 6.3 and 4.7 GHz. At a frequency of 4.7 GHz, WiMax and Wi-Fi operated with a gain of 3.9 dBi. At a frequency of 6.3 GHz, WLAN and Wi-Fi operated with a gain of 3.53 dBi and at a frequency of 7.7 GHz, satellite communication operated with a gain of 3.67 dBi. The antenna obtained maximum gain at a frequency of 4.7 GHz and antenna gain was more at 7.7 GHz. The VSWR was obtained to be within the range of 1 to 2, which is considered desirable. The results conclude that in future this antenna can be used for wide range of applications.

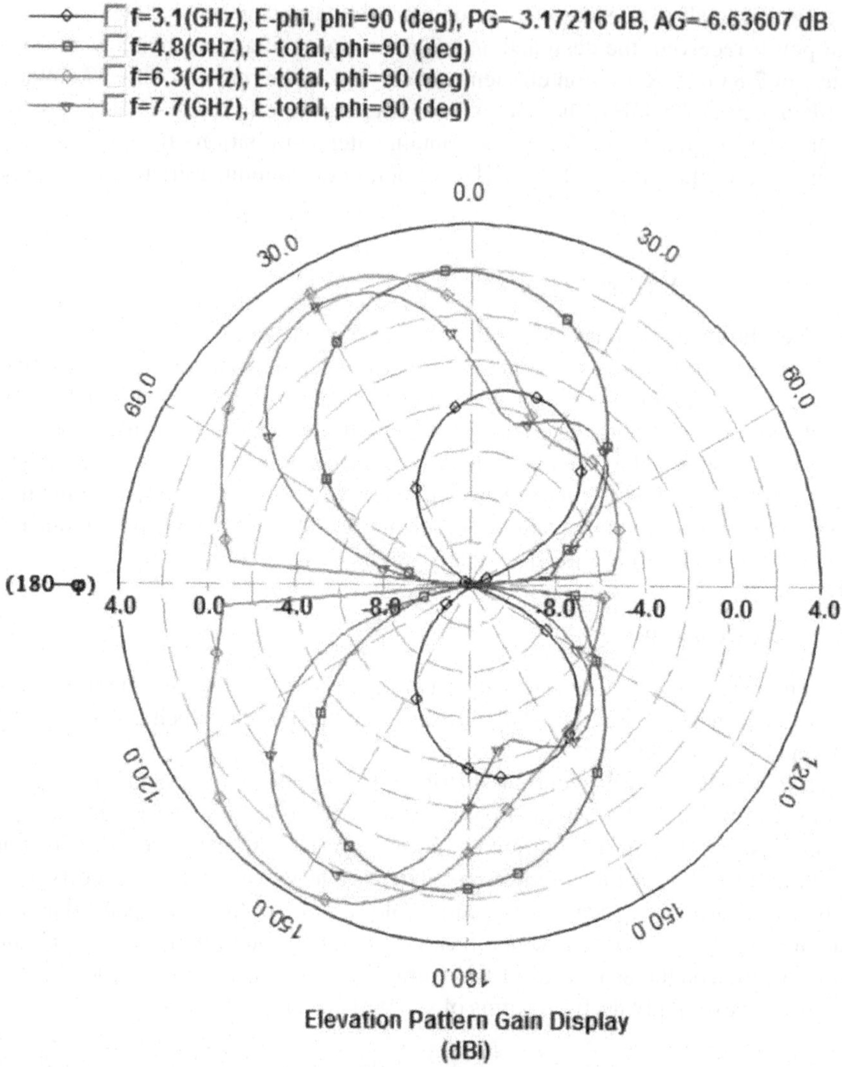

FIGURE 5.6 Two-dimensional radiation pattern.

REFERENCES

[1] M. I. Khan, M. A. Kabir, and A. Hira, "A simple multiband patch antenna for application in wireless communication", *2019 International Conference on Electrical, Computer and Communication Engineering (ECCE),* Cox's Bazar, 2019, pp. 615–618.

[2] A. Mahesh, K. S. Shushrutha, and R. K. Shukla, "Design of a multi-band antennas for wireless communication applications", *2017 IEEE Applied Electromagnetics Conference (AEMC),* Aurangabad, 2018, pp. 1–2.

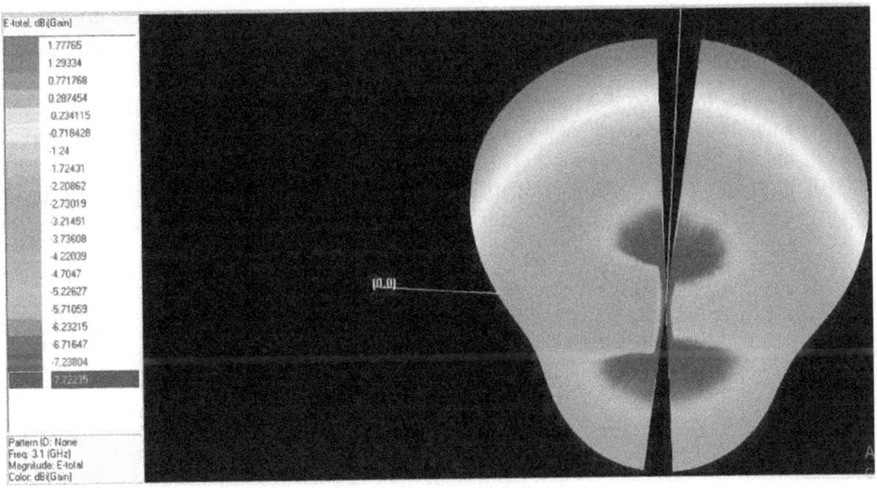

FIGURE 5.7 Three-dimensional radiation pattern for a resonating frequency of 3.1 GHz.

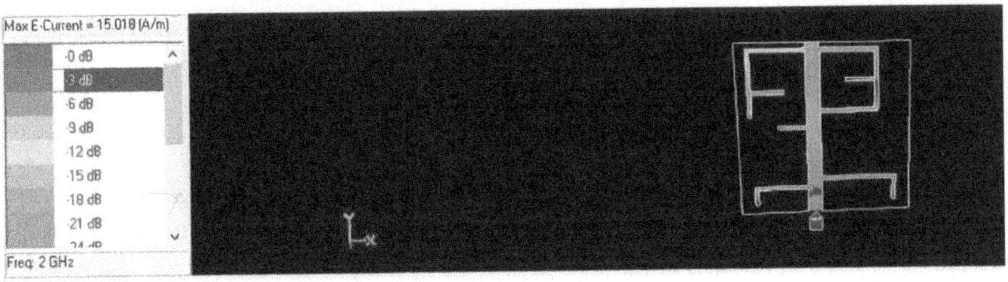

FIGURE 5.8 Distribution of surface current.

[3] A. Asrokin, M. K. A. Rahim, and M. Z. A. A. Aziz, "Dual band microstrip antenna at 2.4 GHz and 5.2 GHz for wireless LAN application," *2018 Asia-Pacific Conference on Applied Electromagnetics,* Johor, 2018, 4 pages.

[4] J. Pei, A. Wang, S. Gao, and W. Leng, "Miniaturized triple-band antenna with a defected ground plane for WLAN/WiMAX applications", *IEEE Antennas Wireless Propagation*, vol. 10, pp. 298–302, 2018.

[5] M. Moosazadeh and S. Kharkovsky, "Compact and small planar monopole antenna with symmetrical L-and U-shaped slots for WLAN/WiMAX applications", *IEEE Antennas Wireless Propagation Letters*, vol. 13, pp. 388–391, 2014.

[6] X. L. Sun, S. W. Cheung, and T. I. Yuk, "Dual-band monopole antenna with frequency tunable feature for WiMAX applications", *IEEE Antennas Wireless Propagation Letters*, vol. 12, pp. 100–103, 2013.

[7] M. Kendre, A. B. Nandgaonkar, P. Nirmal, and S. L. Nalbalwar, "U shaped multiband monopole antenna for spacecraft, WLAN and satellite communication application", *2016 IEEE International Conference on Recent Trends in Electronics, Information Communication Technology (RTEICT)*, Bangalore, 2016, pp. 1538–1532.

[8] A. Kumar, "A compact H-shaped slot triple-band microstrip antenna for WLAN and WiMax applications", *2016 IEEE Annual India Conference (IN-DICON)*, Bangalore, 2016, pp. 1–4.
[9] K. Upadhyay and S. Vijay, "Design simulation of back to back F shape slotted rectangular microstrip patch antenna for satellite communication", *IJARI*, vol. 3, Issue 1, pp. 64–67, 2015.
[10] A. Singh, A. Arya, and S. Sharma, "High gain of C shape slotted microstrip patch antenna for wireless applications", *2019 ISSN International Journal of Applied Engineering Research*, 2019, vol. 7, pp. 201–209, Issue 11.

6 Design of Elliptically Etched Circular and Elliptical Printed Antennas for Dual-Band 5G Mobile Applications

Aqeel Shaik, V. A. Sankar Ponnapalli and Babu P. Vinod

6.1 INTRODUCTION

The telecommunication and mobile industry have been rapidly evolving since the last few decades. For example, various technologies such as global positioning system for mobile communications, wideband code division multiple access, high speed packet access and long-term evolution have found rapid growth. The mobile industry is running towards the advanced fifth generation (5G) and beyond, that is, machine-type communication. These advanced mobile technologies are supposed to congregate the requirements of lower latency, high versatility, large reliability, huge broadcasting data, simply manageable with earlier generations and more connectivity [1–4]. Rapid beamforming is also a necessary 5G parameter that can be achieved with sophisticated printed antennas, and with advanced array antenna systems. Recently various array antenna models were developed to achieve advanced wireless goals. Among these models, fractal arrays are one of the best solutions. Fractal arrays, self-similar array antennas and various design methodologies such as concentric circular and elliptical sub-array generators are discussed in the literature [5–7]. Different printed, fractal-shaped antennas and elliptical-shaped slotted printed antennas have evolved and are discussed in this chapter.

DOI: 10.1201/9781003093558-8

The broadband elliptical monopole microstrip antenna has been designed (see [8, 9]) for the ultra wideband (UWB) and 5G applications. These antennas have been designed for various frequency ranges from 3 to 14 GHz and from 20 to 40 GHz, respectively. Various UWB antennas are discussed in the literature but elliptical related models are only considered in this chapter. A simple printed antenna with square patch and rectangular slots were designed (see [10]) for 5G applications at 28 GHz and with a bandwidth of 716 MHz. A circular patch antenna with multiline slots were developed for radar applications with dual-band frequency ranges from 5.24 to 6.16 GHz and from 7.9 to 11.7 GHz. This antenna has achieved above 6 dB gain for both the frequency ranges [11]. Fractal printed and arrays are also playing a vital role in advanced wireless communication systems such as vehicular, satellites and 5G applications due to their space-filling and multiband capability [12]. A square patch designed with multiple squares and rectangular slots is discussed in Ref. [13] for vehicular communications, and better radiation pattern properties have been achieved in this contribution. Owing to the multiple and repetitive slots, these antennas have achieved multiband behavior and this can happen due to the change in current lengths. In Ref. [14], elliptically etched patch with sectorial disk proposed for dual-band 5G applications and a wideband width of 1.5 GHz are also achieved. The substrate materials do show significant effect on the performance of printed antenna. Basically, FR4 is the most commonly used material, easily available and the cost is also less but the performance of the antenna is moderate. Various other materials are also available to achieve better antenna patterns like RT duroid, Roger and so on. In this research work, Neltec NH9320 with dielectric constant of 3.2 is used for better antenna efficiency [15, 16]. This chapter presents an elliptically etched circular and elliptical microstrip patch antennas for dual-band 5G applications. Three design models are presented in the Section 6.2. Related results of the presented work are discussed in Section 6.3. Finally, conclusions and future scope are drawn in Section 6.4.

6.2 DESIGN MODELS OF THE ELLIPTICALLY ETCHED CIRCULAR AND ELLIPTICAL PRINTED ANTENNAS

The proposed printed antenna was executed at three different levels. In the first level, the basic circular patch antenna was designed. In the second level, the elliptically etched circular printed antenna was designed. Finally, the elliptically etched elliptical printed antenna with a circular slot at the center was designed and the resultant comparison was presented. To study the effect of each element of the patch antenna, a step-by-step construction was followed for better compatible structure and radiation performance. A simple circular patch antenna with a center conductor with a radius of 1.5 mm was constructed on an Neltec NH9320 substrate of thickness 0.762 mm and 10 mm × 10 mm length and width was considered, and this patch was fed with a 50 ohm microstrip line of length 4 mm and width 1 mm designed in the first level and depicted in Figure 6.1.

The design equations of the basic circular patch are presented in Equations 6.1 and 6.2. In the second level, an elliptical slot of major axis of 4.64 mm and minor axis of

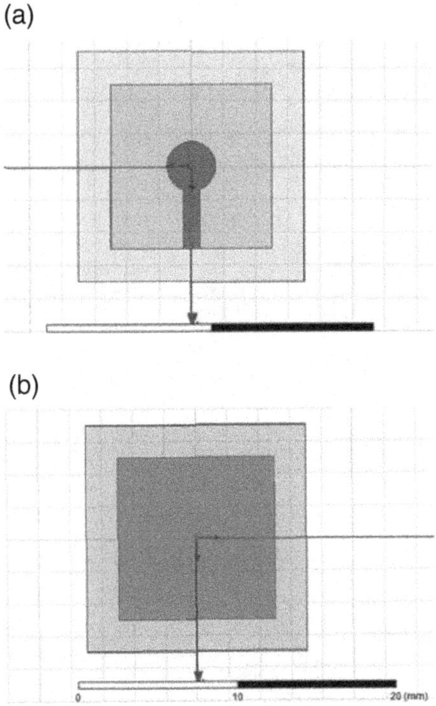

FIGURE 6.1 (a) Basic circular patch top view. (b) Back view of the antenna.

2.076 mm is incorporated into the basic design (Figure 6.1), with the same center and this patch had been fed with a 50 ohm microstrip of length 4 mm and width 1 mm and the resultant printed antenna is shown in Figure 6.2.

$$a = \frac{F}{\left\{1 + \frac{2h}{\pi \epsilon_r F\left[\ln\left(\frac{\pi F}{2h}\right) + 1.7726\right]}\right\}^{1/2}} \quad (6.1)$$

$$F = \frac{8.791 \times 10^9}{f_r \sqrt{\epsilon_r}} \quad (6.2)$$

where a is the radius of the patch, f_r is the operating frequency, h is the height of the substrate and ϵ_r is the dielectric constant. Finally, an elliptically etched elliptical patch with a major axis of 2.32 mm and minor axis of 1.038 mm with a center conductor having a slot radius of 0.5 mm has been constructed on a Neltec NH9320 substrate with a thickness of 0.762 mm, permittivity 3.2, length and width 10 mm × 10 mm.

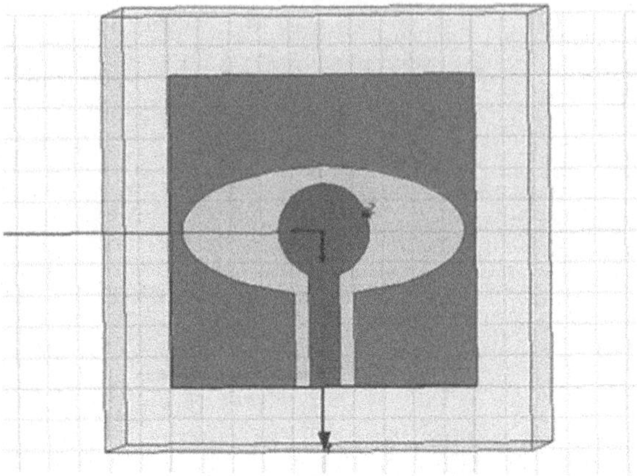

FIGURE 6.2 Elliptically etched circular patch top view.

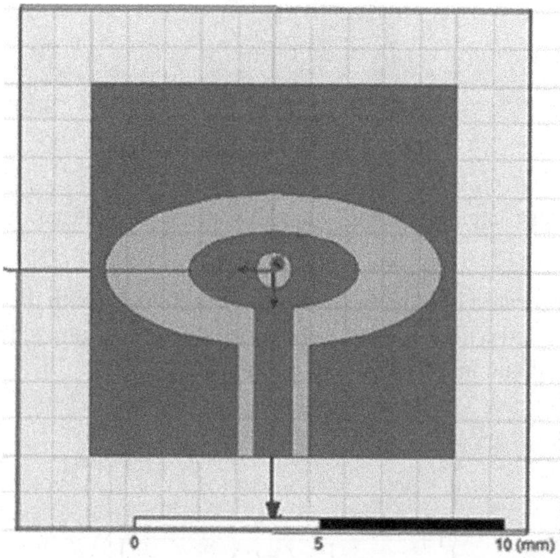

FIGURE 6.3 Top view of elliptically etched elliptical patch with circular slot.

This elliptically etched design has a major axis of 4.64 mm and minor axis of 2.076 mm and this patch is fed with a 50 ohm microstrip with a length of 4 mm and a width of 1 mm as shown in Figure 6.3. With a small slot in the elliptical patch, the return loss is further improved and is exemplified in Chapter 7. Thus, instead of plane surface, a small slot is introduced on the patch. Owing to the concept of systematic etching, the size of the antenna is reduced significantly. Because of this, the presented antenna is more suitable for advanced mobile applications and also for automatic vehicular systems.

6.3 RESULTS AND DISCUSSION OF THE PRESENTED ANTENNAS

The simulated results of basic circular patch, second and third level of presented antennas are depicted in Figures 6.4, 6.5 and 6.6, respectively. The resonant frequencies of the basic circular patch are 26.5 and 36.7 GHz with return loss of −11 and −10 dB, respectively. But for the practical scenarios, the return loss achieved in this case is not enough due to the external factors. Owing to this reason, the antenna is taken into the second level as shown in Figure 6.2. For the elliptically etched circular patch, the resonance frequencies are observed at 28.2 and 38.4 GHz with return loss of −10 and −23.6 dB, respectively, with band of frequencies under resonance. In this case, a wideband width of 3 GHz has achieved a second resonating frequency. Considering practical scenario, the return loss at first resonance frequency is not sufficient due to the external factors and therefore the antenna is further modified as shown in Figure 6.3. For the elliptically etched elliptical patch, the resonance frequencies are observed at 27.9 and 36.7 GHz with return loss of −31 and −29.6 dB, respectively, with band of frequencies under resonance. A high return loss values of −31 and −30 dB with a gain of 7.1 has been observed as shown in Figure 6.6. In this case, notable bandwidths of 1 and 1.2 GHz have been achieved at resonating

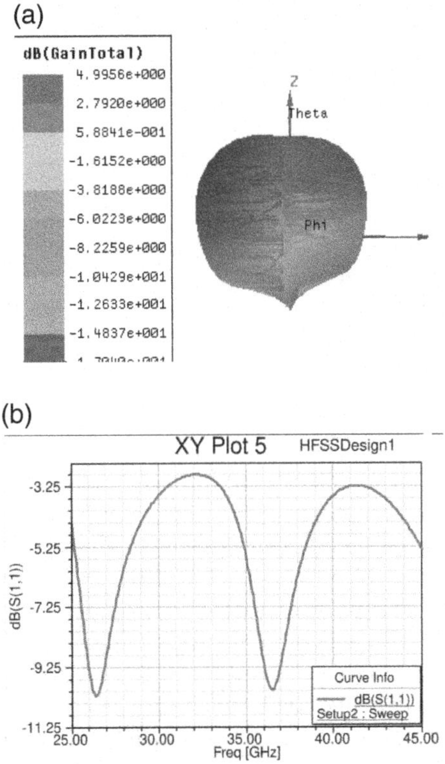

FIGURE 6.4 (a) Radiation plot. (b) Return loss plot of basic circular patch antenna.

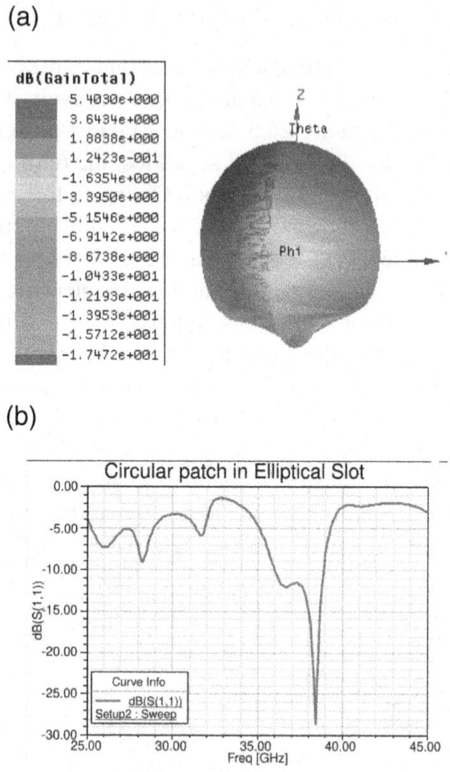

FIGURE 6.5 (a) Radiation plot. (b) Return loss plot of elliptically etched circular patch.

frequencies one and two, respectively. Owing to notable radiation properties observed in the third level design, this antenna is most suitable for the 5G and other advanced wireless applications. Radiation property values of the three antennas are tabulated in Table 6.1.

6.4 CONCLUSIONS

Dual-band elliptically etched circular and elliptical printed antennas for 5G mobile applications were discussed in this chapter. There is an incremental gain observed in the three cases considered in the work. A low return loss value of −10 dB has been achieved in the basic circular patch of resonating frequency 26.5 and 36.5 GHz. A high level return loss of −31 dB has been achieved at 27.9 GHz for third level of antenna design. Design results show center frequency around 28 and 38 GHz, with notable gains, radiation patterns and high bandwidths. The design illustrates a direction beam that makes it a good candidate for 5G and other high frequency applications. A further aim is to build an array with this element to improve the directivity for the application of these frequency ranges.

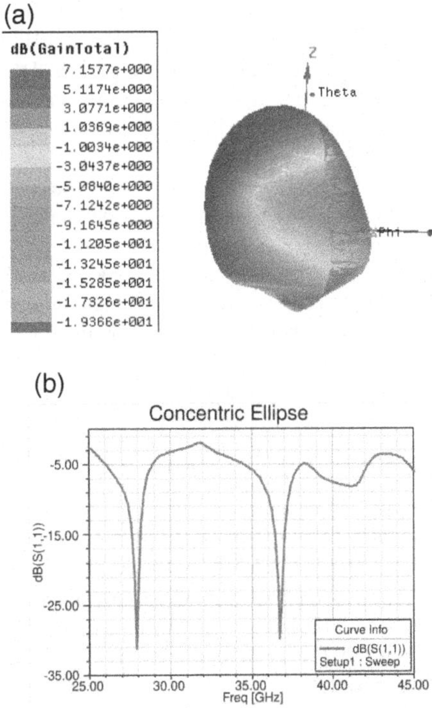

FIGURE 6.6 (a) Radiation plot. (b) Return loss plot of elliptically etched elliptical patch.

TABLE 6.1
Gain and return loss of presented antennas

Antenna type	Return loss (resonance freq. 1)	Return loss (resonance freq. 2)	Gain at resonance	Bandwidth (resonance freq. 1)	Bandwidth (resonance freq. 2)
Basic circular patch	−10 dB (26.5 GHz)	−10 dB (36.5 GHz)	5.00 dB	-	-
Elliptically etched circular patch	−10 dB (28.1 GHz)	−24 dB (38.5 GHz)	5.40 dB	-	3 GHz
Elliptically etched elliptical patch with circular slot	−31 dB (27.9 GHz)	−30 dB (36.7 GHz)		1 GHz	1.2 GHz

REFERENCES

[1] N. T. Le, M. Arif Hossain, A. Islam, D.-Y. Kim, Y.-J. Choi, and Y. M. Jang, "Survey of promising technologies for 5G networks," *Mobile Information Systems*, Vol. 2016, pp. 1–25, 2016. doi: 10.1155/2016/2676589.

[2] Z. Ying, "Antennas in cellular phones for mobile communications," *Proceedings of the IEEE*, Vol. 100, No. 7, pp. 2286–2296, 2012. doi:10.1109/JPROC.2012.2186214.

[3] C.-X. Wang, F. Haider, X. Gao, X.-H. You, Yang Yang, D. Yuan, H. M. Aggoune, H. Haas, S. Fletcher, and E. Hepsaydir, "Cellular architecture and key technologies for 5G wireless communication networks," *IEEE Communications Magazine*, Vol. 52, No. 2, pp. 122–130, 2014.

[4] Y. Wu, S. Zhang, Z. Liu, X. Liu, and J. Li, "An efficient resource allocation for massive MTC in NOMA-OFDMA based cellular networks," *Electronics*, Vol. 9, pp. 1–27, 2020. doi: 10.3390/electronics9050705.

[5] W. Roh, J.-Y. Seol, J. Park, B. Lee, J. Lee, Y. Kim, J. Cho, K. Cheun, and B. Aryanfar, "Millimeterwave beamforming as an enabling technology for 5G cellular communications: theoretical feasibility and prototype results," *Communications Magazine, IEEE*, Vol. 52, No. 2, pp. 106–113, 2014.

[6] V. A. Sankar Ponnapalli and P. V. Y. Jayasree, "Thinning of Sierpinski fractal array antennas using bounded binary fractal-tapering techniques for space and advanced wireless applications," *ICT Express*, Vol. 5, No. 1, pp. 8–11, 2019. doi: 10.1016/j.icte.2017.12.006.

[7] A. Praveena and V. A. S. Ponnapalli, "A review on design aspects of fractal antenna arrays." *2019 International Conference on Computer Communication and Informatics (ICCCI)*, 2019. doi: 10.1109/ICCCI.2019.8821860.

[8] A. Elboushi, O. M. Ahmed, A. R. Sebak, and T. Denidni, "Study of elliptical slot UWB antennas with A 5.0–6.0 GHz band-notch capability." *Progress in Electromagnetics Research C*, Vol. 16, pp. 207–222, 2010. doi:10.2528/PIERC10080905.

[9] O. M. Haraz, "Broadband and 28/38-GHz dual-band printed monopole/elliptical slot ring antennas for the future 5G cellular communications." *Journal of Infrared, Millimeter, and Terahertz Waves*, Vol. 37, No. 4, pp. 308–317, 2016. doi: 10.1007/s10762-016-0252-2.

[10] G. Muhammet Tahir and S. Cihat, "Compact microstrip antenna design for 5G communication in millimeter wave at 28 GHz." *Erzincan University Journal of Science and Technology*, Vol. 12, No. 2, pp. 679–686, 2019. doi: 10.18185/erzifbed.477293.

[11] S. Ghnimi, R. Wali, A. Gharsallh, and T. Razban, "A new design of an S/X dual band circular slot antenna for radar applications." *Journal of Microwave Power and Electromagnetic Energy*, Vol. 47, No. 2, pp. 138–146, 2013. doi: 10.1080/08327823.2013.11689853.

[12] A. Rahim, P. K. Malik, and V. A. Sankar Ponnapalli, "State of the art: A review on vehicular communications, Impact of 5G, Fractal antennas for future communication," *Proceedings of First International Conference on Computing, Communications, and Cyber-Security (IC4S 2019), Lecture Notes in Networks and Systems*, Vol. 121, pp. 3–15, 2020. doi: 10.1007/978-981-15-3369-3_1.

[13] A. Rahim, P. Kumar Mallik, and V. A. Sankar Ponnapalli, "Fractal antenna design for overtaking on highways in 5G vehicular communication ad-hoc networks environment," *International Journal of Engineering and Advanced Technology*, Vol. 9, No. 1S6, pp. 157–160, 2019. doi: 10.35940/ijeat.A1031.1291S619.

[14] N. Kathuria and S. Vashisht, "Dual-band printed slot antenna for the 5G wireless communication network," *2016 International Conference on Wireless Communications, Signal Processing and Networking (WiSPNET)*, 2016. doi: 10.1109/WiSPNET.2016.7566453.

[15] S. D. Mahamine, R. S. Parbat, S. H. Bodake, and M. P. Aher, "Effects of different substrates on rectangular microstrip patch antenna for S-band," *2016 International*

Conference on Automatic Control and Dynamic Optimization Techniques (ICACDOT), 2016. doi: 10.1109/icacdot.2016.7877765.

[16] A. A. Roy, J. M. Mom, and D. T. Kureve, "Effect of dielectric constant on the design of rectangular microstrip antenna," *2013 IEEE International Conference on Emerging & Sustainable Technologies for Power & ICT in a Developing Society (NIGERCON)*, 2013. doi:10.1109/nigercon.2013.6715645.

7 Design and Analyses of Dual-Band Microstrip Patch Antennas for Wireless Communications

D. Singh, A. Thakur, R. Kumar and Geetanjali

7.1 INTRODUCTION

This research work presents dual-band rectangular microstrip patch antennas resonating at X, Ku, and K-bands for wireless satellite communications. Three antenna structures with different configuration techniques are positioned on the surface of the patch mounted upon a grounded substrate made of Rogers RT Duroid 5880. The radiation mechanism of the suggested antenna design is studied, compared, and analyzed with its contemporary structures. Their radiation performance is evaluated in terms of reflection coefficient, −10 dB bandwidth, VSWR, input impedance, radiation efficiency, gain and found satisfactory. These structures find their application in wireless satellite communications. Over the years with the growing advancements in modern

DOI: 10.1201/9781003093558-9

technology, the mode of wireless communication is being utilized in day-to-day activities. It is hard to imagine connectivity without wireless communication, and antennas are the gateway for it. The geometrical parameters and the type of antenna might be different and vary according to their application areas; nevertheless, it is the unique mode of communication for every wireless system. The most popular and coherent structure is the rectangular microstrip patch antenna (MPA). It typically offers distinct advantages over other antennas in term of its planar size, cost-effectiveness, durability, and compactness. Moreover, it is conformal to planar or curved structures with simple fabrication process and potential to work on high-frequency applications [1–5]. They possess many advantages; however, narrow bandwidth, low gain and efficiency, spurious surface wave generation within sufficient end-fire radiations are some of the limitations associated with them [6–11].

In today's communication between many devices like mobiles and laptops, there is a requirement for very thin, dual-band and smart antennas, which can support several applications and higher bandwidth. This can only be achievable by typically using MPA rather than conventional antennas. There is a demand for antennas that can transmit or receive more than one frequency. In some specific applications, dual-band or multiband characteristics are most desirable. Depending on the signal strength and prime location, a dual-band antenna can utilize both selected bands simultaneously or flip between the two bands. The benefits of using a dual-band antenna are its capability to provide superior mobility, and strong and stable wireless connection in some non-reachable locations. To brace the increasing requirement for modern wireless communication, several double-band patch antennas have been suggested [12–15]. Dual-band antenna finds its applications in cellular communication, 5G, LTE, WLAN, Bluetooth, Wi-Fi, and so on, and operates efficiently on mobile communication bands [16–19]. In satellite and radar communications, dual-band or multiband characteristics are most desirable and hence dual-band antenna configurations are typically required at C, Ku, K, and Ka-bands, respectively [20–22]. This can be achieved precisely using various techniques. The most practical technique for achieving multiband or dual-band characteristics is by creating a slot or slit in an antenna. A coupled-slot technique with matching feed position modifies the dominating modes of a patch and perturbs the resonance at other frequencies.

In this proposed work, dual-band antennas are deigned in X, Ku, and K-bands of the frequency spectrum, which are suitable and find their application in fixed satellite services (FSS) and direct broadcast services (DBS). These services are defined by ITU, which is a specialized international agency responsible for matters related to telecommunications. Three different dual-band MPA structures are proposed, designed, and analyzed. The layout of the research work is as follows: Section 7.2 typically discusses modern design and functional analyses of a compact dual-band patch antenna with double slots (DBDS). In Section 7.3, a dual-band antenna with a square ring (DBSR) will be uniquely designed and accurately described. The structural description of a dual-band patch antenna with a possible pair of bend slots (DBBS) will be elaborated in Section 7.4. The radiation performance analyses of all structures have been evaluated in terms of scattering parameters (S_{11}), parametric sweep, input impedance (Z_{11}), voltage standing wave ratio (VSWR), and efficiency (η). The directive gain and realized gain have also been analyzed for dominating electric and magnetic fields in horizontal and vertical planes. The performances of all structures are then compared in Section 7.5. Finally, Section 7.6 concludes the work and provides future recommendations.

7.2 DESIGN OF A DUAL-BAND PATCH ANTENNA WITH DOUBLE SLOTS (DBDS)

In extensive literature, the seven different slot and slit techniques are typically implemented to obtain dual-band behavior for the patch antenna [23–25]. Therefore, in this section we are designing and analyzing the behavior of a dual-band antenna using double slot (DBDS) technique typically operating at 12 and 16.1 GHz. The specific layout of the DBDS patch antenna is illustrated in Figure 7.1.

The front view of a DBDS is provided in Figure 7.1a and the side view with a coaxial feed is shown in Figure 7.1b. For a thin substrate, the coaxial feeding technique has been undoubtedly preferred because of low spurious radiation and considerable ease of impedance matching. In the presented DBDS patch antenna, the radiating component is uniquely positioned at the top of a copper ground plane of dimensions 14.4 mm × 14.4 mm. The structure typically has a considerable length and width of 7.2 mm printed directly on Rogers RTD 5880 Duroid substrate. It has a standard thickness of 1.588 mm. The considerable value of dielectric constant and loss tangent is 2.2 and 0.0009, respectively. RTD Duroid is a high-frequency laminate suitable for carefully extracting stripline and microstrip circuit applications. The substrate is sandwiched between the radiating patch and the conducting ground surface. The parametric details of DBDS antennas are presented in Table 7.1.

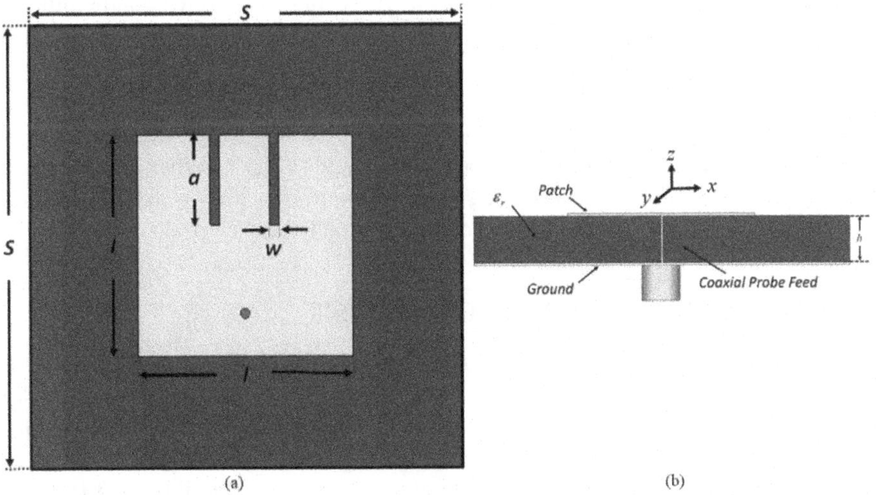

FIGURE 7.1 Dual-band patch antenna with double slots (DBDS): (a) front view and (b) side view (structure 1).

TABLE 7.1
Parameters of the proposed dual-band patch antenna with double slots (DBDS)

Description	Parameters	Dimensions (mm)
Substrate	S	14.4
Length/width of patch	l	7.2
Substrate height	h	1.588
Patch thickness	t	0.035
Slots width	w	0.3
Length of slots	a	3
Spacing between slots	a_1	1

Two slots of equal dimensions are cut down from the top of the patch to achieve dual-band frequency response. The scheming and continuous optimization of the intended geometrical model of the proposed dual-band structure is obtained using parametric analysis. Desired frequency response can only be generated by proper selection and successful extraction of the geometrical parameters, such as considerable length, width, size, and feeding position. Changing antenna design changes its radiation pattern and frequency response. The dual resonance frequency response curve for DBDS (structure 1) is pictured in Figure 7.2(a). A peak magnitude of −41.09 dB is obtained at 12 GHz for the first resonant frequency, whereas the second resonance peak is obtained at 16.1 GHz with the democratic value of reflection coefficient (S11) being −36.49 dB. The value for −10 dB bandwidth of DBDS for first resonant frequency is 540 MHz, which in common is 4.5% of the central resonant frequency. However, the value for −10 dB bandwidth for the second resonant frequency is 96 MHz and its impedance bandwidth is typically just 0.59%.

A parametric sweep is performed to obtain desired resonant peaks by changing the length of slots (a) [Figure 7.2(b)]. It has been observed that with the change in the length of slots, the first resonant frequency remains almost unchanged; however, it profoundly influences and shifts the second resonant frequency peak. The spacing between two slots (a_1) is optimized to achieve desired frequency peaks.

7.3 RADIATION PERFORMANCE ANALYSIS

The radiation performance analyses for VSWR, radiation efficiency (η), and input impedance (Z_{11}) for structure 1 are presented in Figure 7.3. The typical value of the standing wave ratio for DBDS patch antenna operating efficiently at 12 GHz resonance frequency is precisely 1.02. However, it is detected to be 1.06 at 16.01 GHz resonant frequency. This lies within a reasonable and acceptable limit of 1.0 to 2.0 [Figure 7.3(a)]. The input impedance (Z_{11}) simulated response for antenna resonating efficiently at 12 GHz frequency peak is 93.22 Ω; however, for second resonant peak operating at 16.01 GHz, the specific value of Z_{11} is 96.26 Ω [Figure 7.3(b)]. Similarly,

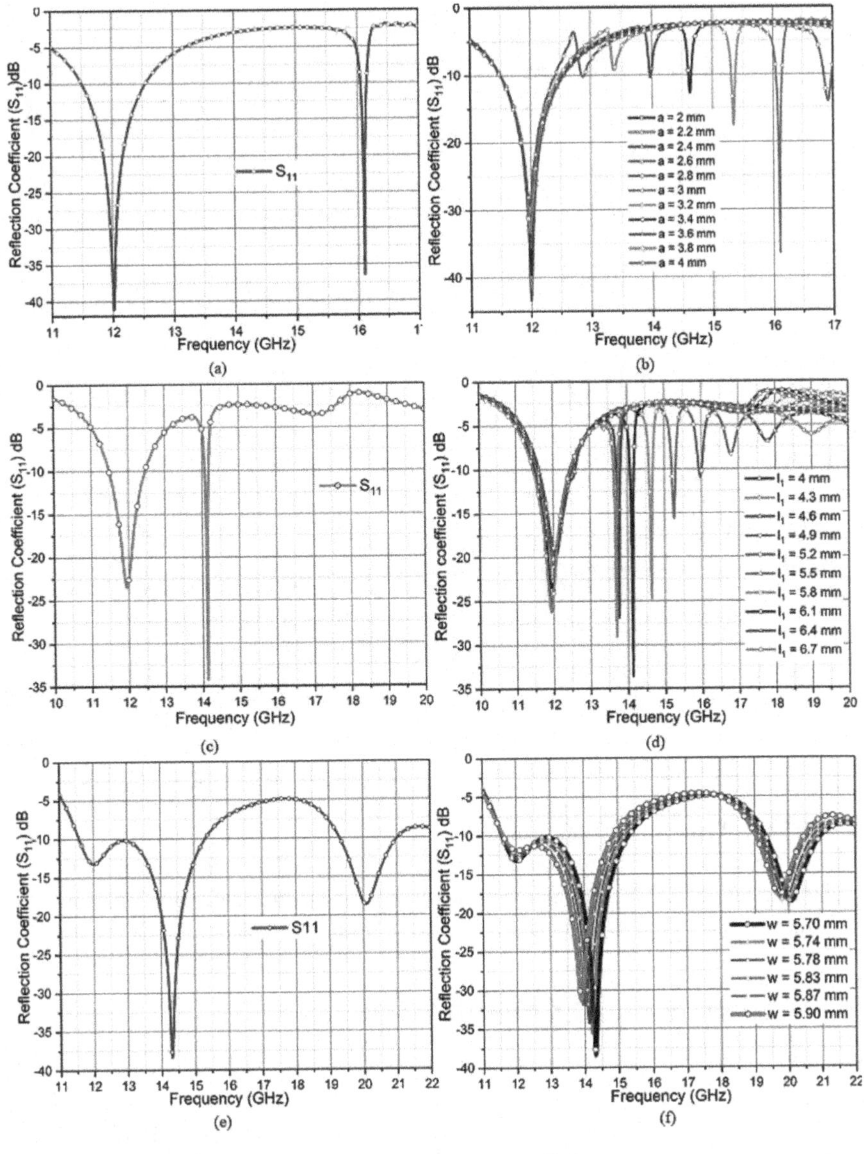

FIGURE 7.2 Dual resonance frequency response curve (S_{11}) at:
Structure 1: (a) 12 GHz and 16.1 GHz (b) with parametric sweep
Structure 2: (c) 11.95 GHz and 14.13 GHz (d) with parametric sweep
Structure 3: (e) 14.31 GHz and 20.07 GHz, and (f) with parametric sweep

in order to determine the value for radiation efficiency, a simulation analysis has been made. It has been observed that for first peak frequency resonating at 12 GHz, the value of η is 98.75%. Whereas, for second resonant frequency operating at 16.1 GHz, a slight decline in η is observed and it corresponds to 73.01% [Figure 7.3(c)].

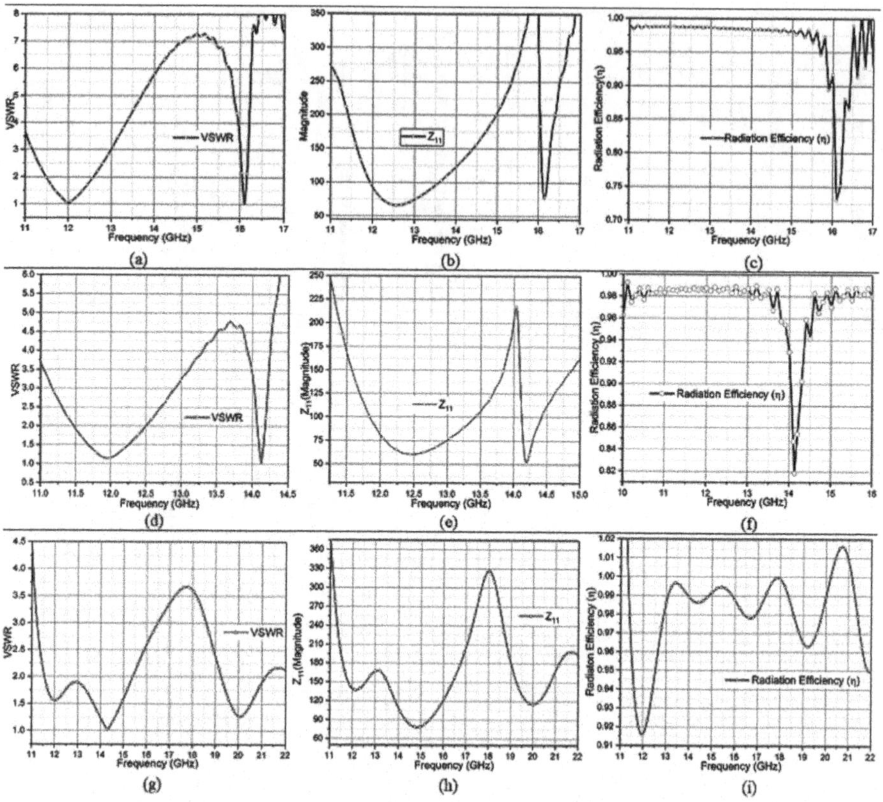

FIGURE 7.3 Radiation performances analyses for:
Structure 1: (a) VSWR, (b) Input Impedance (Z_{11}), and (c) Radiation Efficiency (η)
Structure 2: (d) VSWR, (e) Input Impedance (Z_{11}), and (f) Radiation Efficiency (η)
Structure 3: (g) VSWR, (h) Input Impedance (Z_{11}), and (i) Radiation Efficiency (η)

7.4 DIRECTIVITY AND GAIN ANALYSIS

Simulation analyses for the directivity and gain is done for DBDS antenna (structure 1) for electric and magnetic reference planes (Figure 7.4). At $\phi = 0°$, the directive gain magnitude of main beam is 7.36 dBi at 12 GHz and 6.3 dBi at 16.1 GHz of the resonant frequency; whereas, at $\phi = 90°$, directive gain magnitude of main beam is 7.36 dBi at 12 GHz and 3.35 dBi at 16.1 GHz of resonant frequency, respectively [Figure 7.4(a)]. At $\phi = 0°$, the realized gain magnitude of main beam is 7.29 dBi at 12 GHz and 4.93 dBi at 16.1 GHz of the resonant frequency; whereas, at $\phi = 90°$, realized gain magnitude of main beam is 7.29 dBi at 12 GHz and 1.98 dBi at 16.1 GHz of resonant frequency, respectively [Figure 7.4(b)].

7.5 SURFACE CURRENT ANALYSIS OF ANTENNA

The excited surface current analyses of the DBDS antenna (structure 1) at the two resonant frequencies, namely 12 and 16.1 GHz, are presented in Figure 7.5. Arrows

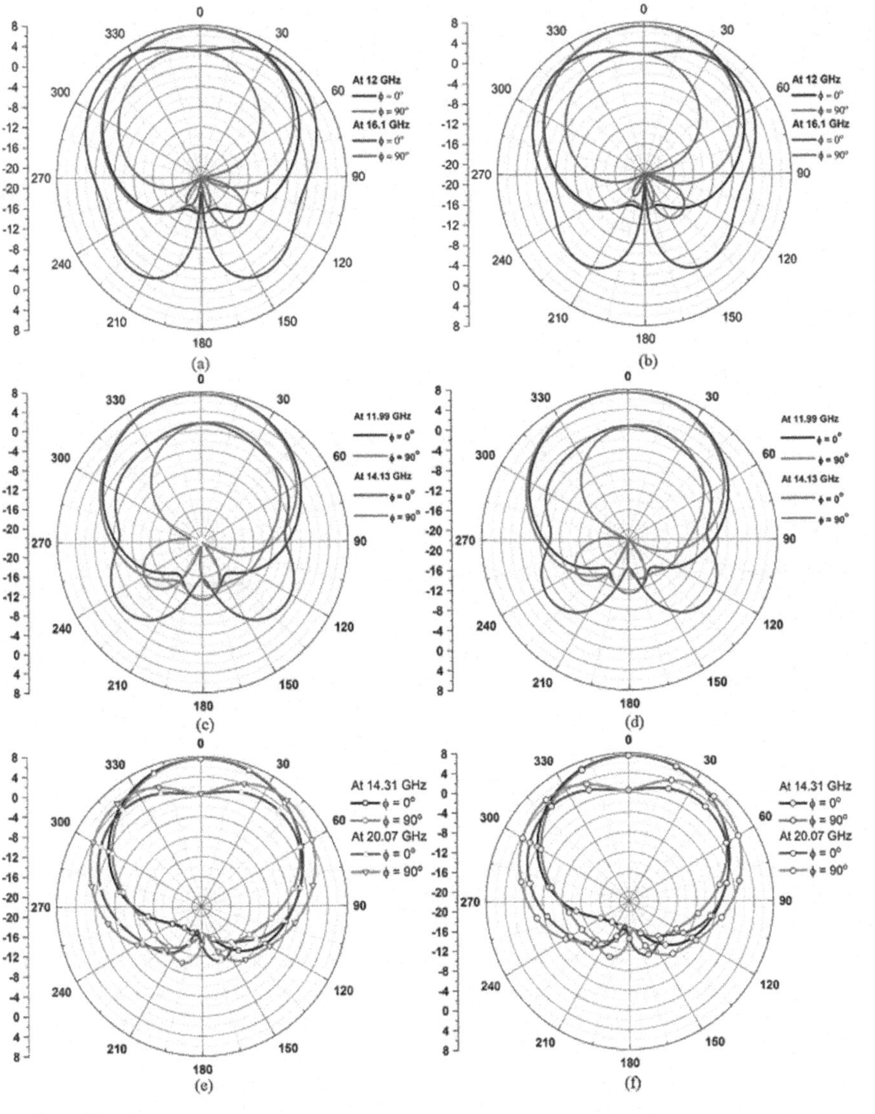

FIGURE 7.4 Analysis of simulated radiation pattern for E-plane and H-plane:
Structure 1: (a) Directivity and (b) Gain
Structure 2: (c) Directivity and (d) Gain
Structure 3: (e) Directivity and (f) Gain

show the current direction and the darker color shows the current intensity on that area of the antenna, while the dark color region shows the null current at that portion. As anticipated, two major current paths have been detected that are the prime causes of antennas having two resonant modes. These two adjoining resonances add together to form the dual-band resonance structure. At f = 12 GHz, the current around the corner of the structure is stronger than the rest of the area. This current is responsible

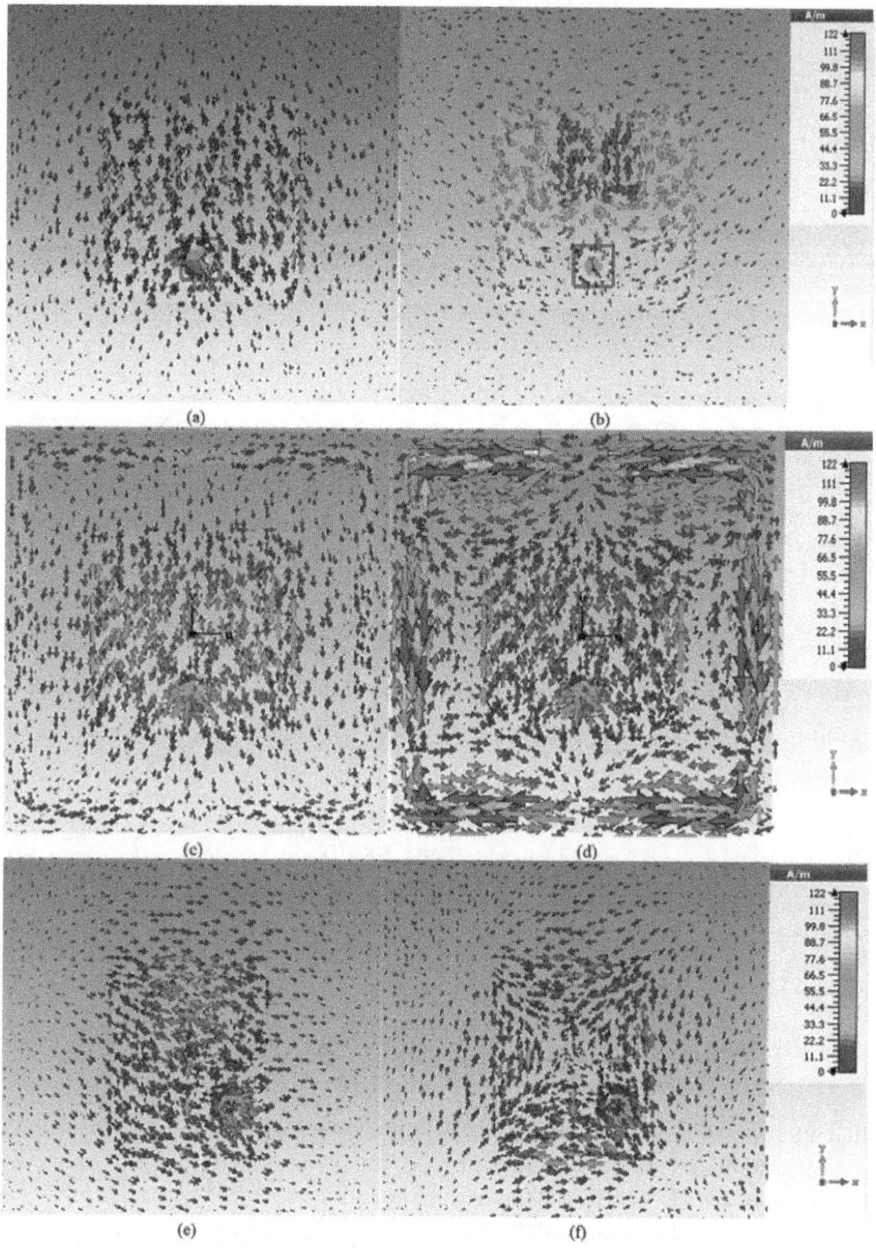

FIGURE 7.5 Distribution of current across the proposed design at:
Structure 1: (a) 12 GHz and (b) 16.1 GHz
Structure 2: (c) 11.95 GHz and (d) 14.13 GHz
Structure 3: (e) 14.31 GHz and (f) 20.07 GHz

for resonance at this particular frequency [Figure 7.5(a)]. However, the current direction at f = 16.1 GHz is more concentrated and dominated around the double slots as compared to the remaining area of the patch. This is clearly visible in Figure 7.5(b) by accumulated arrows around the two slots.

7.6 DESIGN OF A DUAL-BAND PATCH ANTENNA WITH SQUARE RING (DBSR)

Dual-band patch antennas with different square ring configurations are provided in the literature [26–28]. Taking these considerations into account, a dual-band patch antenna modified with a square ring (DBSR) has been designed and analyzed in this section. The DBSR antenna obtained two frequency resonances at 11.95 and 14.13 GHz. The layout of the DBSR microstrip patch antenna is illustrated in Figure 7.6. The exited component is positioned and mounted upon a grounded substrate of dimension

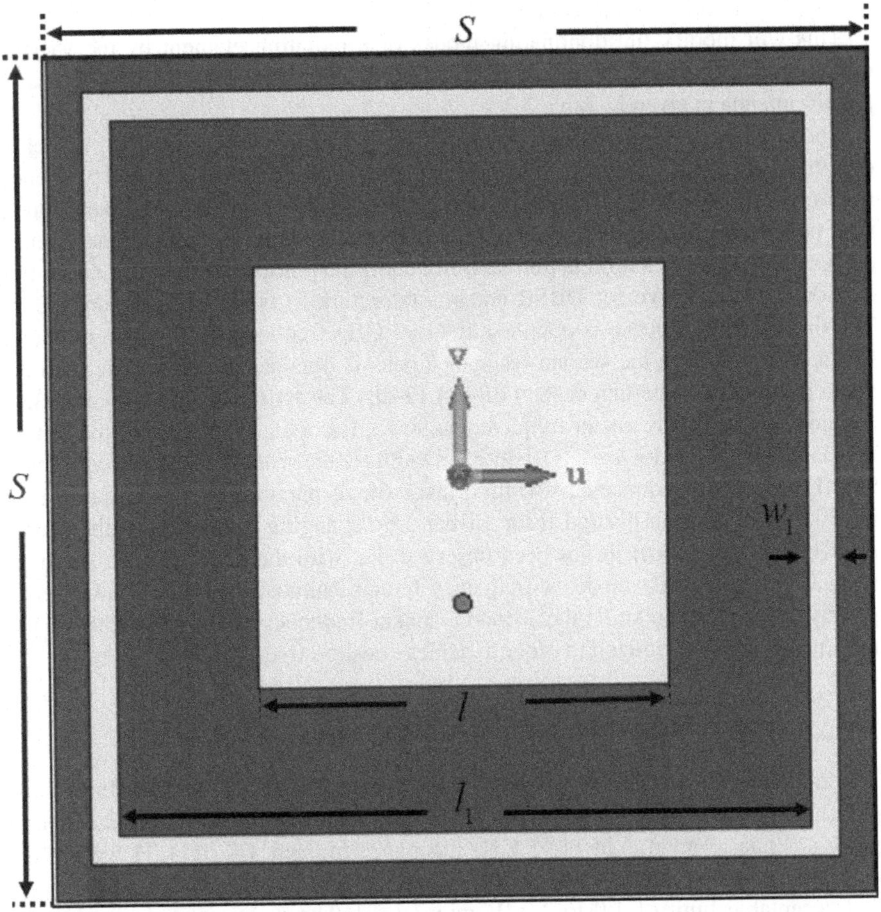

FIGURE 7.6 Dual-band patch antenna with a square ring (DBSR) (structure 2).

TABLE 7.2
Parameters of the dual-band with a square ring patch (DBSR) antenna

Description	Parameters	Dimensions (mm)
Substrate	S	14.4
Length/width of patch	l	7.2
Length of square ring	l_1	6.1
Substrate height	h	1.588
Patch thickness	t	0.035
Slots width	w_1	0.5

14.4 mm × 14.4 mm. The radiating element has a length and width of 7.2 mm printed directly on Rogers RTD 5880 Duroid substrate with a standard thickness of 1.588 mm. A square ring of dimension 6.1 mm × 6.1 mm with a thickness of 0.5 mm is inserted on the patch to obtain dual-band behavior.

This will modify the dominating modes of a radiating element by the varying dimension of a square ring with suitable feed placement. The parameter details of DBSR antenna is given in Table 7.2.

The scheming and continuous optimization of the intended geometrical model of dual-band proposed structure is obtained using parametric analysis. Desired frequency response can only be generated by proper selection and extraction of the geometrical parameters like length, width, size, and feeding position. Changing the antenna design changes its radiation pattern and frequency response. The dual resonance frequency response curve for DBSR antenna (structure 2) is shown in Figure 7.2(c). The first resonance curve is obtained at 11.95 GHz frequency with a magnitude of −23.45 dB; whereas, the second resonance peak is obtained at 14.13 GHz with the value of reflection coefficient (S_{11}) of −34.19 dB. The −10 dB bandwidth of DBSR antenna for the first resonant frequency is 960 MHz, which is 8.03% of the central resonant frequency. However, −10 dB bandwidth for the second resonant frequency is 90 MHz and its impedance bandwidth is just 0.6%. A parametric sweep is performed in order to obtain the desired resonant band by changing the length of the square ring (l_1) [Figure 7.2(d)]. It has been observed that with the change in the length of a square ring, the first resonant frequency remains almost unchanged; however, it greatly influences and shifts the second resonant frequency peak. The width (w_1) of the square ring is optimized in order to achieve desired frequency peaks.

7.7 RADIATION PERFORMANCE ANALYSIS

The radiation performance analyses for VSWR, η, and Z11 for structure 2 are presented in Figure 7.3. The standing wave ratio typical value for DBSR microstrip patch antenna resonating at 11.95 GHz resonance frequency is 1.14. However, it is detected to be 1.04 at 14.13 GHz resonance frequency. This lies within a reasonable and acceptable limit of 1.0 to 2.0 [Figure 7.3(d)]. The Z_{11} simulated response for antenna resonating efficiently at 11.95 GHz is 82.09 Ω. However, for second resonant peak operating at 14.13 GHz, the value of Z_{11} is 91.55 Ω [Figure 7.3(e)]. Similarly, in order to determine the value for radiation efficiency, a simulation analysis has been made. It has been observed that for the first peak frequency resonating at 12 GHz, the value of η is 98.54%, whereas, for second resonant frequency, a minor decline in η is discovered which corresponds to 81.85% [Figure 7.3(f)].

7.7.1 Directivity and Gain Analysis

Simulation analyses for the directivity and gain are made for DBSR antenna (structure 2) for electric and magnetic reference planes and are illustrated in Figure 7.4. At $\phi = 0°$, the directive gain magnitude of main beam is 7.49 dBi at 11.95 GHz and 1.58 dBi at 14.13 GHz of resonant frequency. Whereas at $\phi = 90°$, the directive gain magnitude of main beam is 7.5 dBi at 11.95 GHz and 1.89 dBi at 14.13 GHz of resonant frequency, respectively [Figure 7.4(c)]. At $\phi = 0°$, the realized gain magnitude of main beam is 7.4 dBi at 11.95 GHz and 0.713 dBi at 14.13 GHz of the resonant frequency. Whereas at $\phi = 90°$, realized gain magnitude of main beam is 7.41 dBi at 11.95 GHz and 1.02 dBi at 14.13 GHz of resonant frequency [Figure 7.4(d)].

7.7.2 Surface Current Analysis of Antenna

The excited surface current analysis for the presented DBSR patch antenna (structure 2) at the two resonant frequencies, namely 11.95 and 14.13 GHz, are presented in Figure 7.5. As anticipated, two major current paths have been detected that are the prime causes of antennas with two resonant modes. These two adjoining resonances add together to form the dual-band resonance structure. At f = 11.95 GHz, the current distribution along the vertical direction of radiating patch is stronger as compared to the square ring. This current is responsible for resonance at this particular frequency [Figure 7.5(c)]. However, at f = 14.13 GHz, the maximum current is accumulated and dominating around the square ring as compared to the area of the patch. This is clearly visible through the arrows that are concentrated around the square ring in Figure 7.5(d).

7.8 DESIGN OF A DUAL-BAND ANTENNA WITH BEND SLOTS (DBBS)

The design of a dual-band compact patch antenna with bend slots (DBBS) is analyzed in this section. The dual-band frequency response is found by inserting the antenna structure with bent slots. This structure is observed to resonate at two frequency peaks 14.31 and 20.07 GHz (Figure 7.7). The radiating component is positioned at the upper part of a metallic ground base of dimension 14.4 mm × 14.4 mm. The proposed structure has a length and width of 7.2 mm ×5.7 mm. This is placed on the top of Rogers RTD 5880 Duroid substrate with a standard thickness of 1.588 mm. In literature, this coupled-slot technique is found to have a reduction in the radiating patch size up to about 32% [29–31].

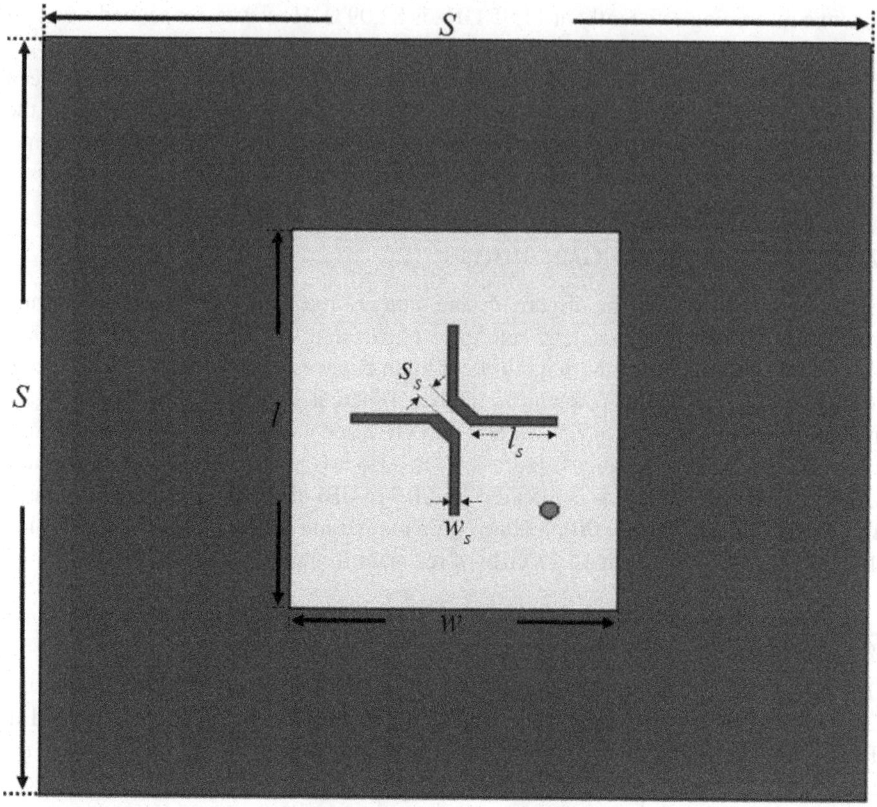

FIGURE 7.7 Dual-band patch antenna with bend slots (DBBS) (structure 3).

This results in a linearly polarized dual-band patch antenna. The length, width, and spacing between the bend slots are adjusted to 3.6, 0.17, and 0.4 mm, respectively. The dimensions of the bend slots are optimized to achieve a sharp double-band response at 14.31 and 20.07 GHz, respectively. The remaining parameters are given in Table 7.3.

The dual resonance frequency response curve for DBBS antenna (structure 3) is shown in Figure 7.2(e). The first resonance curve is obtained at 14.31 GHz with a peak magnitude of −38.36 dB. The second resonance peak is obtained at 20.07 GHz with −18.62 dB as the value of the reflection coefficient (S_{11}). For the first resonant frequency, the value for −10 dB bandwidth of DBBS antenna is 3729 MHz, that is, 26.06% of the central resonant frequency. For the second resonant frequency, the value of −10 dB bandwidth is 1716 MHz and its impedance bandwidth is 8.57%. A parametric sweep is performed to obtain the desired resonant band by changing the width of patch (w) [Figure 7.2(f)]. It is observed that with an increase in the width of copper patch, the dual-band resonant frequency decreases. The value of w is set at 5.7 mm along with the remaining optimized parameters in order to obtain sharp resonance at desired frequencies.

TABLE 7.3
Parameters of the dual-band patch antenna bend slots (DBBS)

Description	Parameters	Dimensions (mm)
Substrate	S	14.4
Length of patch	l	7.2
Width of patch	w	5.7
Length of slots	ls	3.6
Width of slots	Ws	0.17
Spacing between slots	Ss	0.4
Feed position	Pos_x	1.67
	Pos_y	1.7

7.9 RADIATION PERFORMANCE ANALYSIS

The radiation performance analyses for VSWR, η, and Z_{11} for structure 3 are presented in Figure 7.3. The typical standing wave ratio value for DBBS patch antenna operating efficiently at 14.31 GHz resonance frequency is precisely 1.02; however, it is detected to be 1.26 at 20.07 GHz resonance frequency. This lies within a reasonable and acceptable limit of 1.0 to 2.0 as shown in Figure 7.3(g). The Z_{11} simulated response for antenna resonating efficiently at 14.31 GHz is 90.34 Ω; however, for second resonant peak operating at 20.07 GHz, the value of Z_{11} is 115.57 Ω [Figure 7.3(h)]. Similarly, in order to determine the value for radiation efficiency, a simulation analysis is performed. It is found that for the first resonant frequency, the estimated value for η is 98.64% and for the second resonant frequency, a minor decline in η is observed, corresponding to 97.5% [Figure 7.3(i)].

7.9.1 Directivity and Gain Analysis

Simulation analyses for the directivity and gain is performed for DBBS antenna (structure 3) for electric and magnetic reference planes as shown in Figure 7.4. At $\phi = 0°$, the directive gain magnitude of main beam is 7.57 dBi at 14.31 GHz and 3.17 dBi at 20.07 GHz of the resonant frequency. At $\phi = 90°$, the directive gain magnitude of main beam is 7.6 dBi at 14.31 GHz and 5.66 dBi at 20.07 GHz of the resonant frequency [Figure 7.4(e)]. At $\phi = 0°$, the realized gain magnitude of main beam is 7.51 dB at 14.31 GHz and 3.08 dB at 20.07 GHz of the resonant frequency. At $\phi = 90°$, the realized gain magnitude of main beam is 7.54 dB at 14.31 GHz and 5.57 dB at 20.07 GHz of resonant frequency [Figure 7.4(f)].

7.9.2 Surface Current Analysis of Antenna

The excited surface current analysis of the presented DBBS patch antenna (structure 3) at the two resonant frequencies, namely 14.31 and 20.07 GHz, are presented in Figure 7.5. As anticipated, two major current paths have been detected that are the prime causes for the antenna having two resonant modes. These two adjoining resonances add together to form the dual-band resonance structure arising from the resonance of TM_{01} and TM_{10} modes. By suitable feed positioning, it is possible to propagate both the modes simultaneously. At f = 14.31 GHz, the charge is accumulated at the center of the radiating patch. This is responsible for resonance at this particular frequency and clearly visible in Figure 7.5(e). However, at f = 20.07 GHz, maximum current is accumulated and dominating around the corner of the patch as indicated in Figure 7.5(f).

7.10 ANTENNA RADIATION PERFORMANCE COMPARISON

All the three antennas' radiation performance comparison is described in this section. The structures are compared in terms of their resonant frequencies, S_{11}, −10 dB bandwidth, radiation efficiency, and so on and are depicted in Table 7.4.

All the structures are designed using Rogers RTD 5880 Duroid substrate with a standard thickness of 1.588 mm. The dimension of the structures is fixed at 14.4 mm × 14.4 mm. It has been observed that the DBBS antenna structure offers −10 dB bandwidth of 3729 MHz and 1716 MHz for the first and second resonant frequencies, respectively. This bandwidth is much higher and better than the rest of the structures and thus the DBBS offers wideband frequency responses. The value of radiation efficiency for the DBBS antenna is above 98.64% for dual-band frequency curve. This is better when compared to DBDS and DBSP antenna structures. Thus, maximum power is radiated from the antenna which minimizes the energy losses. The value of S_{11} for DBDS antenna is less than −36.49 dB for dual-band which is better than DBBS and DBSP. Thus, the DBDS antenna structure shows a minimum reflection of electromagnetic wave by impedance discontinuity in the transmission medium. A tabular comparison of gain and directivity is given in Table 7.5. From the table, it is clear that the DBBS antenna offers a maximum value of directivity and gain as compared to the remaining structures.

TABLE 7.4
Antenna radiation performance comparison

Structure/performance parameters	DBDS		DBSP		DBBS	
Size	14.4 mm × 14.4 mm		14.4 mm × 14.4 mm		14.4 mm × 14.4 mm	
	1st Freq.	2nd Freq.	1st Freq.	2nd Freq.	1st Freq.	2nd Freq.
Frequency (f_0) GHz	12	16.1	11.95	14.13	14.31	20.07
Reflection coefficient (S_{11}) dB	−41.09	−36.49	−23.45	−34.19	−38.36	−18.62
Bandwidth (−10 dB) MHz	540 4.5%	96 0.6 %	960 8.03%	90 0.6 %	3729 26.06%	1716 8.57%
Total efficiency (η)	98.75	73.01	98.54	81.85	98.64%	97.5%
VSWR	1.02	1.06	1.14	1.04	1.02	1.26
Impedance (Z_{11}) Ω	93.22	96.26	82.09	91.55	90.34	115.57

TABLE 7.5
Comparison of directivity and gain

Performances	Frequency (GHz)	Phi (ϕ)	Directivity (dBi)	Gain (dB)
DBDS	12	$\phi = 0°$	7.36	7.29
		$\phi = 90°$	7.36	7.29
	16.1	$\phi = 0°$	6.3	4.93
		$\phi = 90°$	3.35	1.98
DBSP	11.95	$\phi = 0°$	7.49	7.4
		$\phi = 90°$	7.5	7.41
	14.13	$\phi = 0°$	1.58	0.713
		$\phi = 90°$	1.89	1.02
DBBS	14.31	$\phi = 0°$	7.57	7.51
		$\phi = 90°$	7.6	7.54
	20.07	$\phi = 0°$	3.17	3.08
		$\phi = 90°$	5.66	5.57

7.11 CONCLUSION

This research work presented dual-band rectangular microstrip patch antennas in three different configurations, which are resonating at X, Ku, and K-bands for wireless satellite communications. The benefit of using a dual-band antenna is its capability to provide high mobility and strong and stable wireless connection in some non-reachable locations. Three dual-band antennas have been designed, optimized, and analyzed. The performances of all the structures are compared in terms of reflection coefficient, −10 dB bandwidth, impedance bandwidth, efficiency, VSWR, and Z_{11}. Their directive gain and realized gain performance are also analyzed. It has been observed that DBBS antenna structure offers wideband frequency response with higher radiation efficiency for both resonance peaks as compared to DBDS and DBSP antenna structures, whereas DBDS antenna structure shows a minimum reflection of electromagnetic wave by impedance discontinuity in the transmission medium. This research work concludes that these structures can be preferred and employed in many applications including wireless satellite communications.

REFERENCES

[1] Balanis, C. A. 2016. *Antenna Theory: Analysis and Design*. John Wiley & Sons.

[2] Maci, S. and Gentili, G. B. 1997. Dual-frequency patch antennas. *IEEE Antennas and propagation Magazine*, *39*(6), pp. 13–20.

[3] Papapolymerou, I., Drayton, R. F. and Katehi, L. P. 1998. Micromachined patch antennas. *IEEE Transactions on Antennas and Propagation*, *46*(2), pp. 275–283.

[4] Singh, D. and Srivastava, V. M. 2018. Comparative analyses for RCS of 10 GHz patch antenna using shorted stubs metamaterial absorber. *Journal of Engineering Science and Technology*, 13(11), pp. 3532–3546.

[5] Sharma, A. and Singh, G. 2007. Rectangular microstrip patch antenna design at THz frequency for short-distance wireless communication systems. *Journal of Infrared, Millimeter, and Terahertz Waves*, *30*(1), p. 1.

[6] Sharma, A. and Singh, G. 2008. Design of single pin shorted three-dielectric-layered substrates rectangular patch microstrip antenna for communication systems. *Progress in Electromagnetics Research*, *2*, pp. 157–165.

[7] Singh, D. and Srivastava, V. M. 2018. RCS reduction of patch array using shorted stubs metamaterial absorber. *Journal of Communications*, *13*(12), pp. 702–711.

[8] Garg, R., Bhartia, P., Bahl, I. J. and Ittipiboon, A. 2001. *Microstrip Antenna Design Handbook*. Artech House.

[9] Narayana, M. V., Vikranth, A., Govardhani, I., Nizamuddin, S. K. and Venkatesh, C. 2012. A novel design of a microstrip patch antenna with an end fire radiation for SAR applications. *International Journal of Science and Technology*, *2*(1), pp. 1–4.

[10] Jackson, D. R., Williams, J. T., Bhattacharyya, A. K., Smith, R. L., Buchheit, S. J. and Long, S. A. 1993. Microstrip patch designs that do not excite surface waves. *IEEE Transactions on Antennas and Propagation*, *41*(8), pp. 1026–1037.

[11] Singh, D., Thakur, A. and Srivastava, V. M. 2018. Miniaturization and gain enhancement of microstrip patch antenna using defected ground with EBG. *Journal of Communication*, *13*(12), 730–736.

[12] Hasan, M. M., Faruque, M. R. I. and Islam, M. T. 2018. Dual band metamaterial antenna for LTE/bluetooth/WiMAX system. *Scientific Reports*, *8*(1), pp. 1–17.

[13] Zhu, J. and Eleftheriades, G. V. 2007. Dual-band metamaterial-inspired small monopole antenna for WiFi applications. *Electronics Letters*, *45*(22), pp. 1104–1106.

[14] Abramov, O. J., Nagaev, F. I. and Salo, R., Airgain Inc. 2011. *Dual-band antenna*. U.S. Patent 7,965,242.

[15] Ojaroudi Parchin, N., Al-Yasir, Y. I., Basherlou, H. J. and Abd-Alhameed, R. A. 2020. A closely spaced dual-band MIMO patch antenna with reduced mutual coupling for 4G/5G applications. *Progress in Electromagnetics Research*, *101*, pp. 71–80.

[16] Sharaf, M. H., Zaki, A. I., Hamad, R. K. and Omar, M. M. 2020. A novel dual-band (38/60 GHz) patch antenna for 5G mobile handsets. *Sensors*, *20*(9), p. 2541.

[17] Camacho-Gomez, C., Sanchez-Montero, R., Martinez-Villanueva, D., Lopez-Espi, P. L. and Salcedo-Sanz, S. 2020. Design of a multi-band microstrip textile patch antenna for LTE and 5G services with the CRO-SL ensemble. *Applied Sciences*, *10*(3), p. 1168.

[18] Bainsla, A., Singh, P. and Sahoo, G. 2020. A dual band slotted circular microstrip patch antenna for bluetooth and commercial WLAN applications. *Available at SSRN 3546463*.

[19] Raviteja, G. V., Madhan, B. V., Sree, M. K., Avinash, N. and Surya, P. N. 2020. Gain and bandwidth considerations for microstrip patch antenna employing U and quad L shaped slots with DGS and parasitic elements for WiMax/WiFi applications. *European Journal of Engineering Research and Science*, *5*(3), pp. 327–330.

[20] Viswanadha, K. and Raghava, N. S. 2020. Design and analysis of a dual-polarization multiband oval ring patch antenna with L-stubs and folded meander line for C-band/X-band/Ku-band/K-band communications. *International Journal of Electronics*, Vol. 108, Iss. 4, pp. 647–663.

[21] Gunaram, Sharma, V., Shekhawat, S., Jain, P. and Deegwal, J. K. 2020, May. Coaxial feed dual wideband dual polarized stacked microstrip patch antenna for S and C band communication. In *AIP Conference Proceedings* (Vol. 2220, No. 1, p. 130067). AIP Publishing LLC.

[22] Nakmouche, M. F., Fawzy, D. E., Allam, A. M. M. A., Taher, H. and Sree, M. F. A. 2020, April. Dual band SIW patch antenna based on H-slotted DGS for Ku band application. In *2020 7th International Conference on Electrical and Electronics Engineering (ICEEE)* (pp. 194–197). IEEE.

[23] Motin, M. A., Hasan, M. I., Habib, M. S. and Sheikh, M. I. 2013, May. Design of a modified rectangular patch antenna for quad band application. In *2013 International Conference on Informatics, Electronics and Vision (ICIEV)* (pp. 1–4). IEEE.

[24] Gupta, A. and Gangwar, S. P. 2020. Dual-band modified U-shaped slot antenna with defected ground structure for S-band applications. In Advances in VLSI, Communication, and Signal Processing (pp. 209–220). Springer.

[25] Srivastava, K., Mishra, B., Patel, A. K. and Singh, R. 2020. Circularly polarized defected ground stub-matched triple-band microstrip antenna for C- and X-band applications. *Microwave and Optical Technology Letters*, *62*(10), 1–7.

[26] Pandey, S. K. 2020. Printed CPW-fed dual-band antenna using square closed-ring and square split-ring resonator. *Applied Physics A*, *126*(8), 1–8.

[27] Alkanhal, M. A. and Sheta, A. F. 2007. A novel dual-band reconfigurable square-ring microstrip antenna. *Progress in Electromagnetics Research*, *70*, 337–347.

[28] Hyeon, M., Sung, Y., and Kim, E. 2018. Dual-band circular polarised square-ring patch antenna loaded lumped capacitors. *IET Microwaves, Antennas & Propagation*, *12*(15), 2326–2331.

[29] Wong, K. L. and Yang, K. P. 1998. Compact dual-frequency microstrip antenna with a pair of bent slots. *Electronics Letters*, *34*(3), 225–226.

[30] Maci, S. and Gentili, G. B. 1997. Dual-frequency patch antennas. *IEEE Antennas and Propagation Magazine*, *39*(6), 13–20.

[31] Singh, D. and Srivastava, V. M. 2018. An analysis of RCS for dual-band slotted patch antenna with a thin dielectric using shorted stubs metamaterial absorber. *AEU-International Journal of Electronics and Communications*, *90*, pp. 53–62.

Part III

Multiple-Input Multiple-Output (MIMO) Antenna Design and Its Applications

8 Multiple-Input Multiple-Output Antenna Design and Applications

Anup P. Bhat, Sanjay J. Dhoble and Kishor G. Rewatkar

DOI: 10.1201/9781003093558-11

8.1 INTRODUCTION

The new-generation 5G communication framework is considered a transformation in the wireless communication market where exceptionally high data transfer capacity is required for present-day advanced mobiles. This quick insurgency has driven scientists to build up communication innovation in the programming and instrumentation fields. In addition, in terms of SIMO–MIMO–Massive MIMO, antenna configuration is considered a fundamental application that provides a standardized 5G wireless communication system with different microstrip antenna design and structure to be established and represented. The advancement of innovation for sending and accepting data by means of electromagnetic wave proliferation has come a long way. The creator of this theory does not plan to portray each headway along with the guide since G. Marconi's spearheading test. Instead, the scope of this discussion centres on a platform for various innovations, which is an essential part of both new and up-and-coming portable correspondence frameworks.

With the need to acquire higher information rates in both wireless communication and broadcasting, the existing constraints on expanding traffic in a cell and a growing number of customers, multiple-input multiple-output (MIMO) is a helpful innovation [1]. Multiple antennas are used at the two closures of the connection, which upgrade the transmission execution without additional frequency and boosting assets.

Traditionally, wireless communications were utilized for voice transmission only, small amounts of information could be transmitted using low radio frequency (RF) technology. The majority of information requiring high-rate transmission utilized wired interchanges. Nowadays, complex wireless media applications, supporting phones with associated camera, messaging, navigation, and GPS functionality, have been an expanded feature. Accordingly, there are more prerequisites for wireless fast information transfer, which conventional antennas are not equipped to convey because of multipath and co-channel interference and interference mismatch [2].

These days, wireless access systems have limitations with respect to information rate, nature of administration (QoS), and unpredictive productivity. Numerous applications are concerned, such as wireless LAN with wireless network (WLAN), broadband wireless access (BWA), and versatile radio frameworks (3G, LTE-An) while endeavours have been made to improve plans for current frameworks application. The pattern to build information rates will most likely proceed to reach 1 Gbps ensuring moderate portability in communication.

5G mobile communication is anticipated very shortly with some new promising features [1]. Compared to higher mm-wave 5G bands, such as 38 GHz, 60 GHz, and 73 GHz [2], the 28 GHz band has the advantage of the lowest path loss due to its relatively low atmospheric absorption. Low-profile, compact antennas are the best candidates for incorporation into host devices. In addition, the MIMO antenna array is necessary to achieve higher data rates due to spatial multiplexing. In Ref. [3], a combined 28 GHz beam antenna with two main radiators is reported. Another miniaturized 28 GHz ($4 \times 4 \times 1.34$ mm^3) antenna is recorded in Ref. [4]. In Ref. [5], an inkjet-printed antenna running at 27.75 GHz is recorded for the mobile applications.

The MIMO innovation is another idea to satisfy those requirements without more specifications in the existing system. Since time, frequency reuse, and handling are pushed as far as possible, the space diversity can be improved drastically. The MIMO guideline can be characterized as sending various packets of information on numerous antennas at the same operating frequency. The utilization of numerous operating antennas improves the framework execution. In a perfect world, the channel limit directly develops with the quantity of transmitting (Tx) and receiving (Rx) antennas [1].

This procedure can be seen as a speculation of the space diversity technique with numerous antennas [2]. It predicts a direct wealth in different ways, to misuse free transmission channels between the Tx and Rx antenna exhibits. This transmitting and receiving structure can be demonstrated utilizing a network capacity and off the channel. Data hypothesis shows that two essential instruments should be considered during the time spent for moving information.

The diversity that depends on generating different de-correlated redundant packets of information and multiplexing which depends on the receiving of various packets of data as per the ground layout is explained in Figure 8.1.

These two instruments, assorted different types and multiplexing, battle interference impacts and increment the channel limit individually [3]. The huge improvement in connection of unwavering quality or potentially information rates is anticipated and data hypothesis will come.

The advancement of the technology from the national radio wave phenomenon is considered "multipaths" in which the propagation of popular communication information bounces back from the walls, floors, and other structures approaching towards the antenna. Receiving reflection is based on the multiplied signal at the distinctly placed antenna. The angles and positioning of each antenna receiving the contact separation is defined in the limit of array design. The communication signal intensity strength is the measure using the cross-correlation antenna element to minimize the correlation functions between the antenna element and maximize the capacity. This is the basic application of the memory-based antenna using new base antenna diversity. It is most important because of diversity that it is possible to apply the same antenna for multiple communicating working triple C with that working frequency. Sufficient

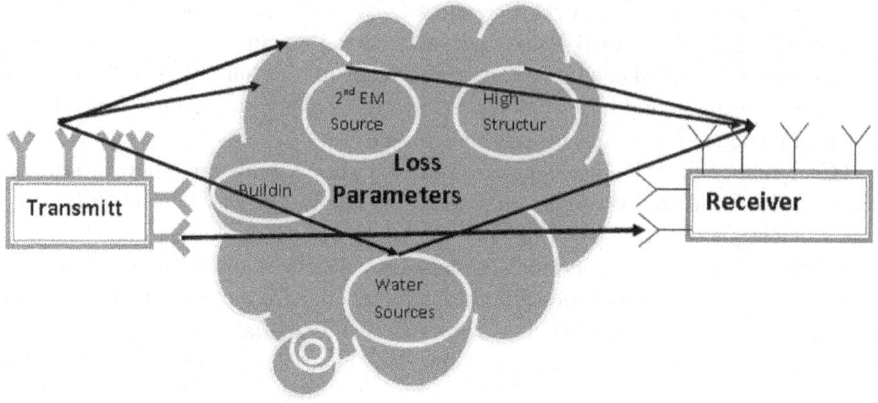

FIGURE 8.1 Spatial diversity different type implementation [10].

power is provided by LTE-based applications. Thus using LTE-based applications we have different designs: the design of LTE-based application comes up from 1.60 GHz so using the frequency variation we are able to design any type of antenna. Secondly, they can give the orientation in a circular fashion called omnidirectional antenna. This makes it more powerful than diversity switching. It improves the communication network with the time or space multiplexing technique and is far more robust. It can be improved using high-quality antennas. These structures are used in addition to antenna design. The design and implementation of new types of MIMO microstrip antennas consist of microstrip feed with coaxial cable or feed with a different feeding method, reporting the s-parameter (S11) in gain enhancement.

In the industry, wireless communication systems are a rapidly advancing and changing system, especially the cellular system. The global mobile phone use in 2019 was 4.70 billion, approaching 4.88 billion in 2020 year end (statista.com). Researchers have investigated the wireless communication system and in this transformation have discovered two trends: the first is communication, meaning transmitting and receiving signals between two active points or ports with the same signal or multipath signal communication. The other corresponding word is wire-free, which means that there is no strong correlation between the two objectives of communication. Because of multiple antenna placements, performance is defined by wireless communication system as a communication channel without a unique connection between the two different centres on bandwidth [4].

The radio spectrum of electromagnetic waves from a transmitter to a beneficiary is described by the accessibility of multipath connectivity because of different phenomena, such as reflection, refraction, dispersion, and diffraction. The presentation of MIMO frameworks is to a great extent reliant on increasing medium coverage and the structure of the antenna array cluster. In this specific situation, both the space-time block coding based transmit diversity and demonstrating of the channel utility are basic in MIMO.

The interest for MIMO throughput and improved customer experience universally attracts buyers, and service provider organizations keep on developing, with better network administrators rapidly improving with extending efficacy within the constraints of government guidelines and wireless system protocols developed commercially. The innovation in improvements to high-speed data transmission have progressed the advance antenna system (AASs), which is a practical choice for existing 4G and 5G versatile systems. AASs empower best-in-class beamforming and various information, different yield (MIMO) strategies that have incredible assets for improving the end-customer experience, limit, and attractiveness. These advances have made great improvements to mm-wave communications within the last 20 years. AAS fundamentally upgrades and organizes execution in both the uplink and the downlink. Finding the most appropriate AAS variations to accomplish execution gains and cost-effectiveness in a particular system arrangement requires comprehension of the attributes of the two AAS and multi-antenna highlights [1–5].

MIMO is probably the current research type of smart antenna innovation to improve the communication execution and performance. The standard norms and the use of spatial multiplexing in MIMO were regularly used in the new system. The principal system was created and functionalized from 2001. The MIMO was applied commercially in association with symmetrical Orthogonal Frequency-Division Multiple Access Technology (MIMO-OFDMA), which maintained both diversity coding and spatial multiplexing [4].

The main driving power behind major developments in wireless innovation has been the need for higher data transfer rates. Nonetheless, the greatest conceivable information pace of a communication framework is generally restricted by channel capacity with the Shannon limit governing the communication validity in the communication channel design [1]. The innovative work of the cutting-edge powerful and flexible system known as the International Mobile Telecommunication (IMT)-Advanced system is also in progress to achieve greater multipoint broadband speeds as well as higher rates over the entire integrated network region [2]. Multi-antenna system with MIMO innovation will play a key role in giving the objective information rate of 1 Gbps with high range efficiency and coverage area.

Mobile Communication Technology has advanced in recent decades with each phase carrying with it new and energizing abilities and administrations to the end customer. Advancement in 4G wireless versatile interchanges has demonstrated that by utilizing various data sources and numerous yields (MIMO), innovation at both the transmitter and beneficiary points, the wireless framework limit and unwavering quality can be improved without the requirement for expanding the power communicated or utilizing more range by using the multipath assorted different types. This is despite a lot of examination that has just been done on the plan of MIMO and other types of handset antennas, the structure of low profile, small impression, and multi-standard (wideband or multiband) diversity different type antennas for handset instruments despite difficulties in implementation.

In the last phases of 4G transmission, there was an influx of advancements of new 5G innovation. The emphasis is on mm-wave communication over 2.4–28 GHz utilizing antenna clusters equipped for giving higher transmission capacity. It is a difficult undertaking for an antenna cluster configuration to accomplish a wide data transmission with small size, great throughput, and straightforward creation procedure. The last part of this centres around building up a number of different novel, smaller, high addition and wide transmission capacity 5G antenna exhibits which are capable of giving wide application without data loss [4–6].

8.2 MIMO TECHNOLOGY

The most widely recognized innovations in radio frequency circuits are microstrip antenna utilization. The 5G principle highlights must be fulfilled in the proposed plan such as double polarization, high data transmission (>1 GHz) and high acceptable increase (>12 dBi) in the bandwidth.

The MIMO technology uses microstrip antenna planners; likewise they use at least two emitting patches for transmission in the framework. In the past few decades, several MIMO-diversity antennas were proposed by the researchers. A couple of these are intended to function in the 3.1–10.6 GHz frequency range, in UWB applications. A few assessments on MIMO antenna placement in a system with two oppositely placed and four physical dimension wise placed transmitting components have been performed. The various placement methodologies are proposed to enhance the isolated particle between the elements of the antenna. Different structures, such as mushroom-moulded, EBG structures and ground plane structures, were suggested to remove the standard association by reducing the ground current between the transmitting components and surface use for broadcasting.

In a miniaturized two-port UWB, connectivity is established among monopole antenna element and the reflected ground plane performs at −26 dB used in UBS dongle. The rated 50 ohm interferences transition is centred at 3.1 to 5.15 GHz. Because of the inadequate range of the available frequency band, as an extra frequency range band of the 5G cell arrangement, the millimetre-wave frequency band known as microwave band has been utilized for communication in a human context, such as microwaves in the kitchen, mobiles and satellite TV [8].

The fundamental objective of 5G is to build an attractive and more extensive system limit at a lower cost. The basic acknowledgment among various examination bunches for future 5G specialized works is that the static customer's information transfer speed is 10 Gb/s, whereas the versatile customer gets a speed of about 1 Gb/s, within structured network and where the urban territory is under 100 Mb/s. The innovation is concentrated to present these high information speed transmission objectives in MIMO. The idea of stretching out multi-customer MIMO to place the number of antennas at a base station is an advance arrangement with mechanical strength which fundamentally builds customer throughput and system limits by permitting beamforming information transmission and reducing interference with other sources. The huge increment in the transmission path inference at high frequencies must be resolved by a higher antenna gain. The gain rise is accomplished by expanding the number of antennas at the transmitting base station. The objective of 5G R&D is still 1 millisecond slower than the 4G gadget deferral and has lower power utilization detailing the bandwidth applicable as shown in Figure 8.2 [3, 6–8].

As of late, the use of millimetre-wave transmission frequencies for mobile cell systems, including high-increase flexible antennas, has been triggered by the

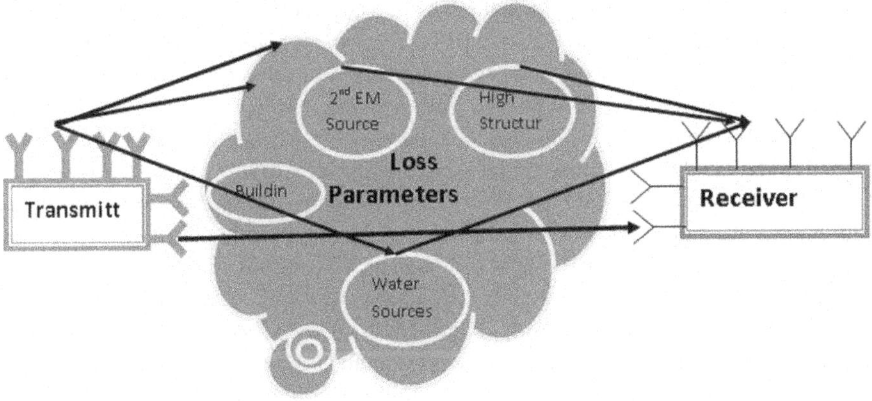

FIGURE 8.2 Spectrum analysis with allocation frequency [3].

interest in fast cell–cell data transformation and the interest in large coverage range. Millimetre-wave bands have pulled in a great deal of consideration due to the enormous measure of accessible data transmission [3].

Separation of <–20 dB is achieved in another diversity antenna [9] by improving the condition of the perfect ground plane by openings with the emanating components. The antenna's working frequency is in consideration scope of 3.1–5 GHz. Both of these antennas will, in any case, cover only the lower UWB band. A different type of antenna covering the entire UWB was structured in Ref. [10], in which stubs are familiar with reducing the shared coupling. At the cost of increased unpredictability and scale of the general antenna system, enhanced disengagement is acquired in Ref. [10].

Ultra-wideband (UWB) in the past known as 'beat radio' is a cutting edge innovation for radiating data over an enormous transfer speed (>500 MHz), promising high information transfer speed with low power utilization. The unlicensed utilization of 3.1–10.6 GHz has been recommended by the Federal Communications Commission for short separation high-speed indoor applications. As of late, International Telecommunication Union Radio Communication Sector (ITU-R) characterized UWB as the high data transfer capacity of the produced signal surpasses the base of either 500 MHz or 20% of the middle frequency govern by the channel limits. The partial data transfer capacity ought to be bigger than 20% of all the transmission [7–8].

As of late, the improvement of media transmission innovation has brought numerous challenges in the plan of cell phones, particularly in the antenna territory. With these days various frequency groups in media transmission frameworks, multiband antenna is mostly utilized. Numerous investigations have reported multi-frequency band antennas for GSM, UMTS and Wi-Fi for cell phone applicability. However, international LTE standard protocol is modified and restructuring starts to introduce new frequency with the security concern. With low frequency utilization, the structure of an antenna may work over a more extended frequency band incorporated with a cell phone structure by an array factor that turns into more of a test antenna at the same time frame. MIMO is an innovation that will be utilized in 5G unique systems and is expected to expand the limit of the framework without utilizing additional data transmission. The limit is explicitly proportional to the number of antennas on the transceiver system. It achieves a higher output with minimal effort but it could extend the intricacy of the system structure [20].

Each antenna in the MIMO framework communicates diverse information streams simultaneously and in equal channels, the sent information is partitioned into sub-information streams or packets, which is sent by various antennas in time interval–based framework while the streams are joined together for giving the primary information as space division multiplexing which can be used to rebuild the limit.

Another space–time coding technique performance is improved for accomplishing a different diversity, type of gain by utilizing different antennas. There are numerous antennas that are structured and applied for the diverse zone, such as antenna at the transmitter, receiver, handset, base station and cell-microcell application.

8.3 IMPLEMENTATION

The MIMO antenna at a 28 GHz frequency with dimensions of 4×4 mm^2 has been manufactured for 5G base station applications, with the ability to be connected with an RF front-end permitting beamforming with the dimensions of 78.5×42 mm^2 with 1.5 GHz working data transmission and an announced feasible increase of 18 dBi. The millimetre-wave band is characterized as part of the electromagnetic spectrum starting from 30 GHz to 300 GHz comparing frequencies with a scope of 10–1 mm wavelength.

Truly, mm-wave frequencies are utilized generally in protection communication, along with radio space science applications. In light of the significant cost and restricted accessibility and availability of electronic instrument communication wavelength, the ongoing progression of single-chip system innovation and the quickly developing applications in communication such as radar, high-end imaging and superior quality video move prerequisites and many more, require advancement of broadband frequency and allied devices especially the incorporation of low power utility wireless system including high-proficiency planar antennas [4].

Transmission capacity can be described as a data rate from a middle frequency or a range from as low as possible to a higher level. The various boundaries and constraints like gain, productivity, interference, correlation are good enough at a specific level. Since antenna attributes are variant from the tilting angle, the line of sight and frequency band are influenced by the centre frequency governing the data transfer speed. Generally, it is indicated independently for every application at the different configuration of antenna arrangement at transmitter and receiver.

Ultra-wideband, formerly recognized as 'Beat radio' is a cutting-edge technology for data transmission over tremendous transmission ability (>500 MHz), offering high data rates with low use of associate network. The Federal Communications Commission has approved the restriction over the use of 3.1–10.6 GHz for short separating indoor applications. The UWB was defined by the International

Telecommunication Union Radio Communication Sector (ITU-R) as the transmission where another data transfer potential of the signal generated exceeds the limit of 500 MHz or 20% of the mean frequency; also the partial transfer speed limit governed by Shannon channel limit in the transmission should be greater than 20% of operational frequency [11].

The bandwidth B_f is characterized as the proportion of data transmission at return loss of less than −10 dB to its centre frequency, which is given by the following equation:

$$B_f = \frac{BW}{f_C} 100\% = \frac{f_h - f_l}{\frac{f_h - f_l}{2}} 100\% = 2\frac{f_h - f_l}{f_h + f_l} 100\% \tag{8.1}$$

where

f_h = upper cut-off frequency

f_l = lower cut-off frequency

On the bandwidth utilization basis there are three modes of utility. In these three modes of communication area, network strength and radiation loss measurements are prime for improving the signal condition as narrowband $B_f < 1\%$ wideband $1\% < B_f <$ 20% ultra-wideband $B_f > 20\%$.

Because, UWB system with centre frequency above 2.5 GHz is needed to have a −10 dB data transmission rate at 500 MHz, while UWB system with frequency centre at 2.5 GHz ought to have a base partial transfer speed of 0.2% of bandwidth. The FCC has approved UWB transmission at high speed can work from 3.1 GHz to 10.6 GHz, with the variable power spectral density (PSD) fulfilling a particular unpleasant appreciation over allocated frequency region governed by the FCC [12].

Versatile terminals require wide operational data transmission bandwidth, and small antennas that are commonly giving restricted data transmission. As clarified in the literature, there is a compromise among data transfer capacity and coordinating level, which can be tackled by finding a diversity trade-off between them.

8.3.1 Need of MIMO

The data transfer capacity can be expanded with the determination of antenna size and effectiveness [13]. Multi-resonating coordinating circuits give a productive method to improve data transfer capacity. Including resonators can double or significantly increase the accessible transfer speed, and yet, the multifaceted nature and misfortunes of the entire framework will increase.

Customarily, wireless exchange was utilized for voice and small information moves, while the vast majority of the high-rate information move items utilized wired communications. The ongoing trends in wireless communication involving interactive media applications – like PDAs having a coordinated camera, messaging capacity and GPS – have been expanded. Thus, there are more prerequisites for wireless information transfer, which is not possible with customary antennas as it is not equipped for conveying as a result of multipath and co-channel interferences [14].

Notwithstanding the necessities of rapid information moves, to keep up certain Quality of Service (QoS), multipath interference impact must be managed. As the transmitted signal travels the longer path to different obstacles on its way to the receiver, the signal is lost as some component is blurred and misshaped. This marvel is called multipath interference. Co-channel interferences to the obstruction are brought about by various coded signals utilizing a similar frequency.

Antenna cluster structures are additionally proposed for upgrading the directivity and accomplishing the necessity to accept the increased number of antennas for the 5G communication system. Bandwidth utility capacity will likewise be accomplished by embedding a phase shifter in transmission line into the previous structure. This proposed alteration expects to include the capacity for inclining the principle radiation as an example to specific bearing at all frequencies.

Consequently, different antennas are to minimize the packet rate just as to improve the quality and limit of wireless transmission channel capacity. This is finished by coordinating the radiation just to the expected bearing and altering the radiation as indicated by the communication traffic condition and signal conditioning. Every one of the numerous antennas is outfitted with a few antennas either in the transmitter or the beneficiary or both. A modern signal processor and coding innovation are the key elements in various antennas.

Cell phones, or wearable instruments, normally expect antenna to be of low profile, small structure and factor [7], great radiation ability, and wide working frequency range, handshaking between communications base stations and organize parts in system within a cell. These days, the physical size of cell phones is operating small and more slender to satisfy customers' appearance prerequisites. Such instruments comprise numerous subsystems to accomplish variable and valuable capacities as explain in Table 8.1. Hence, the antennas ought to be planned as small as could be expected under the circumstances. Small size causes difficulties as different requirements can't be ignored. Additionally, to acquire the structural framework available for communication reality expands business utilization. Cell phone antennas ought to be intended to be effectively produced. Moreover, the antenna configuration is pointless on the off chance that it isn't strong against mechanical harms or can't be mass-created at sensible expenses.

Antennas are a vital part of all versatile communication system. Over the most recent five years or somewhere in the vicinity, the usefulness of the individual segments of inactive antennas has been refined to meet the present higher limit requirement. For instance, cross enraptured transmitting components, feed systems, variable phase shifters for electrical down tilt, the number of antenna exhibits in a base station antenna arrangements has developed impressively without loosening up the RF or mechanical details of the antennas. A few cluster full-scale antennas usually known as Quad-port and Hex-port (4 and 6 port) were the standard in 12" x 15" wide phase. Mobile communication advances experienced a few phases or ages of development and improvement in their exhibition which are distributed in ages summarized in Table 8.1 [19].

8.3.2 The Need of Multiple Antennas

The radio environment is evolving rapidly on resources that are electrically weak. One antenna radio was used in isolation until recently; typically multi-operating

TABLE 8.1
Mobile generation, technological data

Generation	Frequency	Multiplexing	Channel bandwidth	Data speed	Switching	Method
1G	150 Mhz–900 MHz	AMPS	25 Khz	2 kbps	Circuit	Analoge
	800 MHz	1G-AMPS	30 KhZ	10 kbps		
2G	900 MHz–1800 MHz	GSM. TDMA-FDMA	30 KhZ	64 kbps 144 kbps	Circuit Packet	Digital
2.5G	850 MHz–1900MHz	GPRS	30 KhZ			
3G	80–900 Mhz, 1600–1900 MHz	UMTS, WCDMA CDMA2000	500 MHz	2 Mbps	Packet	Digital
4G	2 GhZ–8 Ghz	OFDM LTE WiFi, WiMax	1 GHz	100 Mpbs–1 Gbps	Packet	Digital
5G	2.8 GHz– 300 GHz	WWWW, OFDM	>>1 Gbps	>1 Gbps	Packet	Digital

Source: Ref. [20].

antenna are attached in one antenna system. Today, things are very different: a phone may have four cellular bands, GPS, Bluetooth™, and Wi-Fi. There is typically just one radio used directly. WLAN radios are often present tools. This implies that it is important to have more RF signal receptors and a filtering mechanism. Using just a single antenna to establish diversity or for MIMO applications is also becoming popular for multi-radio systems.

In order to fight the fading, minimize other services and increase the standard reliability in the communication system, antenna diversity is used with WLAN radio. Three kinds of diversity commonly employed with a two antenna arrangement are mostly deployed to establish possible spatial diversity. There will be orthogonally oriented diversity of polarization or there will be different radiation patterns of beams.

The area available for placement of multiple antennas is normally smaller than available for one antenna. There is great restriction of utilizing the space within handsets, so the space available for the antenna is constantly diminishing. Similarly, the size requirements as consideration with availability imposed on the antenna especially the width are continuously decreased with WLAN antennas.

There are filtering mechanisms and allied circuitry for separating the bands when several bands are used. More filtering is essential to separate applicable signals when a whole radio spectrum is received. In addition, antenna switching and demultiplexing are typically required. Typically all those frequencies are collected into an FEM (Front End Module). Losses are implemented by FEMs and more space is taken up on the PCB. The FEM introduces approximately 3 dB of loss on a standard handset and takes up about 100 mm^2 on the PCB.

8.4 DIVERSITY IN MIMO

The spatial diversity is a part of antenna diversity with different strategies in which different antennas are utilized to improve the signal fidelity of a wireless network. Ordinarily, in high density populated territories, there is no line of sight (LOS) between the transmitter and the receiver. Subsequently, the multipath impact happens on the transmission way. In spatial distributed diversity, a few communication antennas are set a way off from each other. Accordingly, if one antenna encounters a blur, another will have a feasible or a reasonable signal [15–18, 22].

A smaller wideband MIMO antenna was given two resonance structures with space and VSWR parameter S11 shows −20 dB working at 5 GHz–10 Ghz anger with max −45 dB at 8 GHz. The second structure with a single component antenna work at 3 GHz–18 GHz range; the gain is −55 dBi at 5.5 GHz as planned at the transmitter station.

A double-band MIMO antenna for 5G mobile has been implemented at 2.6 GHz to 3.6 GHz range. The single-component antenna has a reasonable size with S11 about −15 dB at 2.65 GHz and 3.75 GHz. The MIMO antenna has four components and its S11 shows a profound worth equivalent to −80 dB at 2.8 GHz [10]. A double energized depression supported gap antenna for 5G MIMO applications has been introduced. The activity band is around 30 GHz, the restriction and loss parameter in the form of fossil losses are explained in Figure 8.1.

The combination of the structured antenna is a different shaped circular-rectangular antenna with the tightened opening frequency range. The S11 shows −25 dB on the range from 2.5 GHz to 36 GHz with complex structure hard to create genuinely. A reduced aperture antenna exhibiting permissible application for 5G millimetre-wave massive MIMO with existing communication framework has been introduced. The framework has a diversity bar shaping execution because the space between cluster components meets the prerequisite of half frequency. The activity band of the antenna cluster from 2.5 GHz to 32 GHz is used for the antenna design in 5G. The S11 is −10 dB along with the activity band and its profound estimation is about −35 dB which is from 25 GHz to 30 GHz [6, 10, 16].

On account of base stations in a full-scale cell condition, enormous cells with high antennas, a separation up to ten frequencies is expected to guarantee a low shared interfering connection. Notwithstanding, if there should be an occurrence of handheld instruments, because of the absence of area, half of the frequency is sufficient for the normal outcome. The purpose of this space is typically in the full scale/cell situation. The interference of which is brought about by multipath relationships that have happened close to the zone of the terminal. Along these lines, from the terminal side, in various ways transmitted from more extensive point of communication terminal, subsequently requiring small separations, from the transmitter side, the way edge is generally low. That is the reason for separation in elements required in the array cluster.

In a versatile antenna system, various antennas are utilized both in the transmitting and accepting side of communication connect to streamline the transmission over the channel. An AAS framework will centre its communication vitality towards a beneficiary and will centre its vitality towards the transmitter while accepting. The strategy utilized in the AAS is known as beam framing.

Beam framing empowers directional signal of transmission or gathering without physically guiding the antennas. In this pillar shaping strategy, a few transmitters are separate from each other. They all communicate a similar signal with various phase contrasts and deferral. Subsequently, the obstruction that happened in all the transmitters can be utilized to control a signal to a specific heading. In a versatile antenna system, signals can be scattered all the way around while numerous far off instruments are receiving the data packets. The state of these pillars can be controlled so that the quality of signal between the transmitter and the receiver is consistently greatest. The versatile antenna framework can expand the connection quality by consolidating the impacts of multipath interference just as by correcting diverse information streams received from various antennas.

8.5 DESIGN CHALLENGES

The system traffic keeps on increasing at a tremendous rate and the end-customer execution requests are increasing; these stress the Radio Access Network (RAN) to convey expansion in terms of attractive, limit and end-customer throughput. This requires wireless administrators' systems to convey step-by-step higher acquired limit after some time. Since information utilization is expanding at a quicker rate than relating income, versatile system administrators (MNOs) must develop the RAN in a manner that empowers both a decreased expense for each part and provides the necessary end-customer execution increments. The circumstance is presently directly for the telecom business to make an innovation move to AASs in both uplink (UL) and downlink (DL), and the practicality to assemble AAS cost successfully is upheld by innovation progresses in the reconciliation of baseband, radio and antenna, and a decrease in the computerized preparing cost to perform progressed beamforming and MIMO. Serious issues with MIMO being developed are as follows:

1. *Disconnection*: Mutual coupling between antennas is a primary concern when planning MIMO frameworks. Common coupling does not at this point just influence the antenna effectiveness and disengagement; not exactly −16 dB is required during the working area of the antenna framework.
2. *Bandwidth*: Return mismatch (in dB) ought to be not exactly −10 dB from 3.1 to 10.6 GHz so the interference in transmission capacity covers the total UWB. Concurrent improvement of disengagement, interference in data transmission in a solitary antenna configuration are perhaps the hardest test that exist in the structure of a UWB MIMO antenna framework.
3. *Size*: MIMO has been adjusted to versatile telephones, which are utilized in various innovations, such as WCDMA, WI-Max, WLAN and UWB. To acknowledge the rapid information transmission utility needs a conservative wideband MIMO antenna due to space limitation in wireless instruments. Subsequently, a reduced UWB MIMO antenna framework with low coupling losses among the antennas is utilized for UWB applications.

The new advances have become accessible and therefore oblige a portion of the limit implementations as per needs. The limit necessities will become considerably quicker. In this manner, the administrators need extra to fulfil limit needs. It is generally both causing technological trouble and a costly matter to become the number of locales. The accessibility of destinations is restricted, and the expense and time for procuring another site are not worthless. Subsequently, administrators normally centre around amplifying the utilization of existing locales features before extending the site lattice. AAS is an approach to build Signal to Interference-Noise Ratio (SINR) at the beneficiaries increment and rapid productivity with effectiveness of AAS, and especially cost productivity [5, 13, 31].

Radio engendering is significantly more testing at mm-wave groups in correlation with lower groups in the scope of 1–3 GHz transporter frequencies and 2.5 to 28 GHz for higher transport including satellite communication. The lower group's application using substrates for implantation of MIMO is FR1 (under 6 GHz); the limit upgrade is more fundamental because of the more interference based restricted condition. The higher range, FR2 (over 6 GHz), upgrade is more basic because of all the more testing engendering and way misfortune conditions.

8.6 DIFFERENT TYPES OF ANTENNAS

The MIMO technique uses different types of antennas as per the device specification. Portable antennas work on a wide frequency bandwidth in a small mechanical size. To accomplish the small size and wide frequency band, the following antenna types were examined and analysed.

8.6.1 Dipole and Monopole

The dipole antennas are broadly utilized due to their straightforward structure, simplicity to manufacture in large scale, high effectiveness and omnidirectional radiation design. They are resounding antennas, in which the operational frequency is based upon their length. The most generally utilized one is the half-wave dipole. Commonly, dipoles are narrowband antennas, yet there are compromises that can be used to extend the transmission capacity. A monopole is fundamentally a dipole sliced down the middle taking care of the ground plane. The monopole arrangements are half that of a dipole, and as needs are, the radiated power is half that of a dipole with a similar current, as the radiation happens just on one side of the equator.

Multiple-input multiple-output (MIMO) uses different antennas at the transmitter and the receiver stations. They have a dual capacity to consolidate the technologies of SIMO and MISO. By using Spatial Multiplexing (SM), they can create a capacity-based communication networking. There are several rational points of concern in the MIMO strategy over single-input single-output (SISO) strategies. Spatially assorted different types unbelievably kill the interference; low power is needed in contrast to various procedures in MIMO [21].

The hypothetical limit of the MIMO channel is communicated by the following formula:

$$C = E_H\left[\log_2 \det\left(\mathrm{I}_{mr} + \frac{1}{2} QHH^H\right)\right] \tag{8.2}$$

where $Q = E[xx^H]$ is the input covariance matrix and
$\rho(SNR) = Es/N_0$
Es is the total transmit power
N_0 is the noise power of each antenna

8.6.2 Loop and Slot

Circular antennas are another type of basic wire antennas that can take different structures such as square, round and three-sided. They are flexible enough, generally utilized in short-wave communication applications just as a substitute for monopoles because of their strength against demand of diversity. Circular antennas typically come in two classes – electrically small (periphery $<\lambda/10$) and electrically huge perimeter on the request for λ.

8.6.3 Planar Antennas

Customary dipole or monopole antennas are typically too huge to even consider fitting in the restricted body space of cell phones. Nonetheless, monopoles can undoubtedly be designed to oblige the size necessities by twisting the wire antenna to the structures of transformed L-antennas (ILA) [14]. Supplanting wires with a strip changes the structure to a planar transformed L-antenna (PILA), which brings about more extensive transmission capacity.

8.6.4 Microstrip Antennas

The microstrip antennas are effectively used in current cell phones. The microstrip antenna (MSA) can be imprinted on miniaturized size and adaptable substrates, which offer the close coordination to fit in cell phone space and similar to different mounting surfaces and structures. Another possibility of application in favourable position of microstrip antennas is that they can be coordinated in microwave incorporated circuits, to be off-track structures to abbreviate configuration time and lower configuration costs, and obviously to spare physical space [14].

8.6.5 Bow-Tie Antenna

The radiation example of a bow-tie antenna is like that of the dipole antenna, with vertical polarization, indicated exclusively between the two metal parts spaced at specific geometry. The antenna feed is at the focal point of the antenna, which is the radio positive and negative terminals associated with an antenna. Hypothetically, if the tie antenna is long in both ways, it would have an unbounded transmission capacity because the antenna operates to be identical at all frequencies. When cutting the bow-tie antenna arm to a specific length, the structure can be wideband [19, 10].

8.6.6 Log Periodic Antenna

By including the arms in a log intermittent way, bow-tie antennas' data transfer capacity can be further expanded, including log periodic dipole array (LDAP), log periodic toothed antenna and log periodic trapezoidal antenna with various arms plan techniques [15]. Indeed, even log intermittent antennas can offer wideband working highlights. This sort of structure needs an exact control of the arms sweep and points, which is hard to acknowledge on 5G applications when frequency operating shorter to millimetre extend plus, comparable with bowtie antennas. Log periodic antennas need to execute the feed point in the centre just as baluns in their taking care of the structure.

8.6.7 Leak Wave Antenna

Leak wave antennas (LWA) are collinear antenna clusters arranged sequentially at edges. One of the most significant highlights of break wave antennas is that they change the principle by changing the working frequencies, so various points of information checking can be accomplished. Full wave antennas have numerous comparative highlights as the Franklin antenna cluster. They are broadly applied in numerous communication systems, radar and millimetre-wave imaging systems with minimal effort, a straightforward structure, lightweight, simple to make and incorporated into other microwave circuits [16]. The first leaky wave antenna working data transmission was moderately thin [16] and could be easily incorporated.

8.6.8 Substrate Integrated Waveguide

The substrate integrated waveguide (SIW) innovation is for consolidating the benefits of both the traditional microwave and planar microstrip segments. The SIW can be depicted as a planar structure with the end goal that the top metal layer and the base ground layer are associated utilizing metallic chambers as two occasional columns. Hence, SIW segments have advantages such as small size, less weight and negligible expense since the creation and unwavering quality are improved.

8.7 MICROSTRIP ANTENNA DESIGN

The microstrip antenna (MSA) is also called a fixed antenna and a printed antenna. This kind of antenna is operating in mainstream inside portable applications, since it tends to be printed on the circuit board or conductive plastic surface. Fix antennas are effectively created and have minimal effort. The microstrip antenna is designed using three parts. The first is the metal-generated transmitting-radiating patch placed over the dielectric substrate, which is the second part of the MSA. The last component is the perfect ground plane that is created and suspended under the substrate using a similar metal from the fixed layer. Conversely, the size of the MSA antenna relates to its operational frequency. Particularly for 5G wireless communication applications, this component extends the well-known applications of it. The design of the antenna is based on the frequency conversion to fractional wavelength, as the dimensions of the MSA antenna vary according to the space availability inside the instrument and required performance. The designing of the antenna comes up with the following approach. The rectangular antenna designing steps are considering the substrate performance and its applicability for the den device. The fixed antenna ought to follow these guidelines:

Decide on the substrate that needs to be used and the substrate thickness (h) and dielectric consistent (ε_r). Decide the frequency (f_r) with appropriate bandwidth requirement and selection criteria. Compute the physical dimension of the fixed antenna.

There are also dielectric materials that are basically used as surface for dielectrically charged antennas. The conducting aspect here is that the radiator is used to scale back the physical size, and therefore the dielectric. The ceramic here is the radiating part and is pushed against a small ground plane. Efficiency and resistance to detuning are often fine for these antennas, but typically the bandwidths are modest. Broad ground planes like square, round antenna design on substrate are at quarter wavelength ($\lambda/4$) may provide high gain, greater than 6 dBi, but also narrower bandwidths. The ceramic is again the radiating element, but it is not pushed directly under the antenna against a ground plane. These antennas possess very large bandwidths, but gains are modest, normally 0.5 to 2 dBi, and they are detonated by a point of proximity. The configuration method implemented to first select the substrate and determine the resounding frequencies and then calculate the antenna measurements in order to plan a rectangular fix antenna for the use of the 5G communication network. Using low misfortune Teflon-based RT/duroid 5880 substrates, this antenna is structured. The pre-owned substrate's dielectric is equal to 2.2–5.5 and its thickness is equivalent to 0.381 mm for the 5G antenna development.

8.7.1 The Transmission Line Technique

The rectangular fixed antenna can be referred to in this model as two small originating spaces with displacement separated by width (W), height (h) and length (L). The miniaturized scale strip line is known as a non-homogeneous line of these dielectrics. There are two distinctive dielectrics air and substrate on which the MSA is printed. Due to the enormous measurement of the waves moving through the substrate and the rest of the waves travelling in the air, there is a viable dielectric stable Ereff. This means that the substrate and the air are replaced by Ereff, in a homogeneous medium.

$$\varepsilon_{ref} = \frac{\varepsilon_r + 1}{2} + \frac{\varepsilon_r - 1}{2}\left[1 + 12\frac{h}{W}\right]^{-1/2} \tag{8.3}$$

$$\frac{\Delta L}{h} = 0.4132 \frac{(\varepsilon_r + 0.3)\left(\frac{W}{h} + 0.264\right)}{(\varepsilon_r - 0258.3)\left(\frac{W}{h} + 0.84\right)} \tag{8.4}$$

So the dominant mode will be

$$(f_{r010}) = \frac{1}{2L_{eff}\sqrt{\mu_{eff}}\sqrt{\frac{1}{4}_0\mu_0}} = \frac{v_0}{2(L + \Delta L)\sqrt{\mu_{eff}}} \tag{8.5}$$

8.7.2 Cavity Mode

The rectangular fixed antenna can be referred to in this model as two small originating spaces with displacement separated by width (W), height (h) and length (L). The miniaturized scale strip line is known as a non-homogeneous line of these dielectrics because there are two distinctive dielectrics (air and substrate). Due to the enormous measure of the waves moving through the substrate and the rest of the waves travelling in the air, there is a viable dielectric stable Ɛreff. This means that the substrate and the air are replaced by Ɛreff, a homogeneous medium.

$$(fr) = \frac{1}{2\pi\sqrt{\mu\varepsilon}}\left[\sqrt{\frac{m\pi}{h}} + \sqrt{\frac{n\pi}{h}} + \sqrt{\frac{p\pi}{h}}\right] \quad (8.6)$$

πm = the number of variation of the half-cycle field along the x-direction.
πn = the number of variation of the half-cycle field along the y-direction.
πp = the number of variation of the half-cycle field along the z-direction.

The accompanying table decides the estimation of the dominant mode where the frequency is the least catered with microwave mode of operation for all fix antennas which have a huge enough length and width. The mode of operation is a function of the relation with the L, W, H summarized in Tables 8.2 and 8.3.

8.7.3 Antenna Shape

The rectangular type of design is associated with the L, W, H dimensions, with the specified thickness of dielectric on the same way circular shape and the circular dimension is point multiple of the wavelength. The dimensional detail and designing approach are explained in Equations (8.7) and (8.8), Table 8.4 and Figure 8.3. Also the elements in the antenna have specific applications in achieving the diversity with

TABLE 8.2
Antenna design with dimensional details

Freq GHz	W Um	L um	Effective dielectric Ɛreff	ΔL	Lg um	Wg Um
1.227	97256392	83180172.61	2.160000035	0.841936	83180180	97256401.2
1.575	75767360	64801315.27	2.160000045	0.841936	64801323	75767368.9
1.72	83917190	80971803.68	1.160000022	1.05366	80971811	83917198.8
2.1	56825520	48600986.28	2.16000006	0.841936	48600994	56825528.8
2.4	49722330	42525862.91	2.160000069	0.841936	42525871	49722338.7
2.6	45897536	39254642.64	2.160000074	0.841936	39254650	45897545.1
3	39777864	34020690.2	2.160000086	0.841936	34020698	39777873.7

TABLE 8.3
Mode of operation and resonant frequency calculations

Condition	Mode TM_{mnp}	Resonant frequency
L>W>h	TM010	$(f_r)_{010} = \frac{1}{2L\sqrt{\mu\varepsilon}} = \frac{v_0}{2L\sqrt{\mu\varepsilon}}$
L>W>L/2>h	TM001	$(f_r)_{001} = \frac{1}{2W\sqrt{\mu\varepsilon}} = \frac{v_0}{2L\sqrt{\varepsilon_r}}$
L>L/2>W>h	TM020	$(f_r)_{020} = \frac{1}{L\sqrt{\mu\varepsilon}} = \frac{v_0}{L\sqrt{\varepsilon_r}}$
W>L>h	TM001	$(f_r)_{001} = \frac{1}{2W\sqrt{\mu\varepsilon}} = \frac{v_0}{2W\sqrt{\varepsilon_r}}$
W>W/2>L>h	TM002	$(f_r)_{002} = \frac{1}{W\sqrt{\mu\varepsilon}} = \frac{v_0}{W\sqrt{\varepsilon_r}}$

TABLE 8.4
Standard MIMO circular patch antennas

Legend	Dimension
W	4.60
L	4.43
Cr	1.35
Wg	0.65
G	0.023
Ws	0.95

AASs. Concerning the roundabout fix antenna, given the operating frequency of 2.5 GHz to 28 GHz, its sweep can be determined as

$$F = \frac{8.791 \times 10^9}{f_r\sqrt{\varepsilon_r}} \tag{8.7}$$

$$C_r = \frac{F}{\left[2.7726 + \frac{2h}{\pi \varepsilon r F}(\ln)\frac{\pi F}{[2h]}\right]\mu^{1/2}} \tag{8.8}$$

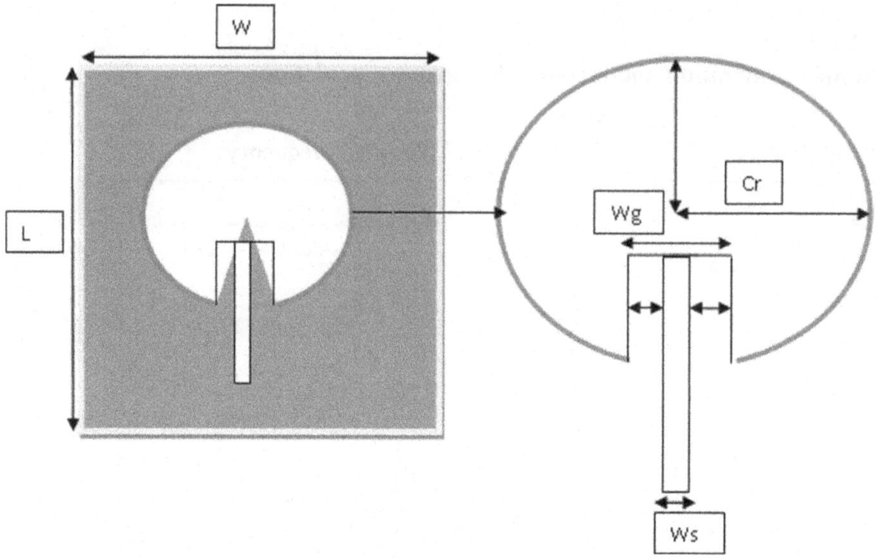

FIGURE 8.3 Circular path antenna designing.

8.7.4 Single Component Antenna without Spaces

The arrangement is centred around the 28 GHz single-band antenna with space added to the fix to operate in other bands (38 GHz). In the structure, a hole-coupled feed line is used to increase the speed of data transfer in antennas as shown in Figure 8.4 and Table 8.5.

8.7.5 H-Shaped Single Component

An H-shaped position was made on the fix to conduct double-band operation for the miniaturized scale strip fix antenna for the 5G wireless communication network. The H-shaped opening's external components were basically identical to the 5.5 GHz to 38 GHz fixed antenna elements structured on a similar substrate.

8.8 DIVERSITY TECHNOLOGY IN MOBILE COMMUNICATIONS

Disadvantages of the low-frequency multi-reconfigurable antennas are: thin transmission capacity, greater rigidity, effectiveness and low segregation of element. Moreover, the transporter collection is a pattern to improve the range use effectiveness. Because of the MIMO reconfigurable antenna, a methodology of antenna arrangement in different types of frameworks is proposed to illuminate the downside of the low-frequency reconfigurable antenna which provide considerable improvement in traffics. Initially, two reconfigurable resonators are associated with one another through a basic divider.

Points of interest of utilizing numerous antennas, either toward one side or on the two closures of a correspondence channel, can be seen from a few viewpoints.

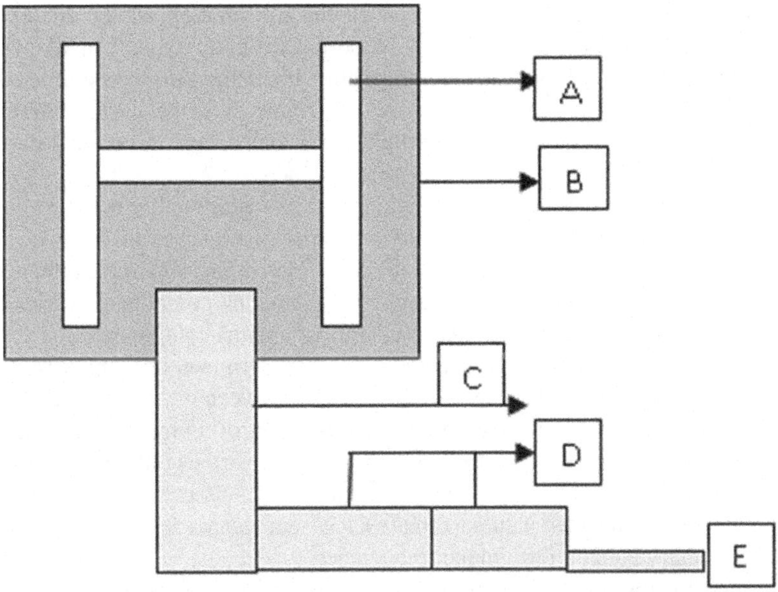

FIGURE 8.4 H-shaped microstrip antenna.

TABLE 8.5
Dimensional details of antennas

Physical dimension	Dimension in mm
W	4.15
L	3.8
A	0.054
B	0.060
C	2.10
D	1.54
E	0.0616
Y	0.85
Z	1.65

How these points of interest work relies upon the applied space–time multiplexing diversity [25, 33]. The indications of directional radiation can be formed by using antennas for spatially-particular transmission or potentially receiving signal, leading to cluster benefit and improved quality of communication. It can also add to the relief of interference in a multi-customer cell framework, e.g. the base station (BS) antenna displays the shape of an invalid towards the course of the meddling service provider in the cluster design.

Another simple process of space–time planning is a diversity-based selection of antennas. On account of RX recuing antenna, collecting the signal from various antennas might result in interference. Given the information on the channel, the received signal can be joined usefully. Assorted different types can likewise be abused at the TX side without channel information by methods for space-time coding. Antennas assorted with different types of procedures have been concentrated for quite a few years, and one current spotlight is on their usage in minimal customer terminals [34].

Space diversity is the most widely recognized and presumably the least difficult technology for accomplishing a diversity principal of branches in space or spatial diversity types. Utilizing two isolated antennas isolated far from one another means that, the phase change signal showing up at the antenna components varies at the interface of the two antennas. The base separating antenna components at a versatile terminal is typically about 0.5 frequency, which is determined utilizing zero-request Bessel work, to obtain a sufficient low relationship between the interference signal.

Frequency diversity is depicted as the transmission of a signal at least two distinctive frequencies transmitted, to such an extent that various radiating signals with different autonomous interference can be accomplished. It is an expensive component because of the multifaceted nature to create a few sent signals and to combine signals at the beneficiary but at a few unique frequencies.

Signals showing up at antennas originate from various points or bearings. Being autonomous in their interference varieties, these signals can be utilized for edge or precise diversity of different types. At a portable terminal, point diversity of different types can be accomplished by utilizing two omnidirectional antennas that operate as parasitic components to one another. The radiation examples can be exchanged to deal with the gathering of signals at various boundaries.

Time and multipath diversity varieties are connected systems that are no doubt applied in the advanced transmission. Time-assorted different types are accomplished by communicating a solitary part of data dully at various time stretches. Varieties of these redundancies of the signals will go free. Multipath mixed signal different types are utilizing time-assorted different types in multipath situations. The monotonous signal is sent to the collector in various voyaging ways [23].

Transmission of one polarization signal gives the chance of accepting two symmetrical polarization signals with variations in the correlated functions. This is moderately useful because a number of small-sized antennas can be utilized.

8.9 ANTENNA ARRAYS

The radiation of single-component antennas is not moderately wide. This implies they have generally low directivity. For upgrading the directivity, the components of the single-component antennas can be developed. Another approach to upgrading the directivity is by gathering transmitting components in an appropriate electrical and mathematical setup to shape the antenna cluster. The exhibit components are generally indistinguishable. This isn't essential, yet it is more straightforward and useful for the creation and planning of MIMO development [24].

8.9.1 Two-Component Cluster

The antenna under scrutiny is a combination of standardized small dipoles located along the z-hub, all outfield radiation from the two factors, expecting no coupling between the components, equivalent to the whole of the two, and it is given in the y–z plane by:

$$Et = a_\theta\, j\eta \frac{lI_0}{4\pi}\left\{\frac{e^{-j\left[kr1-\frac{\beta}{2}\right]}}{r1}\cos\theta 1+\frac{e^{-j\left[kr1-\frac{\beta}{2}\right]}}{r1}\cos\theta 2\right\} \tag{8.9}$$

where ß is the excitation difference in phase between the elements. The magnitude excitation of the radiators is identical

$$I_1 = I_0 e^{-j\left(\frac{\beta}{2}\right)} \tag{8.10a}$$

$$I_1 = I_0 e^{+j\left(\frac{\beta}{2}\right)} \tag{8.10b}$$

For sampling the total electrical field

$$r_1 = r - \frac{d}{2}\cos\theta \tag{8.10c}$$

$$r_1 = r + \frac{d}{2}\cos\theta \tag{8.10d}$$

For amplitude variations

$$r_1 = r_2 = r$$

So, the total electric field becomes:

$$Et = a_\theta\, j\eta \frac{lI_0 e^{-jkr}}{4\pi r}\cos\theta\left\{2\cos\left[\frac{1}{2}\left(kd\cos\theta+\beta\right)\right]\right\} \tag{8.10e}$$

The large field of the array product is equal to the independent average field of a single element at the starting point, which is commonly referred to as the Array Factor, increased by a factor. Subsequently, the exhibit factor is given for the two-element cluster of consistent sufficiency by:

$$AF = 2\cos\left[2\cos\left[\frac{1}{2}\left(kd\cos\theta+\beta\right)\right]\right] \tag{8.11}$$

By fluctuating the detachment d or potentially the phase ß between the components, the qualities of the exhibit factor and all outfield of the cluster can be controlled.

8.9.2 Four-Component Antenna Cluster with Fixed Bar Controlling

The bearing of the antenna radiation design is additionally significant in the antenna structure, where the 5G system requires antennas with a guiding capacity. Higher directivity implies smaller radiation, so it must have the option to control the primary pillar bearing. There are strong characteristics of guiding radiation components of the antenna. The 5G system can be accomplished by numerous techniques that rely upon altering the phase or the extent of the span of the information signal feed to the antenna with the LHP and RHP phase shifter arrangement as shown in Figure 8.5.

This will alter the directivity and radiation efficiency of the primary structure to an ideal impact as shown by the phase distinction between the indicators taking care of the antennas. The phase layout relocated on the four-component antenna exhibit one on the right side and another on the left side, which were included in the phase shifter. These two structures can be achieved by the right to control fixed frames, where the principle barrier can be directed to the θ bearing and the other structure [15].

8.9.3 N-Component Uniform Straight Exhibit

A uniform cluster is a different type of component with indistinguishable great characteristics and each with a dynamic phase placement. Expect that N isotropic components have common amplitudes, however, each succeeding component has a ß dynamic phase. The current in the previous component drives the current of the next

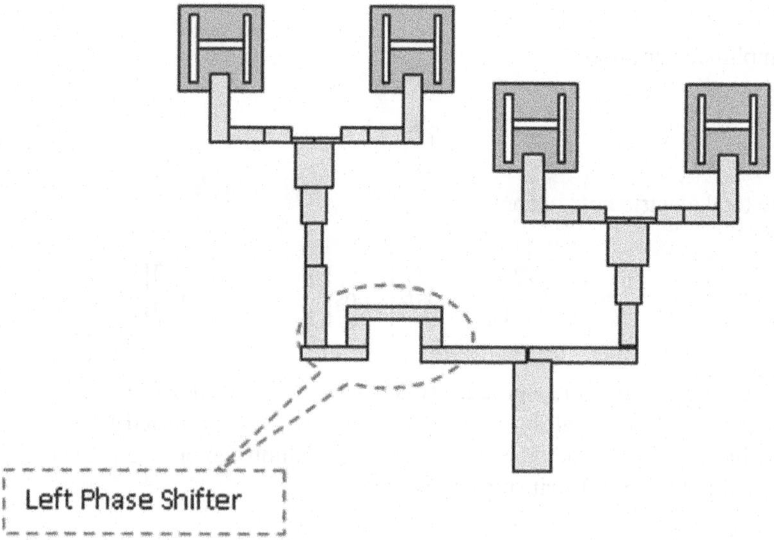

FIGURE 8.5 4 × 2 array arrangement with polarization arrangements in the phase shifter.

component. All of the outfield can be shaped by duplicating the exhibit factor by the field of a separate component [24]. This is the example of multiplication rule and it applies for varieties of common components. They exhibit multiplication factors for the far-field that can be framed by:

$$AF = \sum_{n=1}^{N} e^{j(n-1)\varphi} \tag{8.12}$$

$$\varphi = kdcos\theta + \beta$$

8.9.4 Structuring

MIMO innovation is an alternate way to multiple-input multiple-output innovation. This innovation is utilized in a wireless communication system for expanding the information rate and the framework's ability. The MIMO innovation additionally improves the nature of administrations of the communication system. The transmission capacity of the antenna helps the wireless system for transmitting enormous information at high rates. Likewise, the improvement in common coupling between nearby antennas helps in planning a successful MIMO framework [15].

The currents distributed by the four short patches are fed in phase opposition (the suggested antenna radiates in left hand circular polarization) to achieve circular polarization, while dual polarization is implemented by exciting each antenna pair in antiphase with the same magnitude. The same concept is introduced on an FR4 PCB as a rectangular ground plane below and a quarter wavelength slot antennas inserted diagonally on the back side of the PCB, structured with two printed F antennas on the top layer. To achieve better separation between the F antennas and to be used as a radiator, the quarter wavelength slots are mounted, raising the resonance at the desired frequency.

Increased directivity is important for the antenna in which the antennas are structured to transmit and receive signals promptly. The single-component fix antenna has relatively large emission architecture but subsequently reduces extension and low directivity. The need of a significant rise in directivity and coverage area in the 5G system will not be met from the single-component fix antenna.

The improvement in the throughput of system using an antenna cluster is planned, which enhance the utility of microstrip fix antenna. The exhibit of transmitter would be a range of single-antenna components that are linked together to function as a common radiated signal using the antenna. Increasing the number of antenna dipole causes the rise in gain of the antenna and hence the directivity drastically increases. The recognized antenna cluster is not important in the form of associate indistinguishable components together, but it is favoured for easier creation.

8.9.5 Integrated Antenna

Heritage macrocell design had huge radio frequency (RF) in covers at the base of the important part. Coaxial links ran up the important part from the radio to the antenna.

With the consideration of important part-mounted instruments, tower-mounted amplifier was utilized to rearrange the RF prerequisites of the radio at the base of the important part with the compromise of extra power. Further refining this method, the wireless radio head (RRH) is set on the important part that handles the RF capacity of the radio. Along these lines, improving the gear in the asylum and lessening RF misfortunes from long link lengths, the compromise is more perplexing gear and thus more power requests on the important part.

The advanced part of the radio, i.e. the baseband unit (BBU), is situated at the base of the transmitter part. The connection between the RRH and the BBU is linked with the optical fibre that decreases the tower stacking contrasted and the first coaxial link runs up to the specific part. This partition of RF and baseband practically takes into account a more diminished impression on the important part off and in the safe house just as improving system execution [24–26].

The establishment of intricacy originates with MIMO, the number of ports on an antenna and the radio increment for each band. This infers the number of links that interface the radios to the antennas increments also. This expansion includes multifaceted nature since the radio port definition must match the correct ports on the antenna. Even though the links are a lot shorter than those that ran up the important part, the misfortunes are as yet critical and if one of the links falls flat, the band won't work appropriately.

The name incorporated antenna/radio has been utilized in the past to portray when dynamic instruments are planned legitimately into the antenna. The benefit of a coordinated antenna/radio dispenses with the requirement for multiple cables between the radio and antenna, improving the establishment and including unwavering quality by lessening part tallies. Presently just power, baseband and control data are required by the antenna framework.

8.9.6 Phase Shifters

A phase move module indicated in above diagram is a microwave organize module which gives a controllable phase move of the radio frequency signal. It is utilized in phased clusters. Applications remember controlling the overall period of every component for a phase cluster antenna in RADAR or steerable interchanges interface and cancelation circles utilized in high linearity intensifiers.

8.9.7 Fractal Antenna

The fractal antenna placement ideas are utilized in different buildings and science disciplines such as arithmetic, geology, physiology and so forth. Fractal shapes utilize the electrodynamics conception of wave travelling mechanisms for the improvement of the printed antenna performance, as extremely powerful tools. The fractal antenna has an important commitment in the present testing, antenna building new research area [1]. Fractalization of the antenna can be accomplished through different dual structures and shapes such as Koch bend, Sierpinski and other deterministic or non-deterministic fractal shapes. Distinctive kinds of other fractal structure talked about different applications [2, 3].

Fractalization has been accomplished through numerically downsizing the component of a structure redundantly. Thus, it is exceptionally valuable in accomplishing multiband/multifunctional qualities [6–8]. Presently multipath interference is a significant problem in communication. To moderate improvement in multipath interference, MIMO technology has been used. The channel limit has been upgraded by MIMO without extra transmission capacity. The high information rate is just conceivable with MIMO. Fractal cluster for 5G MIMO applications presented a good solution for the high data rate [9–10].

The antenna exhibit is structured on an FR4 substrate of thickness 1.58 mm with dielectric coefficient $\mathcal{E} = 4.4$ and tangent loss factor of 0.017 because FR4 has safe properties. After all, FR4 has flame-resistant properties and good strength [14]. The changed antenna components, 1 × 2 Sierpinski fractal antenna exhibit and 2 × 2 Sierpinski antenna cluster with two arrangements is created with better execution as far as an addition. The following are the designing steps:

(a) Conventional triangular patch antenna (CTPA) where the components of the fixed antenna have been determined from the numerical articulations [15].

Calculation of side length

$$\varphi = kdcos\theta + \beta \tag{8.13}$$

The compelling dielectric consistent of the triangular microstrip fix antenna

$$a = \frac{2c}{3f_r\sqrt{3\varepsilon_r}} \tag{8.14}$$

The compelling side length of fix (L)

$$\varepsilon_{eff} = \frac{1}{2}(\varepsilon_r + 1) + \frac{1}{4}\frac{(\varepsilon_r - 1)}{\sqrt{1 + 12\frac{h}{a}}} \tag{8.15}$$

At that point, a fundamental fix antenna is adjusted as a filleted corner to adjust the parasitic reactance helpful for the perfect ground achievement. The reflection coefficient, radiation resistance, electrical length have been improved by cutting the opening of round shape or rectangular shape at the corner or at centre in the feed line. A round slice is likewise to get the oddity fit as a fiddle. These oddities served to improve the execution of fundamental fix antenna. Altered three-sided fractal fix antenna with adjusted corner cut in the feed is shown in Figure 8.6.

8.10 DESIGN OF HANDSETS

An end-fire phased antenna has been introduced. The single-component antenna has a leaf bow-tie shape and has served the 3.5–28 GHz, 10.5–38 GHz frequency groups. Its S11 is −20 dB to −15 dB at 26 GHz and −30 dB to 17 dB at 35 GHz having the

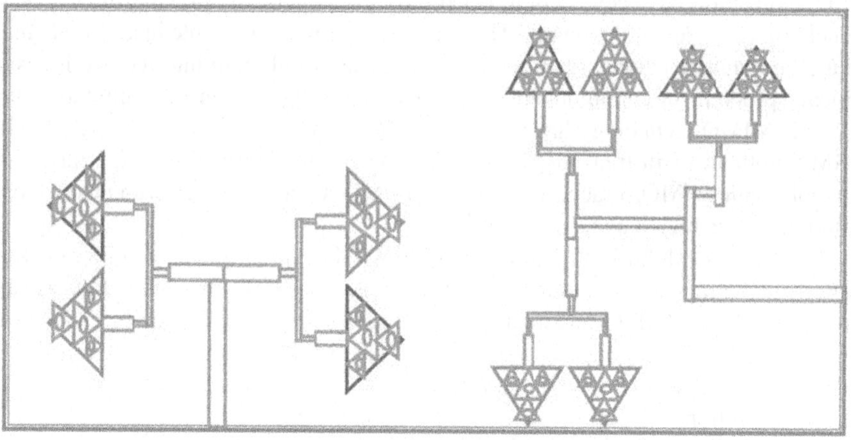

FIGURE 8.6 Handset arrangement of antenna for betterment in throughput [5].

size of leaf-moulded bio-tie shape in the direct phase cluster. The double bandwidth MIMO antenna with the structure for 5G versatile handsets has been introduced in communication. The antenna comprises eight components situated on the symmetrical casing corner of the substrate. The activity groups are 3.4 GHz–3.6 GHz and 4.55 GHz–4.75 GHz. Its S11 is –35 dB at the main operational frequency range and –25dB at the subsequent one.

A double-band antenna exhibits with polarization and far directing ability highlights has been introduced by Mahmoud and Montaser. Its activity groups are at 28 GHz and 38 GHz. The single component comprises three copper layers where the centre layer has a T-shape to link each substrate layer have a gap for opening. This single component was utilized in a 12 component cluster for a versatile handset where they were separated similarly on top, both sides of the portable device [19].

A fast audit on the components of the cutting edge versatile PDAs has been done as shown in Table 8.6. Likewise, in the wake of auditing, some of the earlier literature on the antenna configuration field; the elements of handset configuration have been picked as 110 × 55 mm^2 dimensionally used in the mobile handset.

The different antenna documentation on MIMO framework is considered as innovative and applicable technology for use in 5G. The 4G compact framework is used to improve the information speed with some limitation in terms of base antenna placement and handset antenna alignment [5, 18, 19]. As described before, the current flexible handset continues to function at mm-wave at the current sub-3 GHz band because 5G can run at mm-wave. This means that more MIMO antennas should be incorporated to cover different advances and ranges inside the handset body. The antenna is huge, compared with the limited space within the handset. By using a standalone structure as proposed two structures [14, 21–23]. This limitation can be partially mediated by constructing a MIMO consisting of antenna components that

TABLE 8.6
Modern advanced mobile phone antenna sizes

MIMO Mobile	Width	Length
Galaxy S7	7	14.2
Galaxy S7 edge	7.3	15.1
Galaxy S8	6.8	14.9
Galaxy S8 plus	6.7	16
iPhone 6S	6.7	13.8
iPhone SES	5.9	12.4
iPhone 7S	6.7	13.8
iPhone 7S plus	7.8	15.8
iPhone 8	6.7	13.8
iPhone 8 plus	7.8	15.8
Pixel 2	7.0	14.6
Pixel 2 XL	7.7	15.8
One Plus 3T	7.5	15.3
One Plus 5T	7.5	15.6
LG G6	7.2	14.9
LG V30	7.5	15.2
BL V Dash	6.3	12.4
HTC U11	7.6	15.4

cover cross-groups for 5G and 4G. Towards this end, in the 5G handset industry, MIMO 5G/4G is one of the key variables subsequently applicable for the service provider and customer with better quality experience.

The utilization of data transmission for versatile and cell systems has been expanding at a phenomenal rate, with numerous rising applications. The anticipated increase in portable system use requires a high information speed. The restricted limit of the 4G versatile system is not required to fulfil the anticipated high information request by 2020. To meet this prerequisite, another 5G MIMO-based versatile system has been proposed [4]. To arrive at high limit requests, 5G requires a wide frequency transfer data rate, which is not accessible in the sub-3 GHz frequency range. One of the potential frequency band possibilities for 5G mm-wave centred around 10.5–28 GHz extended till 38 GHz [2, 11].

A convoluted test is the incorporation of a mm-wave blade directing structure into a cell phone handset, with the thought of limited accessible space. The necessary mix with other segments of the instrument and the multifaceted circuit nature of the full guiding framework of the pillar. In addition, the predicting fields strength from the antenna are unambiguously consumed by the surface layers of the human body at mm-wave. Consequently, it is important to use the guide to reduce the EM radiation organized through the body of the customer. Subsequently, maintaining a clear interaction with the transceiver station at a high data rate is a key factor in building up a complete beam steering system for portable handsets, while simultaneously decreasing the EMF penetration to the human.

By consolidating two sub-arrays, the 3D pillar directing capacity is accomplished. For the presentation of the concept of the beam controlling ability, a complete circuit is produced from the 2D guiding model. The full beam controlling circuit comprises a two-layer dipole antenna cluster at the different sides of the PCB, two Wilkinson power dividers, two advanced phase shifters that use a novel technique of controlling the p–I–n diode operation in the circuits, microstrip-to-coplanar strip line baluns and a novel two-layer microstrip feed power divider area. In addition, a tuneable power divider for 5G-directed circuit shaping has been added additionally to enhance the execution of guiding structure. The tuneable power divider is utilized to tune an antenna cluster. To wrap things up, it is proposed to take care of a tuneable circuit to approve the ability to use the proposed tuneable power divider to smother the side-projection level of the antenna radiation design [26].

Most of the time, four-component antenna clusters in the handset strategy is implemented as described. The structure has four elements in which two of them are placed on the upper flat side of the handset and the other pairs are placed on the correct vertical side corner of the handset with acceptable separations between the patches' feed line and the receiving line. The antenna scattering provided the double polarization spotlight to the framework where the vertical polarization was served by the patch on the even size and the flat polarization was served by the side patches. In addition, the proposed structure underpins the innovation of MIMO and achieves double-band operation, an adequate increase (>10 dBi) and ample ability for data transmission (>1 GHz). Subsequently, the proposed antenna configuration for the 5G communication system is a good possibility for handsets.

The ports are named as port 1 and port 2. The shared coupling repeats the radiation results between them from −17 to −30 dB. In this structure, the separations between the MIMO setup are reasonable and all the re-enacted shared coupling results are satisfactory and better than −30 dB as seen from the charts.

A dipole antenna sub-array with 4×2 mm^2 dimensions with metamaterial to coplanar strip line baluns, with a 50 Ω microstrip line as feed line are used several times in the framework. It has a to-centre dispersion at a position of 4.9 mm between adjacent dipoles, precisely 0.45 λ at 28 GHz and 0.35 λ at 36 GHz. The system is constructed using two three-layer substrates where the top and base layers are used for ground layer insulation.

The dipole antennas with dimensions $4 \times 1 \times 2$ mm^3 are arranged as shown in Figure 8.7 and are used in each of the top and base layers to reduce the interference and to provide the feeders with the interface between the antenna monitor and microstrip-to-coplanar strip line baluns [62]. Each part of the antenna is made of a printed dipole using a shortened ground plane as a reflector plane. The two-dipole sub-cluster technique empowers the structure in the H-(yz) plane to direct the pillar, while the 4×1 antenna course of action in each layer empowers the ability to guide the beam in the E – (xy) plane. The consolidation of the two planes therefore gives 3-3-D beamsteering capacity [28].

Therefore, one of the most critical variables for ensuring the protection of rising 5G advances is the reduction of EMF penetrations to the human head. One of the main favourable circumstances of the proposed structure is to assess, by directing

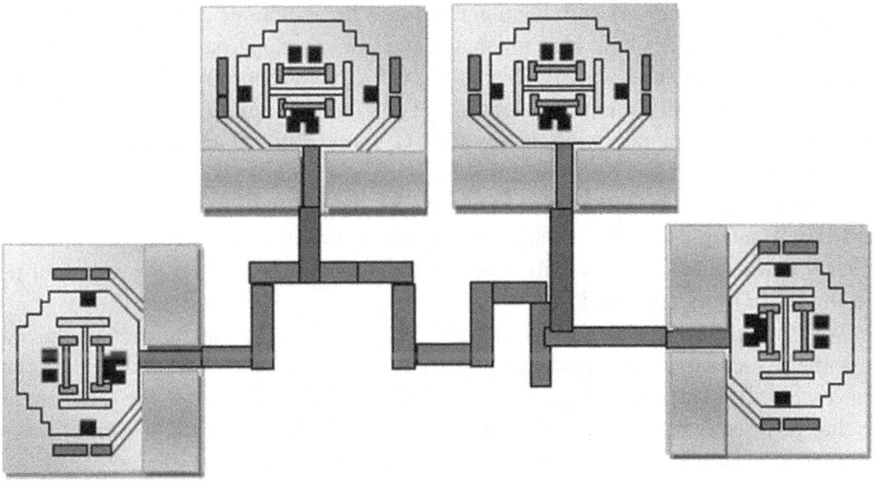

FIGURE 8.7 Mobile antenna with LHCP and RHCP utilizing antenna design.

the beam to the base station, the previously stated EM penetrations to the head of the customer.

Using the specific absorption rate (SAR), the EM level at low frequency is evaluated for advances, including 3G and 4G. In any event, according to the Federal Communications Commission (FCC) and the International Commission on Non-Ionizing Radiation Safety (ICNIRP) [13], the guideline for the implementation of EM was transformed from a SAR cut-off to control thickness (PD) at mm-wave.

A LOESS smooth procedure in MATLAB was utilized to eliminate the commotion from the estimated radiation, coverage, gain and associated dimensional design. The deliberate outcomes are in acceptable concurrence with the re-enacted qualities. The contrasts between them at this high frequency could be brought about by the resilience in the dielectric consistent of the substrate, the misalignment of the plan layers, particularly because of the utilization of plastic screws to join the layers together and conceivable manufacture mistakes [27].

The designing of MIMO is based on the perception and service utility. The device simulation is done using the HFSS antenna simulator and CTS studio. The radiation parameter and probable methods of the performance prediction are visualized.

The primary component in the antenna configuration is circular polarization (CP) to moderate geometrical and multipath disturbances. In addition, this element is correct to the ongoing flexible framework of the 5G. Surprisingly, contact between device to device (D2D) [17] and two CP radiators will suffer from severe effect of extraordinarily low efficacy of polarization if the two transforming methods are set in reverse ways. CP antennas with tilting modes that can be switched between right-hand circular polarization (RHCP) and left-hand circular polarization (LHCP) should be used to identify this problem.

Electronic components, p-I-n diodes, are one of the most commonly recognized component used at high frequency to identify reconfigurable circuitous polarization antennas as they may certainly be integrated into the structures of the antennas. The accessible p-I-n diodes have high mismatch and low coordination for 5G applications at millimetre-wave (mm-wave), which was professed in reporting to limit the use of reconfigured p-I-n diode development in CP antennas at mm-wave. Two new polarization tuneable antenna architectures for 5G applications at mm-wave have been defined [28].

Second, a single fix antenna is generated with the utility to switch between LHCP and RHCP using a straightforward feed circuit. There is a four-component antenna display that can switch its operation between two pivoting modes using only four p-I-n diodes (RHCP and LHCP).

With equal abundance signals and 90° phase delay, the patches are energized in either clockwise, e.g. right-hand (RH), or anticlockwise, e.g. left-hand (LH), succession. It is important to take care of the structures that can provide ±90° phase exchange to understand a reconfigurable CP display. The proposed structural management consists of a branch-line coupler, two arms and two p-I-n diodes.

8.11 PLAN OF BASE STATIONS

A conservative millimetre-wave huge MIMO has been introduced. Its S11 is −17 dB at 28 GHz and −28 dB at 38 GHz. At 28/38 GHz, the increased esteem is 12 dBi at each band. Antenna components are circulated in space for huge MIMO base station engineering with a range of 25 mm. Absolute bars checking of 360° is accomplished by 12 exchanged components. A tapered frustum exhibit antenna of multi-polarization has been introduced. 5G double groups had been served by the base station (28, 38 GHz). Its 32-component exhibit antenna had been circulated in cone-like frustum design and it accomplished 8.17 dBi gain. The triple band and multiband MIMO antennas are provided with a frequency scope of a 28–38 GHz two-port base station and an exchanged bar guiding component has been introduced [21].

The primary thought of this innovation is to send or get distinctive information streams by utilizing different antennas at a similar transporter signal without more power. At the point when the antenna is communicating its electromagnetic wave (information stream), the wave takes various ways as a result of dissipating conditions (multipath proliferation). Thus, in a basic way, we can characterize MIMO as utilizing more than one antenna simultaneously.

The MIMO limit improvement that can be provided by the intrasite multiarea participation is concentrated in a deliberate urban condition. The outcomes show that higher than 40% limit improvement is accomplished at the part edge district, comparative with that of the single-segment connect with no participation.

Throughput is expanded to the customer by exploiting MIMO and transporter conglomeration with a high port check antenna. Moreover, a solitary or pair of the high port tally antenna can diminish the number of antennas walled in areas. In a perfect world, the important part limitations will be maintained by keeping the weight and wind heap of the new antennas to the equivalent or not exactly of the antennas they dislodge. Not all high port tally antenna can keep this last guarantee.

Huge MIMO has appeared to improve the organized limit by making the MIMO innovation one stride further with the joining and utilization of more than eight radio

T/R (communicate/receive) modules inside an antenna. Some random range antennas will remain a similar estimate or become genuinely bigger with the expansion of massive MIMO. Notwithstanding expanded size, weight additionally increments because of the incorporated hardware. RF mismatch and power inter modulation (PIM) are vastly improved because of the RF front-end being coordinated into the antenna. Nonetheless, the power utilization and warmth dissemination of the huge MIMO antenna expands as the quantity of T/R increases.

Base station antennas with various structures are proposed in late writings. Here, the creator intends to structure BS antenna with 4 × 4 mm^2 MIMO double enraptured antennas. These antennas improve the transmission (base station) and recipient (portable terminals) executions. The base station comprises two kinds of antennas, one is arranged destinations and the second one is the endorser site antenna. System locales give the channel highlight point (PTP) and supporter site is used to move information from highlight multipoint (PMP). Sector antenna endorser destinations sector antenna with strong developments for small scale base station applications whose frequency band is 45 MHz. Here, the utilizing one is versatile beam framing and TDD procedure.

The double dumbbell exhibit is new proficient as a MIMO base station antenna. [8] So, structured base station of MIMO antenna with the help of a double dumbbell antenna. Subscribers are at present requesting higher data rates to help applications such as web access, video and games.

The previously introduced four-component antenna exhibit is currently used to build an octagonal crystal structure base station to support the 5G communication system by achieving MIMO to enhance the nature of the communication jointly. On all the eight sides of the base stations [14–17], the propagation of the antennas is indistinguishable.

There are twelve 4-component antenna exhibits on either side of the base station where the three 4-component antenna plans (without phase shifter, with right phase shifter and with left phase shifter) were used. To achieve fixed beam control capability for the proposed 5G station, the exhibits with phase shifters are embedded in the structure.

The cuts on the vertical edges on the sides of the base station are really intended to allow the coaxial link connectors to interface. Double polarization is known to include where the three plans are mounted on the four edges of the side of the base station. Set on the vertical edges, the antenna displays allow the base station to work with flat polarization and the even edges allow it to work with vertical polarization. The two-way phase shifter displays allow the basic beams to be inclined separately at the edges of 9 and −9° at 28 GHz and allow the main bars to be inclined separately at points of 10 and −10°degrees at 38 GHz. Due to indistinguishable findings for displays with the same separation between ports, not all mutual coupling outcomes are shown in the diagrams.

8.12 MIMO CELLULAR SYSTEMS

Until now, the postulation has just acquainted point-to-point MIMO system, i.e. single-dual user situation. The cell systems consist of a multi-selection situation which ought to be tended to. For instance, the downlink instance of the phone framework can be concentrated by numerous RX antennas considering that are spatially disseminated on geological zone, every one of which speaks to one customer with one RX antenna [4, 6, 11, 30].

Space division multiple access (SDMA) is a term recognized in radio access organizing innovation that utilizes multi-dimensions and shaped antenna systems to permit more than one customer service in a subsequent cell to speak with the BS on a similar frequency with a similar time frame [8]. The multi-element antenna framework utilized at the BS can recognize among the customers by methods for their spatial marks. It can likewise frame a pillar for a committed customer to improve the connection quality, just as an invalid to different customers to stifle co-channel interferences.

In this way, to effectively enhance end-customer execution, the aforementioned space–time planning techniques, such as beamforming, different types of diversities and spatial multiplexing, can also be used. In the case of the different access channel (MAC), i.e. the uplink case, the locale limit with either free or joint interpretation to the receiver (i.e. the BS) is equivalent in terms of the point limit highlight structure with no channel information to the transmitter. This is because cooperation at the end of the receiver is conceivable. Nonetheless, in order to fully identify the limit, the aggregate prices for all possible customer blends (along with their associated MAC channels) need to be calculated for this situation.

The limit locale is specified by the individual paces of customers 1 and 2 (i.e. R1 and R2) in a two-customer situation, just as the entire pace of the two customers, R. Similarly, each customer's SNR relies on its own accessible communication capacity, and no sharing of sending power can occur among customers. A pre-coding technique known as filthy surface coding (DPC) [23] is found late in the transmission channel (BC) to reach the downlink limit. A curious duality between the uplink and the downlink maximum exhibits also exists [33].

It is becoming increasingly necessary to achieve higher important part data rates alongside the development of the cell system, but in addition to higher rates across the entire attractive region. In particular, a significant objective for the LTE-Advanced standard is to further enhance cell edge throughput execution.

In fact, this MIMO principle, known as facilitated multipoint (CoMP) transmission and selection, was introduced in LTE-Advanced as an applicant approach to enhance the framework efficiency and attractiveness. It is possible to update the CoMP structure between a few BSs (interstice) or within a few divisions of a solitary BS (intrasite).

Utilizing progressed space–time preparing procedures, the multi-element antenna system of numerous parts in the cells can collaborate to alleviate the interferences and further improve the range efficiency. So far, inter-site helpful MIMO has gotten significant consideration in the writings [25–27]. Utilizing generally isolated circulated antenna systems (DASs) with improved limit execution has been accomplished with inter-site CoMP because of the diminished relationship, expanded SNR on account of more continuous LOS situations and shadowing diversity types. Inter-site collaboration is additionally known to upgrade the position of the compound channel lattice, subsequently offering a superior limit execution.

The phone site antennas utilized in the second era (2G) wireless systems are inactive and regularly comprise of an intermittent straight cluster that transmits a static directional fan-pillar design that is more extensive in azimuth and smaller in the rise. The exhibit is developed with directly captivated radiators that are taken care by a corporate feed, which ordinarily utilizes some type of transmission line innovation, for example, cajole links or a printed circuit microstrip. The corporate feed arrangement provides a structured abundance/phase excitation conveyance to the components, shaping an example that is fixed in its azimuth and height with static qualities for gain, bar width, side flaps and invalid file.

The aloof antenna is advanced to join variable phase shifters in 3G systems to empower the variable electrical tilt of the radiation design. A few usages have a committed phase shifter for each component, while others may share a phase shifter among at least two components, with the trade-off of cost versus execution. The fan bar is directed by shifting the phase circulation along with the radiators. This change can be performed either locally at the antenna or distantly by incorporating a Wireless Electrical Tilt (RET) actuator to the antenna. The dividing of the antenna components is commonly 0.7–0.9λ (frequency) to diminish evaluating flaps as the pillar is guided to the plan tilt limit, ordinarily 10°–12° beneath the skyline. Since that time, the base stations have obliged this capacity into their plans to permit far off tilt alterations through the operations support system (OSS).

Practically, all the base station antennas have double symmetrical captivated clusters that reuse the antenna opening. Each exhibit can be inclined freely with committed RET actuators, or at least two clusters are inclined together, contingent upon usage. Note that these exhibits are intended to work electrically autonomous of one another, thus the dividing between every segment ought to preferably be sufficiently enormous to give at least 30 dB port-to-port segregation. For late plans with two or even four next to each other exhibits, the disconnection may be diminished to 28 or even 25 dB.

Detached antennas are applied in the radio system where earlier arranging and connection spending forecasts have been performed for transmitting. In this way, the antenna's area above territory, azimuth bearing and tilt are indicated for execution. Once introduced, the antenna's electrical tilt can be distantly changed following advance of the ideal cell edge attractive and handover limits.

Rather than the aloof antenna's static nature, a functioning antenna is intended to frame differing design qualities at some random occurrence. In LTE, needs are controlled by CRS (for TM4) or CSI-RS (for TM9) estimations of the air interface with UEs; for instance, an expansive fan pillar for broadcasting to the division, a smaller bar traffic beam to be guided appropriately over a scope of azimuth and height edges or an example with a profound invalid to be aimed at an interferer can be initiated. These versatile modes can all the more ideally give the most ideal RF condition centred on the user(s) over the static example of aloof antennas.

Gigantic MIMO is tied with RF chains, while not including additionally emanating components. Gigantic alludes to the bigger number of handsets (TRx) in the base station antenna cluster. The more prominent number of TRx the antenna is furnished with, the more degrees of opportunity to change the radiation example of the sent signal dependent on where the beneficiary is found. Another more significant component is that there are different RF chains with the goal that few customers can be spatially multiplexed while having various antennas. For this, at any rate, the same number of RF inputs are required as there are customers. Each can get the full cluster gain and the computerized preceding can be utilized to keep away from customer interferences. Operators with range constraints or other system densification expected to help numerous 5G use cases will then again investigate the mm-wave-based wireless communication system [17].

This inadequacy features the basic requirement for beamforming plans that help beat the serious misfortune in signal quality. Its usage, with devoted advanced baseband and RF affix per antenna to control both adequacy and period of the sent signal over all antennas of an enormous exhibit, while accomplishes the alluring directional addition, likewise turns out to be seriously cost restrictive. The equipment cost and power utilization levels related to mm-wave blended signal circuits, including information converters, dishearten their utilization in incredible numbers rather than low/mid band advanced beamforming handset plans.

Another trademark intrinsic to the high-way misfortune attached to mm-wave engendering is the impact of dissipating or low spatial selectivity. Antenna components for the mm-wave system are as of now firmly pressed with small inter-element separations. From the accepting side, the last infers that various adaptations of a similar signal showing up at every antenna have firmly corresponded. Channel attributes displaying low scarcity (amazingly connected channel system) is one motivation behind why mixture beamforming execution can move towards that of its advanced variation, despite actualizing lower unpredictability preceding. Beam switching is the cycle by which the UE and base station monitors the best bar to communicate unicast information on and to guarantee that the UE changes to this best bar for effective consistent information gathering.

The cycle of pillar exchanging includes the following: the UE can be requested by the base station to perform beam quality (for instance, RSRP) estimations on the bars that are noticeable to it. These estimations depend on the SS blocks communicated intermittently by the base station, the gear to get and send information to and from the base station. The bar pair has been chosen by a cycle of beam acquisition as well as beam tracking. This cycle is not ensured to yield the best blending since it is a period and asset compelled measure. The refining system comprises the base station giving to the UE reference signals in the DL (called CSI-RS) relating to a lot of pillars, so UE can perform additionally refined determination of the bar pair dependent on estimations performed on those signals. The arrangement of beams for which the base station gives these signals involves a plan and use case, yet two wide cases can be recognized. One is the place of the intention to refine the Tx part of the bar pair and the second is the place of the reason for existing to refine the Rx divide.

8.13 WIDEBAND 4G COMMUNICATION

The numerous info different yield (MIMO) antenna framework is viewed as a key innovation for the up and coming fifth era (5G) and the current fourth era (4G) versatile antenna framework to improve information rate [14, 12, 30, 21]. In addition, for the MIMO antenna framework, 5G requires a wide transfer speed to amplify the information rate. However, the current accessible transmission capacity at sub-3 GHz is clogged [21]; along these lines, numerous universal associations are doled out of the millimetre-wave band, which incorporates 28, 37 and 39 GHz, as a potential frequency band for 5G [22, 32].

Along these lines, current versatile terminals need to help 5G at mm-wave band notwithstanding. In any case, versatile terminals have restricted space, and it is best not to include numerous antennas. Accordingly, the proposal is to build a MIMO antenna framework that has antenna components covering various groups of 5G and 4G by utilizing a solitary structure as opposed to two structures.

The use of multiband antenna components in the MIMO framework is a key factor for diminishing the antenna impression inside the portable terminal [14, 21–23]. Nonetheless, the reconciliation of 5G and 4G antennas in a similar structure produces numerous difficulties as far as antenna size, detachment between antennas at various groups and antenna gain are concerned. Each MIMO component comprises an opening in the ground plane and two microstrips taking care of ports in the top layer; a low pass channel is additionally incorporated into the structure for segregation improvement. By appropriately configuring the antenna feed from one port, the antenna can work in two diverse working modes. One of the taking care of ports compares to a tightened space antenna with end-fire radiation for 5G, though the other taking care of port relates to an open-finished space antenna with omnidirectional radiation at 4G. MIMO execution factors are likewise determined and talked about. Second, a smaller two-component 5G/4G MIMO antenna framework for portable terminals is introduced. It works for both low microwave region and mm-wave region of wavelength utilizing a solitary space structure as shown in Figure 8.8.

8.13.1 Multiband 5G Antenna Structure

It comprises a tightened opening and microstrip feed line [30]. The size of the single component with the ground plane is 70 × 77.8 mm^2 ($L \times W$), the smaller edge of the opening closures with a $\lambda/4$ round stub. The measurement of the stub is $d2$-4. The tightened opening is taken care of by a roundabout division microstrip line with the distance across. The round part and a 50 Ω feed transmission line are connected through an interference coordinating transformer that has an absolute length as in Ref. [31]. The roundabout stub of the opening and roundabout area of the feed line add to the arrangement of a broadband interference coordinating transfer speed [10, 12].

A boundary study is led to bring ventured tightening into the opening. Four case studies have been concentrated to give the last shape to the opening utilizing various shapes. It is obvious from the reflection coefficient for the single tightened opening that the interference coordinating is poor, though the three-phase tightened space has more extensive data transfer capacity.

Thus, the best possible choice and tuning of the shapes can give more extensive interference transmission capacity. The last tightened space covers the 25–40 GHz band. The primary beam is coordinated towards the negative x-pivot in an end-fire radiation design with the greatest addition of 8.14 dBi at 38 GHz.

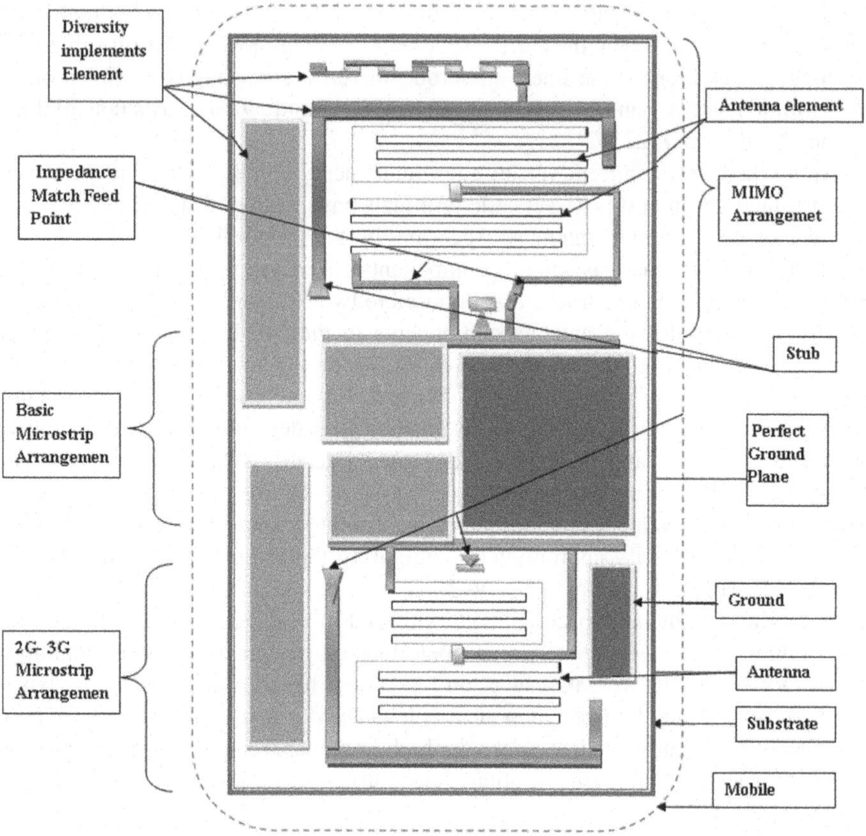

FIGURE 8.8 Mobile antenna placements inside the handset.

8.13.2 MIMO in Cancer Recognition

Breast disease is one of the most well-known terminal illnesses among women globally. The expression "breast disease" characterizes a dangerous tumour that has been created from cells in the breast. With time, carcinogenic cells can possess solid breast tissues and advance into the underarm lymph hub. When they get into the lymph hubs, they have a pathway into different parts of the body. There are four phases of malignant growth in the breast. The breast disease phase states how much the malignancy cells have spread past the first tumour. Odds of fix and endurance diminish as the phase of malignant growth increments from 4h. The fourth phase disease spreads past the breast and to different parts of the body, most normally it goes to the lungs, bones and additionally spreads to the cerebrum [34].

Subsequently, the early diagnosis of malignant growth in the breast is crucial for the quick fix and endurance of patients. Microwave imaging (MI) to distinguish breast tumour is a promising strategy to conquer the issues with strategies that have minimal effort, non-ionizing radiations, high affectability in recognizing tumours and give solace to tolerance, utilizing a microstrip fix antenna for tumour discovery. Microwave imaging is partitioned into two classes as tomography and radar-based microwave imaging. The microwave imaging (MWI) framework comprises a transmitter microwave framework to produce flags within the breast and a collector to distinguish those back-scattered signals after they collaborate with the breast. A few adjustments in the ground plane and opening on the transmitting patch make reproduction results better.

Five altered antenna structures are mimicked with the 3D breast model. Electric field, attractive field and flow thickness boundaries show a perceivable contrast for recreation of the proposed antenna with the breast model having no tumour and with a breast model comprising the tumour. A tumour sweep of 20 mm was considered for re-enactments.

The structure of a microstrip fix antenna for breast malignancy discovery has been audited. Microwave imaging is found as a diversity option over other breast screening procedures. The human body's introduction to microwaves is innocuous. Microwave imaging is a more secure and less expensive strategy to recognize breast tumours when contrasted with other existing breast screening strategies [34–35].

8.13.3 Bluetooth MIMO

A significant thought in the compact MIMO antenna configuration is to decrease the relationship between the numerous components, specifically common coupling and electromagnetic cooperations that exist in the middle of various components [2].

While planning an antenna for Bluetooth instruments, a wideband antenna in wireless correspondence frequency must be planned. The components of the proposed antenna are chosen according to the physical elements of the Bluetooth gadget, Wi-Fi and WLAN 802.11 frequency groups. To meet the necessities, size decrease and wide data transmission are turning out to be the significant plan requirements for structuring viable instruments.

The plan involves two indistinguishable "dumbell" modelled radiators that were taken by coaxial test-taking method [6] with streamlined feed; position of length and width of opening measurements are Lsl = 28.32 mm and W = 18.18 mm. The area of feed point is chosen to get the best interferences matching. The primary bit of leeway by utilizing coaxial test feeds that the internal conductor can be situated at any favoured area is to have a superior interferences coordinating as shown in Table 8.7.

Length of antenna

$$L \approx 0.49\lambda_d = 0.49\frac{\lambda_o}{\sqrt{\varepsilon_r}} \tag{8.16}$$

TABLE 8.7
Antenna design with different dimensions

Freq GHz	W um	L um	Effective dielectric Ɛreff	ΔL um	Lg um	Wg um
12.27	97.256	83.18	2.1635	0.841	83.180180	97.2
15.75	75.767	64.801	2.1645	0.841	64.801323	75767368.9
17.2	83.917	80.971	1.1622	1.0536	80.971811	83917198.8
2.5	56.825	48.600	2.166	0.841	48.600994	56825528.8
6.4	48.722	42.525	2.1669	0.841	42.525871	49722338.7
8.6	45.897	38.254	2.1674	0.841	38.254650	45897545.1
12.27	38.777	34.0202	2.1686	0.8416	34.020698	39777873.7

Patch width

$$Width = \frac{c}{2f_r}\sqrt{\frac{2}{\varepsilon_r + 1}} \tag{8.17}$$

Effective dielectric

$$\varepsilon r = \frac{\varepsilon r + 1}{2} + \frac{\varepsilon r - 1}{2\sqrt{1 + \left(\frac{12h}{w}\right)}} \tag{8.18}$$

Extension of the length (ΔL)

$$\Delta L = 0.412h \frac{\left[\varepsilon_{reff} + 0.3\right]\left[{}^{w}\!/_{h} + 0.264\right]}{\left[\varepsilon_{reff} - 0.258\right]\left[{}^{w}\!/_{h} + 0.8\right]} \tag{8.19}$$

Interference of microstrip

$$Zo = \frac{120\pi}{\sqrt{\ } \varepsilon reff(1.393 + \left(\frac{W}{h}\right) + \frac{2}{3}\ln\left(\frac{w}{h} + 1.444\right)} \tag{8.20}$$

Resonant dominant frequency

$$fr = \frac{1}{2L\sqrt{\varepsilon r}\sqrt{\varepsilon \mu}} = \frac{v}{2L\sqrt{\varepsilon r}} \tag{8.21}$$

$$fr = \frac{v}{2L\sqrt{\varepsilon reff\left(L + 2\Delta Leff\right)}} \tag{8.22}$$

To increase the effective electrical duration of the patch longer than its physical length, the fringing effect is used. The resonance state, therefore, depends on L. Therefore, the effective length of the patch is given by:

$$Leff = Lp + 2\Delta L \tag{8.23}$$

The actual length of the patch (L_p)

$$Lp = \frac{C}{2f0\sqrt{\epsilon_o eff}} - 2\Delta L \tag{8.24}$$

Substrate length (L_{sb})

$$L_{sb} = 12h + L_p \tag{8.25}$$

Width of the substrate (W_{sb})

$$W_{sb} = 12h + W \tag{8.26}$$

Length of slot (L_{sl})

$$L_{sl} = \frac{L_p}{\epsilon\, eff} \tag{8.27}$$

Width of slot(W_{sl})

$$W_{sl} = \frac{W}{2} \tag{8.28}$$

8.13.4 Antenna PDA

The dimensions of the ceramic rink are 7.1 mm wide, 14.65 mm long, 4.6 mm high and thus the relative permittivity of the ceramic, Ɛr, is 2.2–40. Where the microstrip runs under it, the ring is metallic. This ensures good contact and allows the PCB and thus the ceramic ring to create a stable solder joint. With an overhang of 2.15 mm, the microstrip feed line is 3.2 mm long. With εr of 2.33, the substrate is 1.6 mm thick. The microstrip's middle offset is 1.5 mm from the puck's centre. The bottom plane, because of the puck, is at the same distance. Extensive bandwidth has been shown to give this design: 4.8 GHz to 6.0 GHz at a return loss of −10 dB. Further changes were made to the antenna configuration and the size was then reduced to 11 mm in length, 4.8 mm in width and 3.2 mm in height as shown in Table 8.8.

8.13.5 Designing Laptops

The OEMs and ODMs of laptop communication like Wi-Fi, Bluetooth require the antenna to be placed in the bottom of the laptop device, the antennas be placed inside

TABLE 8.8
Antenna design with parameters

S. No.	Parameter	Dimension
1	Length of patch	28.85
2	Width of patch	36.10
3	Length of ground	48.34
4	Width of ground	55.35
5	Length of substrate	48.65
6	Width of substrate	55.15
7	Length of the slot	15.66
8	Width of slot	8.012
9	Feedpoint X	14.55
10	Feedpoint Y	18.34

the cover. The laptop test rig was made from an FR4 double-sided PCB. Solder joints were continuous between the parts of FR4. After passing through a small opening, the connecting cables were RG405 semi-rigid, tack soldered to most of the board and the FR4. The outer section of the wire was soldered to the ground of the antenna PCB and the inner part was soldered to the track at the top of the antenna PCB. All the four coaxial feed cables were of identical length, thereby ensuring that the losses were equal in all. The cables and connectors protruding from the top of the board should be at an equal distance from the last grounding point, the cable was meandered.

8.13.6 EBG

Electromagnetic bandgap (EBG) materials are periodic structured materials that, when interacting with electromagnetic waves, exhibit useful features such as frequency stop-bands, pass-bands and bandgaps. By adding periodic disturbances such as sized holes, dielectric rods and patterns in waveguides and PCB substrates, EBG structures are created. Many EBG structures are recorded as ground plane holes, mushroom-like EBG, compact uniplanar EBG (UC-EBG) [4–6]. EBG structures have a good range of applications when coupled with antennas. These include image rejection, enhancement of gain, surface wave suppression, antenna pattern ripple suppression and antenna bandwidth. The low-profile antenna positioned above the EBG structure results in complex interactions between the structure of the antenna and the EBG. The wideband of the EBG structure must provide good return loss and radiation efficiency for the antenna in order to ensure an efficient application of the EBG structure to these low-profile antennas.

8.13.7 Photonic Bandgap

The conventional MIMO terminals equipped with multiple antennas use space diversity assisted by co-polarized linear arrays. This arrangement is appropriate for the scenario at base station, but not for a compact mobile/device terminal. Compact arrays for MIMO wireless communication terminals are subject to extensive research and development efforts with new design of antenna. The basic criteria for this technique are that the antennas must be diverse, i.e. while they are closely spaced, they need to be able to transmit numerous signals. The use of multiple antennas for MIMO communication systems on a portable terminal (laptop, PDA or handset) remains a challenge at the moment. For this, there are two main explanations. First, the placement of multiple antennas in a small handset is difficult to incorporate. Second, the output of the handset antenna is affected by the inter-antenna reflection, impedance variation, operational temperature and coupling that forms the cross-correlation of the envelope and coupling to biological tissues within the user's head and/or hand.

Two designs were proposed for the MIMO laptop solution – the size and orientation of the folded loop on the bottom, side panel and screen layout based on the differences. Mechanically, the folded loop was not rigid enough and was therefore designed to add a brace (fibreglass GRP $\varepsilon r = 3.99$). Not only does the fibreglass sustain the folded ring, but it also allows the antenna size to scale back. The laptop lid was made of 0.4 mm thick copper and the folded loop was made of 0.2 mm thick copper.

For the first MIMO laptop, the antenna is $13.2 \times 5.5 \times 2.9$ mm^3 in dimension; thus, the dimension of the laptop is $270 \times 210 \times 10$ mm^3. For the second MIMO laptop design with its antenna features, the antenna is $10 \times 5.5 \times 6.5$ mm^3 in dimension; thus, the dimension of the laptop is $270 \times 210 \times 10$ mm^3. The size and orientation of the antenna elements in the two designs are different.

A ceramic slab ($\mathcal{E}r = 6$) is packed into the folded ring. The handset is made of 0.4 mm thick copper and therefore the folded loop is made of 0.2 mm thick copper. The handset has dimensions of 40 mm × 100 mm × 10 mm while the folded loop has dimensions of $12.5 \times 5.5 \times 4$ mm.

8.13.8 UWB MIMO Antenna

A UWB MIMO antenna with a bandwidth from 3.1 to 10.6 GHz with a compact size of 26×40 mm^2 comprises two 50 Ω microstrip-fed planar-monopoly (PM) elements that are placed perpendicular to each other to provide good insulation between the two input ports. Antennas are designed on a 0.8 mm thick ROG5880, RO4350B substrate with a permittivity of 3.5 and a loss tangent of 0.004. A small rectangular slot was cut on the upper fringe of the bottom plane for better matching at high frequencies. Two long ground stubs were added, positioned in parallel with the corresponding PM, to extend insulation and impedance bandwidth. To satisfy the needs of the UWB operation, these structures were placed (stub 1 generated a resonance at 3.5 GHz for S11 and stub 2 on S22 for 4 GHz). In addition, the short ground strip is placed to scale back the mutual coupling between the elements and strengthen the isolation. To minimize the effects of multipath fading, different radiation patterns were used to scale back the risk of error, where two PMs were located perpendicular to each other. Therefore, two distinct radiation patterns are used to obtain signals from different directions, gaining pattern diversity, as noted in Ref. [32].

Two antennas with different space and polarization were built in to their results. The critical element consists of a ring monopole printed on a RO3003 (0.75 mm thick and r = 3) coffee loss substrate fed by a 50 Ohm microstrip line. The antenna was also studied with the technique of CPW feeding and philtre addiction, proving to be a much better solution.

Two separate antennas were designed to operate at a frequency of 2.45 GHz (ISM band), on an FR4 substrate of εr = 4.6 and 1.6 mm thickness, suitable for body-centred and wireless communications, respectively. The researcher uses an electromagnetic bandgap (EBG) to obtain beneficial electromagnetic properties that cannot be seen in natural materials. A microstrip line [25, 26] follows the feeding technique used in both the cases. To increase the reflection coefficient at lower frequencies, circular disc monopole (CDM) with a CPW feeding using FR4 substrate, as well as with 50 Ohm impedance matched with the coaxial feed, and two notches cut on the circular disc patch, as show in Ref. [27], are used as the stimulating solution for UWB applications.

8.14 CONCLUSION

It is a very difficult job to build compact and effective MIMO antennas. While antennas are considered to be a critical component of the overall radio system, system-related studies typically simplify them. As a consequence, an antenna with high efficiency and omnidirectional radiation is considered good enough, at least on the user terminal side of a SISO system. When it comes to designing MIMO antennas, both these advantages are still extremely desirable. High efficiency is critical because even with advanced baseband signal processing techniques, the power loss due to low-efficiency antennas cannot be compensated for. The omnidirectional pattern of radiation is often preferred by the complex world that the consumer would actually experience [36].

These guidelines are not appropriate, however, for the design of powerful MIMO antennas. For instance, space–time processing enables the parallel transmission of data streams in the case of spatial multiplexing schemes. Effective MIMO antennas should also be capable of giving the device the ability to access these orthogonal channels. One basic requirement is that the correlation should be kept relatively low between the various TX–RX channels.

REFERENCES

[1] L. J. Vora, (2019), Evolution of mobile generation technology: 1G to 5G and review of upcoming wireless technology 5G, *International Journal of Modern Trends in Engineering and Research*, 29, 13–36.

[2] M. Abu Saada, T. F. Skaik, and R. Alhalabi, (2017), Design of efficient microstrip linear antenna array for 5G communications systems. *International Conference on Promising Electronic Technologies (ICPET)*, 2017, 43–47. DOI: 10.1109/ICPET.2017.14

[3] Modi, Hardik & Gohil, Asvin & Patel, Shobhit. (2013). 5G technology of mobile communication: A survey. 2013 International Conference on Intelligent Systems and Signal Processing, ISSP 2013. 10.1109/ISSP.2013.6526920.

[4] First 5G NR specs approved, [Online]. Available: www.3gpp.org/news-events/3gpp-news/1929-nsa nr 5g.
[5] Hubregt J. Visser, (2005), *Array and Phased Array Antenna Basics.* ISBN: 978-0-470-87117-1.
[6] T. H. Elhabbash, (2019), Design of MIMO antenna for future 5G communication systems, Master of Electrical Engineering, Thesis.
[7] A. P. Bhat, (2019), Development of smart instrumentation for nanostructured antenna and solar cell, RTMNU, Nagpur Thesis.
[8] M. A. Al-Tarifi, M. S. Sharawi, and A. Shamim, (2018). Massive MIMO antenna system for 5G base stations with directive ports and switched beamsteering capabilities. *IET Microwaves, Antennas & Propagation, 12*(10), 1709–1718.
[9] C.-X. Wang, F. Haider, X. Gao, X.-H. You, Y. Yang, D. Yuan, H. M. Aggoune, H. Haas, S. Fletcher, and E. Hepsaydir, (2014) Cellular architecture and key technologies for 5G wireless communication networks. *IEEE Communications Magazine*, 52(2), 122–130.
[10] T. A. Milligan, (2015) *Modern Antenna Design*, 2nd ed. Wiley-IEEE Press, ISBN: 978-0-471-45776-3.
[11] K. R. Mahmoud and A. M. Montaser, (2018). Synthesis of multi-polarised upside conical frustum array antenna for 5G mm-wave base station at 28/38 GHz. *IET Microwaves, Antennas & Propagation*, 12(9), 1559–1568.
[12] K. Klionovski, A. Shamim, and M. S. Sharawi, (2017), 5G antenna array with wide-angle beam steering and dual linear polarization, *2017 IEEE International Symposium on Antennas and Propagation & USNC/URSI National Radio Science Meeting*, pp. 1469–1470.
[13] T. Varum, A. Ramos, and J. N. Matos, (2018), Planar microstrip series-fed array for 5G applications with beamforming capabilities, *2018 IEEE MTT-S International Microwave Workshop Series on 5G Hardware and System Technologies (IMWS-5G)*, pp. 1–3.
[14] X. Shi, M. Zhang, S. Xu, D. Liu, H. Wen, and J. Wang, (2017). Dual-band 8-element MIMO antenna with a short neutral line for 5G mobile handset. In *2017 11th European Conference on Antennas and Propagation (EUCAP)*, pp. 3140–3142. IEEE.
[15] Z. U. Khan, Q. H. Abbasi, A. Belenguer, T. H. Loh, and A. Alomainy, (2018), Hybrid antenna module concept for 28 GHz 5G beamsteering cellular devices, *IEEE MTT-S International Microwave Workshop Series on 5G Hardware and System Technologies (IMWS-5G)*, pp. 1–3.
[16] X. Zhu, J. L. Zhang, T. Cui, and Z.q. Zheng, (2018), A miniaturized dielectric-resonator phased antenna array with 3D-coverage for 5G mobile terminals, *IEEE 5G World Forum (5GWF)*, pp. 343 –346.
[17] J. Bang and J. Choi, (2018), A SAR reduced mm-wave beam-steerable array antenna with the dual-mode operation for fully metal-covered 5G cellular handsets, *IEEE Antennas, and Wireless Propagation Letters*, 17(6), 1118 –1122.
[18] S. F. Jilani and A. Alomainy, (2017), Millimeter-wave conformal antenna array for 5G wireless applications, *2017 IEEE International Symposium on Antennas and Propagation &USNC/ URSI National Radio Science Meeting*, pp. 1439–1440.
[19] S. F. Jilani and A. Alomainy, (2017), A multiband millimeter-wave 2-D array based on enhanced Franklin antenna for 5G wireless systems, *IEEE Antennas, and Wireless Propagation Letters*, 16, 2983–2986.
[20] W. C. Y. Lee, (2014), Mobile Cellular Telecommunications: Analog and Digital Systems, 2nd Ed., Tata McGraw Hill.

[21] M. Stanley, Y. Huang, T. Loh, Q. Xu, H. Y. Wang, and H. Zhou, (2017), A high gain steerable millimeter-wave Antenna array for 5G smartphone applications, *2017 11th European Conference on Antennas and Propagation (EUCAP)*, pp. 1311–1314.

[22] S. S. Zhu, H. W. Liu, Z. J. Chen, and P. Wen, (2018), A compact gain-enhanced Vivaldi antenna array with suppressed mutual coupling for 5G mm-wave application, *IEEE Antennas and Wireless Propagation Letters*, 17(5), 776– 779.

[23] J. Chen and Q. Zhang, (2017), High scanning-rate periodic leak-wave antennas using complementary microstrip-slotline stubs, *Sixth Asia-Pacific Conference on Antennas and Propagation (APCAP), 2017.*

[24] H. Jin, W. Che, K. S. Chin, G. Shen, W. Yang and Q. Xue, (2017), 60-GHz LTCC differential-fed patch antenna array with high gain by using soft-surface structures. *IEEE Transactions on Antennas and Propagation*, 65(1) 206–216.

[25] S. Liao and Q. Xue, (2017), Dual polarized planar aperture antenna on LTCC for 60-GHz antenna in-package applications. *IEEE Transactions on Antennas and Propagation*, 65(1), 63–70.

[26] D. Psychoudakis and A. Foroozesh, (2017), Small millimeter wave printed antenna arrays for 5G applications. *2017 IEEE International Symposium on Antennas and Propagation & USNC/URSI National Radio Science Meeting*, pp. 1805–806.

[27] Najam A. I., Y. Duroc, and S. Tedjni, (2011), UWB-MIMO antenna with novel stub structure. *Progress In Electromagnetics Research C*, 19(3), 245–257.

[28] P. Hartley, (2017). *Gimme 5: What to Expect from 5G Wireless Networks | FreshMR.* (Marketstrategies.com). Retrieved from www.marketstrategies.com/blog/2015/03/gimme-5-what-to-expect-from-5gwireless-networks/Antenna-Theory.com- Fractional Bandwidth. (2017). Retrieved Feb 28, 2017, from www.antenna-theory.com/definitions/fractionalBW.php.

[29] A. Agarwal, (2017). *The 5th Generation Mobile Wireless Networks- Key Concepts, Network Architecture and Challenges* (Pubs.sciepub.com). Retrieved from http://pubs.sciepub.com/ajeee/3/2/1/.

[30] M. Prasanna K., (2013) Designal and analysis of two port mimo antennas with wideband isolation, A Master thesis NIT, Rourkela-769008.

[31] R. Tian, (2011). *Designal and Evaluation of Compact Multi-antennas for Efficient MIMO Communications.*

[32] M. Kaur and S. Goyalm, (2020), Microstrip patch antenna designal for early breast cancer detection. *International Journal of Recent Technology and Engineering*, 8(6), March 2020.

[33] D. Srinivasan and M. Gopalakrishnan, (2019), Breast cancer detection using adaptable textile antenna designal. *Journal of Medical Systems* 43, 177.

[34] T. V. P. P. Venkatesh and D. B. N. Sivakumar, (2020) Designal of I-shaped dual C-slotted rectangular microstrip patch antenna (I-DCSRMPA) for breast cancer tumor detection. *Cluster Computing*, 8, 123–130.

[35] S. Shawalil, K. N. Abdul Rani, and H. A. Rahim, (2019), 2.45 GHz wearable rectenna array designal for microwave energy harvesting, *Indonesian Journal of Electrical Engineering and Computer Science*, 14(2), 677–687.

[36] P. Jaglan, R. Dass, and M. Duhan, (2019) Breast cancer detection techniques: Issues and challenges. *Journal of the Institution of Engineers: Series B*, 100(4), 379–386.

9 A Survey of Fifth-Generation Cellular Communications Using MIMO for IoT Applications

Sanket A. Nirmal and Richa Chandel

9.1 INTRODUCTION

Nowadays, good internet connectivity with high speed data is in substantial demand for wireless communications, for good economic growth, and for digitalization of cities across the globe. The Internet of Things (IoT) has changed the world by using various types of sensors and makes the world do a wide range of communications by using smart technologies (Figure 9.1).

The advancement in mobile communication in each generation enhances the data rate and the quality of services as observed previously in 2G, in which we dealt with GSM. 3G was then introduced that used UMTS wherein high speed and faster downloading were possible. 3G was the first mobile standard; later 3.5G, 3.75G, 4G, Long Term Evolution-Advanced (LTE-A) standards were introduced. Though 2G is taken as the first digital standard used in the cellular systems for voice communication with low data rate, the introduction of 3G and 4G has improved the data rate with reduced delay and enhanced network such that data, voice, and video can be accessed at any time and at any place from 1 to 2 Mbps, and in 4G it is from 50 to 100 Mbps. In the coming years, it is expected that 100 billion subscribers will be connected to the global network; however, the requirement is high bandwidth and low latency as the number of users are increasing and in order to achieve higher data rate,

DOI: 10.1201/9781003093558-12

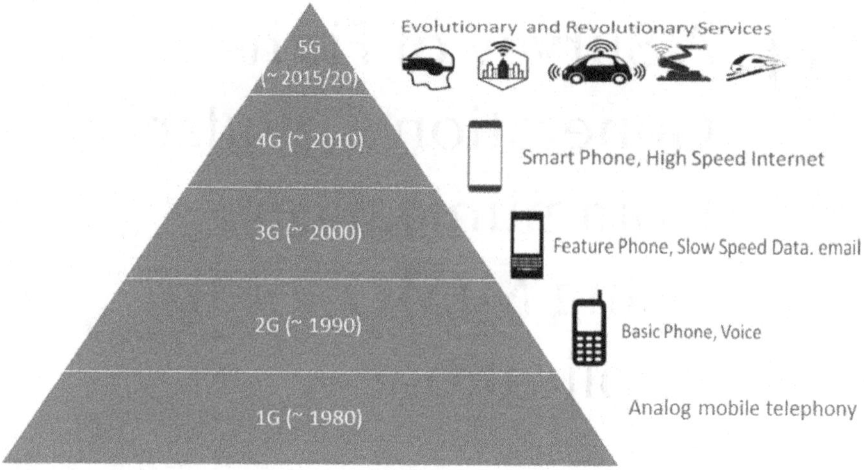

FIGURE 9.1 Evolution of various generations of cellular communications.

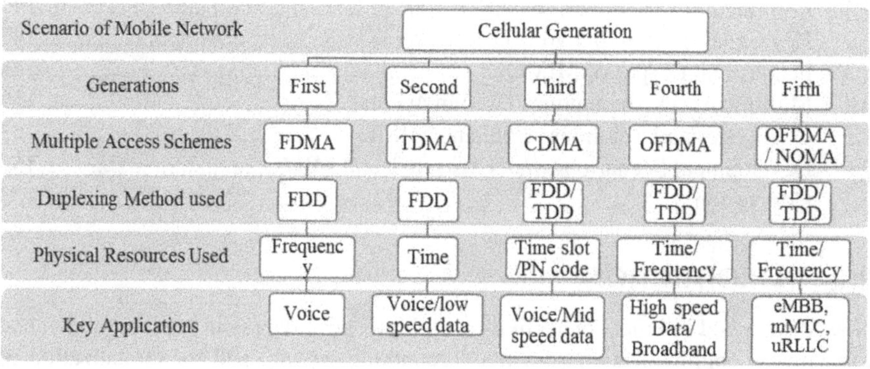

FIGURE 9.2 Cellular generations along with their parametric changes.

a mobile network like 5G will fulfil the need as expected by the researchers. Also 5G is scalable with good connectivity and increase in data transmission rate for mobile networks [1].

The multiple access schemes used from the first generation to the fifth generation are shown in Figure 9.2. The development of actual methods in each generation is discussed in this section. Orthogonal frequency division multiple access (OFDMA) uses orthogonality, which means user accessibility does not interfere each other when sufficient numbers of resources are available but code division multiple access (CDMA) and non-orthogonal multiple access (NOMA) follow non-orthogonal behaviour. In frequency division multiple access (FDMA) and time division multiple access (TDMA) techniques, frequency and time are distributed among users in terms of channels and time slots so that they can be utilized by a numbers of users. In case of CDMA, codes are allocated to users so that users can communicate on available spectrum at the same time. In OFDMA, there are groups of subcarriers in which users are associated with several channels of frequency at various time slots. The OFDMA technique is used to achieve sufficient spectral efficiency in which spectrum overlaps due to subcarriers results into transmission of more data compared to FDMA on the same bandwidth [2, 3]. In the 5G development, NOMA technique is used that consists of power domain and code domain where time slot intervals are flexible and scalable. Also full duplex methods for communication, i.e., time division duplex (TDD) and frequency division duplex (FDD), are emphasized and also the physical resources assigned to users and the key applications of various mobile generations. The main purpose of introducing 5G is IoT as it requires high throughput and high spectrum for its applications like remote access control and various qualities of services. As 5G and IoT are associated with MIMO, millimetre wave communications, and cloud-based platform, it is possible to control the machine in real time using a mobile device; therefore, device to device and machine to machine approaches were introduced. To achieve high data rate in terms of Gigabit-per second (Gbps) and wider spectrum, 5G along with MIMO is the essential need in today's framework of cellular communications. MIMO is the multiple-input multiple-output antenna that can perform the spatial multiplexing and can improve the channel capacity by using massive MIMO, as it involves large numbers of antennas to communicate with a base station with the same frequency and at the same time. Some of the key features of 5G are shown in Figure 9.3. The purpose of introducing 5G is high throughput and high spectrum for its applications such as remote access control and various quality of services. As 5G is associated with mm wave communication and cloud based platform it is possible to control the machine in real time using mobile device using an approach such as device to device, machine to machine. In order to achieve high data rate in terms of Gigabit-per second (Gbps) and wider spectrum 5G along with MIMO is primarily needed in today's framework of cellular communication.

9.2 KEY CAPABILITIES OF 5G WELL-DEFINED BY ITU-R

By 2015, researchers working with ITU-R devolved several key capabilities (Figure 9.1) in which three scaling parameters were formed: low, medium, and high. ITU has provided IMT 2020 standard requirements for 5G, i.e., enhanced Mobile Broadband (eMBB), ultra-Reliable Low Latency Communications (uRLLC), and massive Machine Type Communications (mMTC) [4]. These are explained as follows:

1. *Enhanced Mobile Broadband (eMBB)*: Its main purpose is to provide dynamic internet speed and adaptive delivery in real time and it also supports services such as virtual and augmented realities as shown in Figures 9.3 and 9.4.
2. *Ultra-Reliable Low Latency Communications (uRLLC)*: The requirement of the latency period should be very less, typically one millisecond and below. It has opportunities in the field of automation, smart transport, eHealth, and public safety.

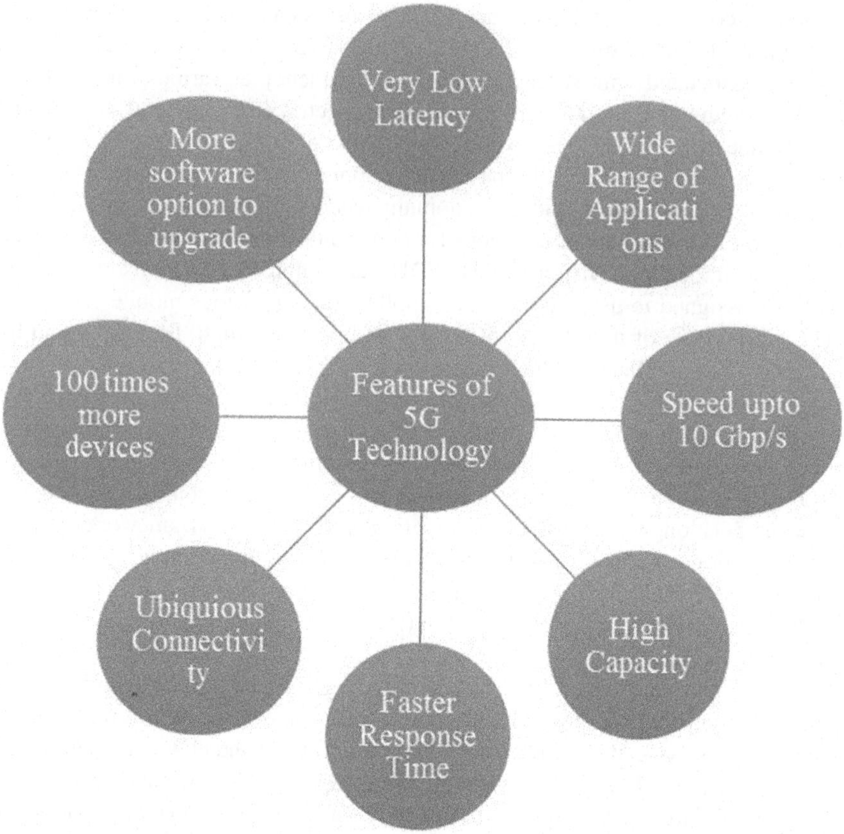

FIGURE 9.3 Key features of 5G.

3. *Massive Machine Type Communications (mMTC)*: It also has an immense number of Internet of Things (IoT) enabled devices. It offers many services such as smart cities, smart power grids, and smart-farms.

9.3 FEATURES AND TYPES OF MIMOs

The foremost thought is to use a large number of antennas at base stations to concurrently work for several independent terminals through spatial multiplexing. Across the world, the cellular communication technology uses multiple-input multiple-output (MIMO) system on an extensive scale. The MIMO technology comprises of numerous antennas on both the ends of communication link so as to boost the performance of MIMO along with reliability.

The are three types of multiple-input multiple-outputs: massive MIMO (MaMi), multiuser MIMO (MU-MIMO), and distributed phased array MIMO (DPA-MIMO) [5]. Massive MIMO is being used in 5G as it involves large numbers of antennas

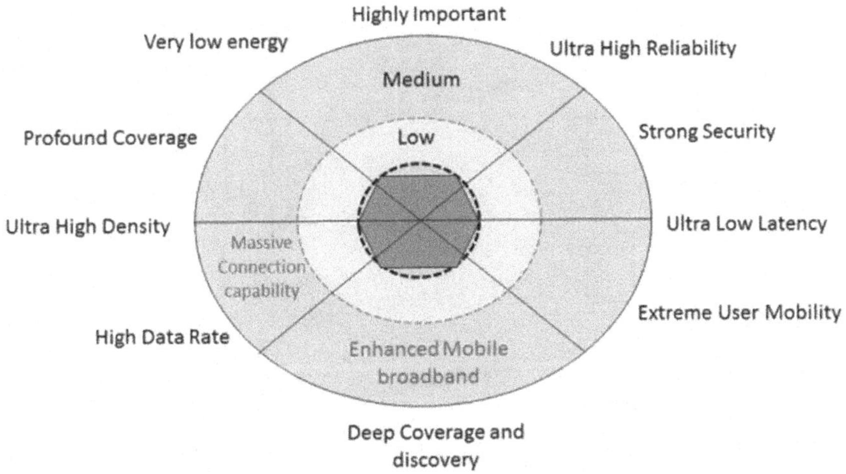

FIGURE 9.4 Important key aspects of the fifth generation given in the ITU Model [4–5].

at the base station and it has functions that can work with many terminals using the principle of spatial multiplexing. MIMO has wide capability at both the ends by using multiple antennas at both transmitter and receiver, using which its performance indices can be improved, enhancing its reliability. It is able to work effectively up to 6 GHz frequency range in LTE and by arranging more number of array antenna we can design a wide frequency range system [6]. Currently, 4G uses MIMO in all its transmissions. Within the limited bandwidth and power levels, the data rate can be increased by the use of multiple antennas. In the supreme situation, within a multipath environment, the channel capacity rate is directly proportional to the number of transmitter and receiver antennas, as shown in Figures 9.5 and 9.6

$$C = B\log_2\left(1 + M \times N \times SNR\right)$$

where C denotes the channel capacity rate in bits/sec, B denotes the bandwidth in Hz, and the number of transmitter antennas and receiver antennas are denoted by M and N, respectively. SNR is the signal to noise power ratio.

The performance metrics of MIMO antenna structures and several current MIMO antenna designs and their characteristics along with improvement methods in consumer wireless electronics makes MIMO of real importance [7].

- i Spatial multiplexing: The goal of spatial multiplexing is to split high data rate stream into independent low data rate stream before transmitting. This process enhances the data transfer rate.
- ii Diversity: Due to multipath propagation, the signals face attenuation, phase delay, and time delay which leads to multipath fading. MIMO antenna helps to combat multipath fading in rich scattering environment.

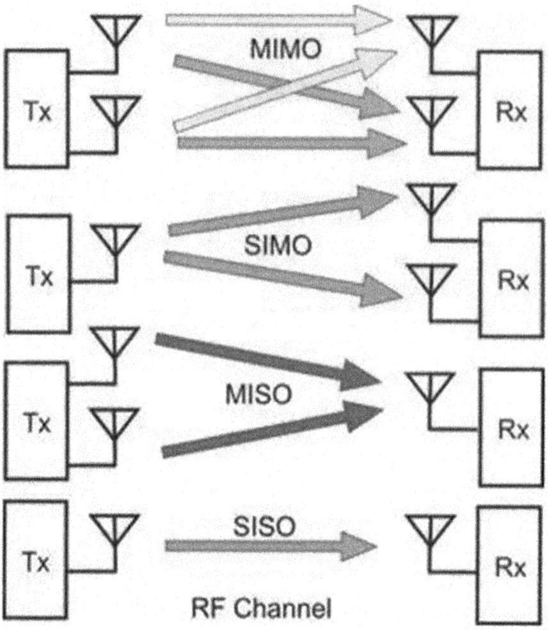

FIGURE 9.5 Different MIMO configurations.

In the past few decades, the MIMO has gained evolutionary and revolutionary importance in the wireless communication era which is all about digital communication. In the last 50 years, there has been tremendous transformation due to researchers working hard to strive for excellent research in the field of MIMO and on several other wireless technologies. During transmission and reception, importance has been given to resolve issues such as signal interference, insufficient spectrum allocation, multipath transmission, and so on. MIMO has transformed the entire issues into the realistic solution and allows us to transmit and receive higher data rate with low latency. It increases the range of communication, available bandwidth, and allows spatial multiplexing diversity. Also we have acquired advantages in channel capacity, enhanced signal to noise ratio and hence increase in signal immunity.

9.4 MULTIPLE ACCESS TECHNIQUES USED IN 4G AND 5G

Increasing the bandwidth efficiency is one of the key concerns in fourth- and fifth-generation cellular communications in order to improve the information rate so that data can be transmitted without interference on a given bandwidth. In LTE-Advanced (4G), this bandwidth efficiency depends on the orthogonal frequency division multiplexing (OFDM) that can handle peak to average power ratio (PAPR). OFDM has several issues; FBMC versus universal filtered multi-carrier (UFMC) versus OFDM modulation schemes are compared in Ref. [8]. 5G introduced NOMA as a

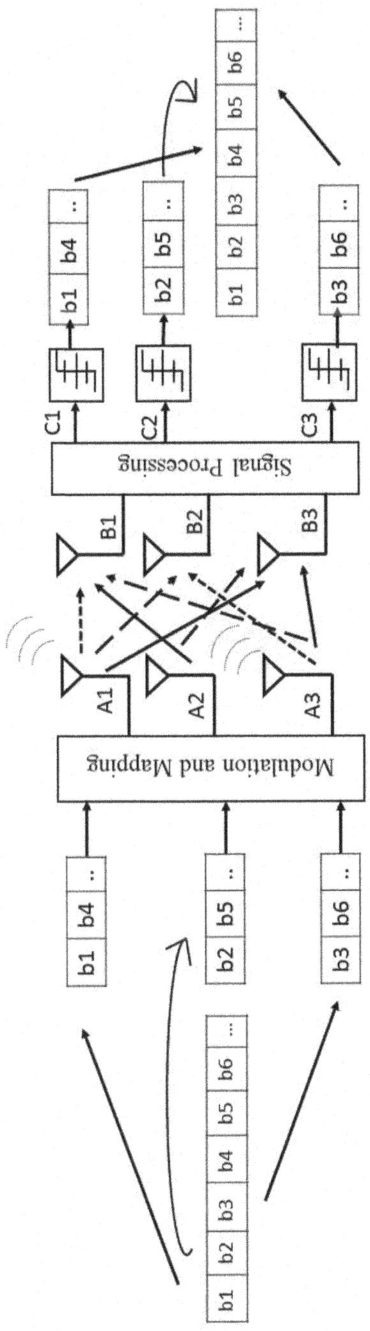

FIGURE 9.6 Data stream transmitted through spatial multiplexing.

new multiple access technique that has two domains, namely, power domain and code domain.

In NOMA, there are many antennas at transmitter end and receiver end, as it has the ability to serve many users at the same time and frequency. Power domain perform multiplexing using power and code domain perform multiplexing following multiplexed code. In NOMA, SIC (successive interference cancellation) mechanism is used at the receiver end, as shown in Figures 9.7 and 9.8. Power domain in NOMA gives high channel gain, which means if any weak channel found it will assign more gain and power so the limited spectrum resources can be used efficiently and can compare both the domains with ease [9].

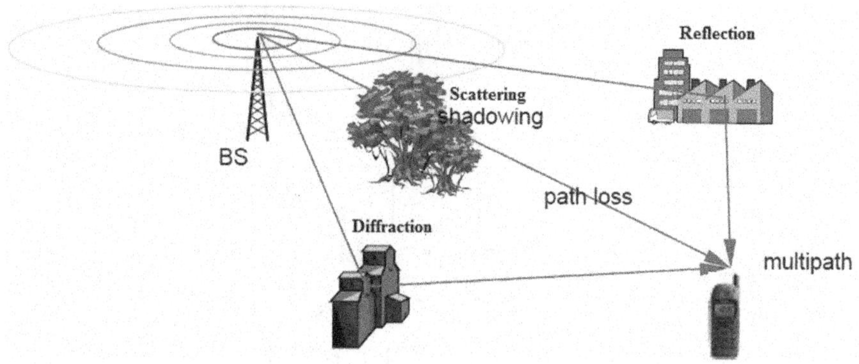

FIGURE 9.7 Multipath diversity.

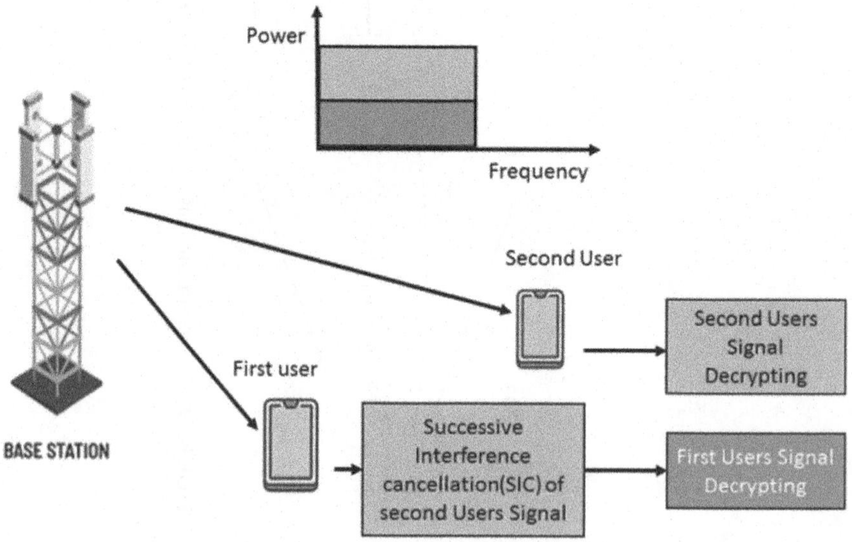

FIGURE 9.8 Downlink NOMA system with one base station and two users.

9.5 CHALLENGES AND SOLUTIONS OF 5G AND MIMO

I Data rate must be 10 times greater than 4G.

II Latency interval should be minimum.

III Bandwidth requirement – using MIMO antenna, it is desirable to have mm wave technologies and by the use of cognitive radio, high spectrum efficiency can be achieved to utilize both the approved and unrestricted spectrum for the user which can fulfil the bandwidth requirement.

IV To achieve good isolation in MIMO configuration of dimension 4 × 4, 8 × 8, and so on, there is limitation of size and space in mobile terminals and it is also stated methods to improve isolation is presented by researchers. In order to solve the issues related to structure density and isolation, MIMO antenna uses appropriate techniques for 5G-enabled mobile devices [9, 10].

9.6 DEVELOPMENT SCENARIO TOWARDS 5G – IoT AND MIMO

The 5G technology becomes important for the global organization of the IoT devices. The 5G evolution in cellular communication with high data rate for the purpose of transmission with very low latency can be used in 5G–IoT nodes and various cloud-based applications such as artificial intelligence, data handling, estimation, machine and deep learning [11]. In 5G, there is improvement of packet data rate compared to 4G; due to this, the delay that is required from one end to the other in order to deliver the package is improved, hence 5G is considered a better performer than 4G, which reduces interference greatly [12].

Researchers investigated several cases to indicate 4G, 5G, MIMO, massive MIMO, and narrowband IoT in which the new NB-IoT wireless technology will emerge in present and upcoming networks. Access networks will be denser in massive MIMO than existing 4G technology and will offer a larger capacity [13].

9.7 CONCLUSION

In this chapter, we highlighted the various aspects of 5G with Internet of Things technology and the use of MIMO antenna that changes the range of communication up to several GHz frequency by using NOMA. The spatial multiplexing is used in both the single user and the multiuser as it supports high data rate. The emerging 5G is required and this technology must be operational as early as possible as per steering report. In this chapter we have also seen various MIMO configurations, multiple access schemes used in 5G, and focussed on the capabilities of 5G defined by ITU and IMT 2020 with 3GPP standards. This review aims to provide researchers relevant information to 5G-IoT and MIMO and its challenges along with possible solutions for the current cellular communication era.

ACRONYMS

Acronym	Definition
1G	First generation
2G	Second generation

3G	Third generation
3GPP	3rd Generation Partnership Project
4G	Fourth Generation
5G	Fifth Generation
5G-PPP	Fifth-Generation 5G Public Private Partnership
IMT	International Mobile Telecommunications
IMT-2000	International Mobile Telecommunications 2000
IMT-A	International Mobile Telecommunications-Advanced
ITU	International Telecommunication Union
ITU-R	International Telecommunications Union – Radio communication
LTE	Long-Term Evolution
LTE-A	Long-Term Evolution-Advanced
MIMO	Multiple-Input Multiple-Output
NOMA	Non-Orthogonal Multiple Access
PAPR	Peak to Average Power Ratio
SIC	Successive Interference Cancellation
UMTS	Universal Mobile Telecommunication System

REFERENCES

[1] L. Chettri and R. Bera, "A comprehensive survey on Internet of Things (IoT) toward 5G wireless systems," *IEEE Internet Things J.*, vol. 7, no. 1, pp. 16–32, 2020, doi: 10.1109/JIOT.2019.2948888.

[2] S. Asif, *Next Generation Mobile Communications Ecosystem: Technology Management for Mobile Communications*. Wiley Inc., UK. 2011.

[3] R. N. Mitra and D. P. Agrawal, "5G mobile technology: A survey," *ICT Express*, vol. 1, no. 3, pp. 132–137, 2015, doi: 10.1016/j.icte.2016.01.003.

[4] Rec. ITU-R M.2083-0, "IMT vision – framework and overall objectives of the future development of IMT for 2020 and beyond," Sep. 2015.

[5] A. Bhardwaj, "5G for military communications," *Procedia Comput. Sci.*, vol. 171, no. 2019, pp. 2665–2674, 2020, doi: 10.1016/j.procs.2020.04.289.

[6] Y. Huo, X. Dong, and W. Xu, "5G cellular user equipment: From theory to practical hardware design," *IEEE Access*, vol. 5, pp. 13992–14010, 2017.

[7] F. Vook, A. Ghosh, and T. A. Thomas, "Massive MIMO communications towards 5G: Applications", *Requirements and Candidate Technologies*. John Wiley & Sons, Ltd, p. 1, United States, 2017.

[8] M. S. Sharawi, "Printed multi-band MIMO antenna systems and their performance metrics [wireless corner]," *IEEE Antennas Propagation. Magazine*, vol. 55, no. 5, pp. 218–232, 2013, doi: 10.1109/MAP.2013.6735522.

[9] Z. Xiao, L. Zhu, J. Choi, P. Xia and X. Xia, "Joint power allocation and beamforming for non-orthogonal multiple access (NOMA) in 5G millimeter wave communications," *IEEE Transactions on Wireless Communications*, vol. 17, no. 5, pp. 2961–2974, May 2018.

[10] J. G. Andrews et al., "What will 5G be?," *IEEE J. Sel. Areas Communication*, vol. 32, no. 6, pp. 1065–1082, Jun. 2014.

[11] A. Zhao and Z. Ren, "Size reduction of self-isolated antenna MIMO antenna system for 5G mobile phone applications," *IEEE Antennas Wireless Propagation Lett.*, vol. 18, no. 1, pp. 152–156, Jan. 2019.

[12] K. Shafique, B. A. Khawaja, F. Sabir, S. Qazi, and M. Mustaqim, "Internet of Things (IoT) for next-generation smart systems: A review of current challenges, future trends and prospects for emerging 5G-IoT Scenarios," *IEEE Access*, vol. 8, pp. 23022–23040, 2020, doi: 10.1109/ACCESS.2020.2970118.

[13] S. B. Ramteke, A. Y. Deshmukh, and K. N. Dekate, "A review on design and analysis of 5G mobile communication MIMO system with OFDM". *IEEE Conference Record # 42487; IEEE Explore*, ISBN: 978-1-5386-0965-1 2018.

10 Design Considerations in MIMO Antennas for Next-Generation 5G Wireless Communications

V. Dinesh, J. Vijayalakshmi, M. Suresh and A. Kabeel

This chapter depicts the opportunities and challenges involved in MIMO antenna design for the portable applications in smart cities. The concept of smart city reveals a broad opening on the existing challenges in urban infrastructure and internetworking capacity.

10.1 INTRODUCTION

The fifth-generation (5G) wireless communication frequency bands will cover shorter distance and will lead to dead zones. This requires signal repeaters at frequent locations to alleviate signal degradation. The concept of MIMO system in wireless communication introduces various possibilities of efficient utilization of existing channels in the near future. The implementation of multiple transmitter and receiver antennas using MIMO technology improves the usage of spectral efficiency for signal transmission over multiple spatial channels. Innovation in 5G wireless system is to improve the transmission data rate that leads to the implementation of MIMO antennas in the handheld portable smart devices. The physical design of MIMO antennas within the smallest smart devices is the design tradeoff. The innovation in miniaturization of

DOI: 10.1201/9781003093558-13

antenna dimensions drastically reduces the bandwidth of operation because of the fact known as antenna Q limit. In addition, reducing the space between the antennas introduces mutual coupling between the radiators; this might degrade the MIMO performance. Hence, this chapter discusses the various models of MIMO antennas proposed in recent papers, which describe the tradeoff in the antenna design and the performance characteristics of the MIMO system. The improved isolation within the MIMO antenna increases the antenna efficiency and gain. Various methods of improving the antenna isolation are defective ground structure (DGS), neutralization line (NL) technique, decoupling network (DN) and parasitic element (PE) technique using metamaterial (MTM) to enhance the isolation of MIMO antennas.

10.2 ISOLATION ENHANCEMENT TECHNIQUES

The miniaturization of MIMO antenna by reducing the space between the antenna elements introduces the mutual coupling between the elements. The different antenna isolation enhancement techniques improve the MIMO antenna performance [1].

10.2.1 Decoupling Network Technique

The implementation of decoupling network techniques in MIMO antenna structures is simple. The varieties of isolation structures, i.e. L-shaped, T-shaped and I-shaped, are discussed. The above-specified decoupling structures degrade the mutual coupling and thereby increase the ohmic losses. In general, the implementation of any coupling element cancels the existing mutual coupling effect, which gradually increases the antenna isolation.

The addition of two individual transmission lines (TL) between the strongly coupled antennas changes the trans-admittance from complex to pure imaginary at the antenna inputs. The generated imaginary value of trans-admittance is reduced by adding a shunt reactive component. Finally, a better impedance matching is achieved by attaching a lumped element circuit at the input ports. The excited current on the antenna surface is analyzed from even and odd modes that describe the radiation characteristics of an antenna array [2].

The isolation enhancement between the two closely packed antennas is obtained by placing a coupling element separating the two radiators. This method of placing an element introduces field cancellation between the antennas and thus enhances isolation. An isolation of more than 30 dB is achieved between the antennas radiators where both the planar inverted F antenna (PIFA) share the common ground plane [3]. The addition of multiple sections of transmission lines as a decoupling network improves the isolation of about 23 dB between the antenna elements. The fabricated antenna operates in the LTE band 13 that results in the envelope correlation coefficient that is less than 0.4 [4].

A hybrid fractals structure is achieved with the combination of two different fractal geometries. The insertion of a T-shaped strip and the etched rectangular slot in the upper layer of the ground plane improves the isolation and impedance matching between the radiators. A fractional bandwidth of 14% in band 1 from 1.65 GHz to

1.9 GHz and 80% in band 2 from 2.68 GHz to 6.25 GHz over the impedance matching is achieved [5]. An implementation of tunable decoupling network between the closely packed antenna elements minimizes the mutual coupling. The variation in decoupling network parameters is achieved by tuning the input–output couplings [6].

The microstrip lines and L-shaped strips are inserted adjacent to the antenna elements and short circuited with ground plane. The strip lines and the L-shaped lines are capacitive coupled with each other, which behave as resonators and minimize the coupling among the input ports. It results in performance improvement of more than 13 dB in isolation and an ECC of less than 0.3 [7]. In the designed 4 × 4 MIMO antenna structure, each antenna set consists of one PIFA antenna element and one coupling element. This coupling element acts as a loop antenna that effectively decreases the induced current flow. It results in performance improvement of more than 17 dB in isolation and an ECC of less than 0.03 [8]. Two C-shaped monopoles and a decoupling element result in dual band operation. Thus, there is a decrease in the coupling between the antenna feed which is a minimum of 13 dB and the ECC is lower than 0.05 [9] as shown in Figure 10.1. The printed T-shaped decoupling element in the monopole antennas structure varies with the ground plane surface current distribution as shown in Figure 10.2.

A planar helix-type antenna array is designed for frequency band of mobile applications. The resonance frequency of the antenna is tuned by varying the length of the helix lines. The suspended strip line within the antenna minimizes the mutual coupling among the antennas. The values of estimated inductance and the capacitance of suspended lines affect the isolation bandwidth. The increase in series inductance leads to better isolation; however, only the improvement in series capacitance results in improved isolation bandwidth [10].

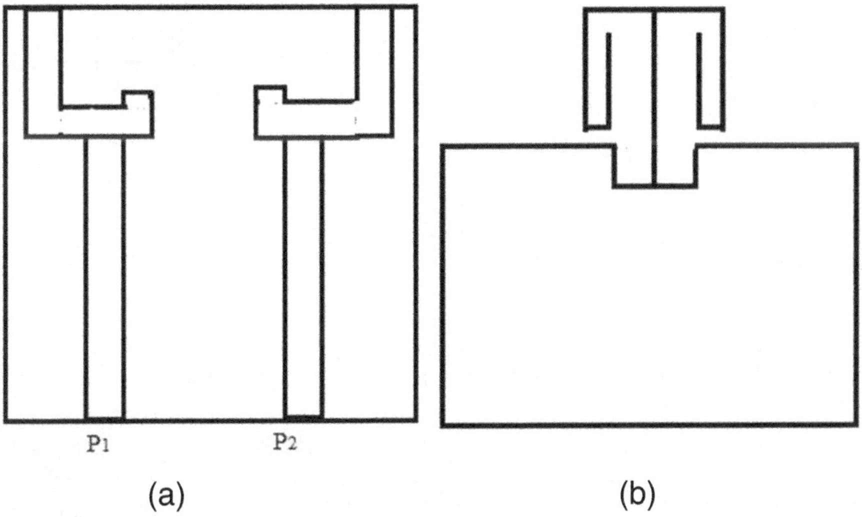

FIGURE 10.1 MIMO antenna with dual-band (a) top layer and (b) bottom decoupling network [9].

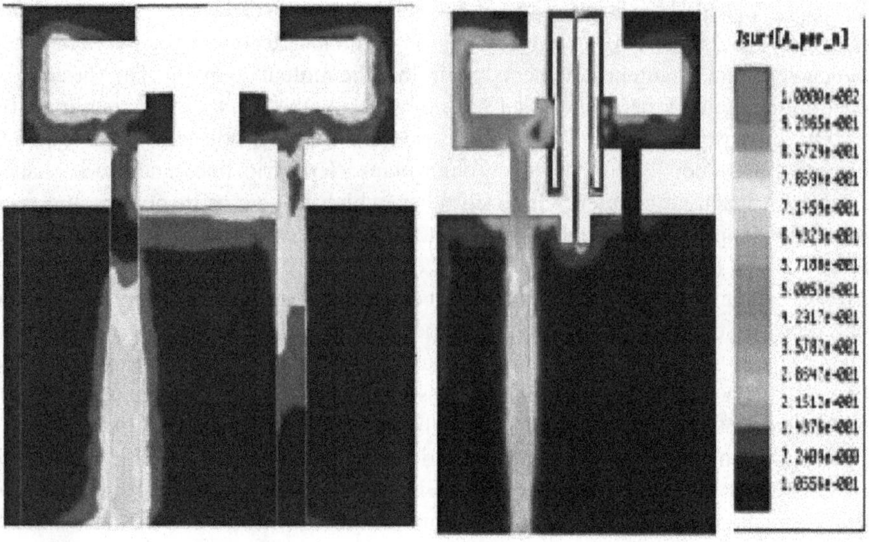

(a) (b)

FIGURE 10.2 Distribution of surface current at 2.4 GHz (a) top patch and (b) bottom ground plane [9].

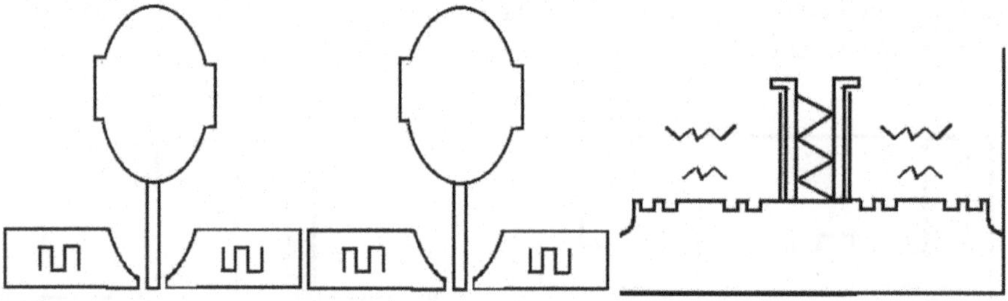

(a) (b)

FIGURE 10.3 MIMO antenna (a) 1 × 2 patch antenna and (b) sine curve-based T-shaped decoupling network [12].

The decoupling technique in the MIMO antenna array is introduced by adding a shunt element between the radiators. This shunt element decreases the mutual coupling between the antennas and improves the radiation performance of ECC providing better gain [11]. The structure of sine curve-based T-shaped decoupling network suppresses the induced current in the antenna, which exists in simple MIMO structure without decoupling network (Figure 10.3). This decoupling network improves the isolation and results in ECC of less than 0.53 and isolation of 20 dB [12].

10.2.2 Parasitic Element Technique

The introduction of parasitic elements (PE) such as dual-slot decoupling structure is implemented to minimize the mutual coupling among the elements [13]. The meander slot between the antenna structures minimizes the reverse coupling and improves the efficiency and bandwidth of the antenna [14]. For the portable smart phone devices, the planar inverted F antenna (PIFA) is the most preferred structure. The PIFA antenna operates for multiband frequencies of 2.4 GHz, 3.4 GHz, and 5.5 GHz for WLAN, WiMax, and HIPERLAN bands, respectively. The isolation between the MIMO antennas is increased by connecting two non-radiating strips between the antenna and ground plane [15]. Different structures of PE are used to control the mutual coupling effects, such as decrease in gain and variation in radiation patterns. An isolation of about 10 dB and 1.2 dBi gain with high radiation efficiency is achieved [16]. The PE placed horizontally at the bottom of the patch antenna decreases the mutual coupling between antennas. The PE results in high gain and optimum isolation for the MIMO antenna design as shown in Figures 10.4 and 10.5 [17].

Lowering the surface current improves the isolation between the antenna elements. The placement of PE with the antenna radiators increases the mutual coupling. Hence it results in an isolation of 30 dB as shown in Figure 10.6 [18].

10.2.3 Neutralization Line Technique

Recently several shapes of antenna designs, slots, defective ground structure (DGS) are incorporated to operate at required resonant frequency. The major tradeoff in implementing different shapes along with the antenna radiators varies the impedance of the antenna. Hence, the issue of impedance matching is minimized by this neutralization line (NL) technique. The selection of metal dimensions such as width and length, the location of metal element in the MIMO antenna geometry is a challenging

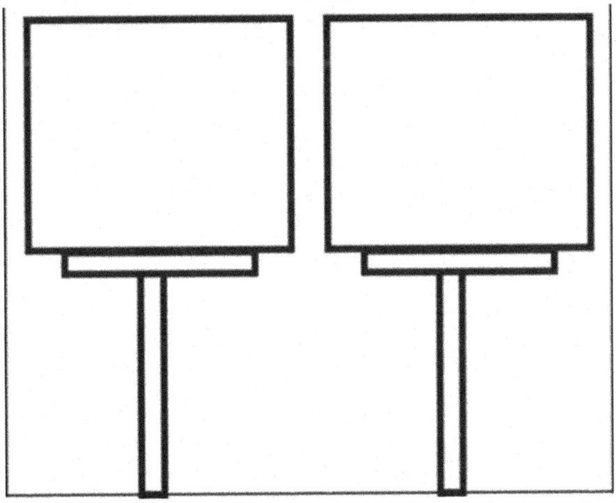

FIGURE 10.4 MIMO antenna of rectangular structure [17].

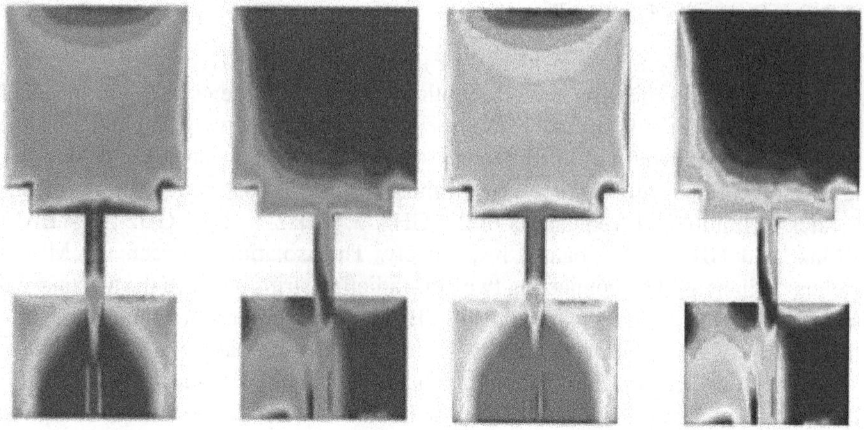

(a) (b)

FIGURE 10.5 Surface current distribution of antenna at 8 GHz (a) without parasitic elements and (b) with parasitic elements [17].

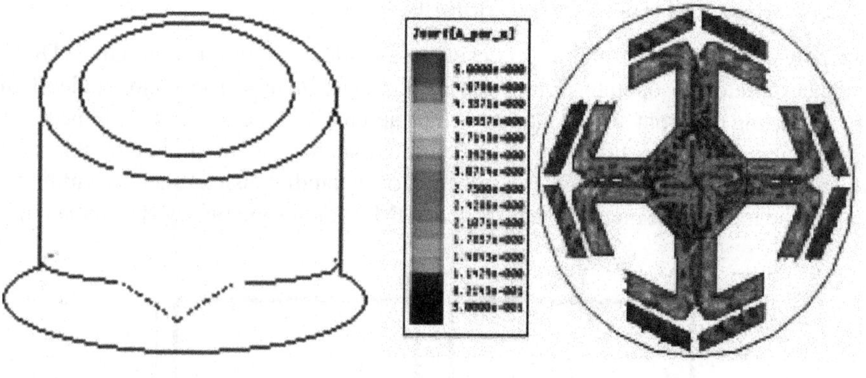

(a) (b)

FIGURE 10.6 Antenna element (a) side view and (b) surface current at 2.2 GHz on the patch with parasitic element [18].

method. The implementation of NL is easy; however, there is a shift in frequency in every design. The MIMO antenna with metal element results in radiation pattern which has a low ECC value of 0.006, gain of 2.10 dBi, and 19 dB inter-port isolation as shown in Figure 10.7 [19]. The design of MIMO antenna for mobile handset applications requires proper impedance matching for all operating frequency band. The suspended line along NL gives a better isolation of 12 dB, gain of 0.9 dBi, and ECC of 0.1 [20]. The F-shaped monopole antenna with ground plane operates at a

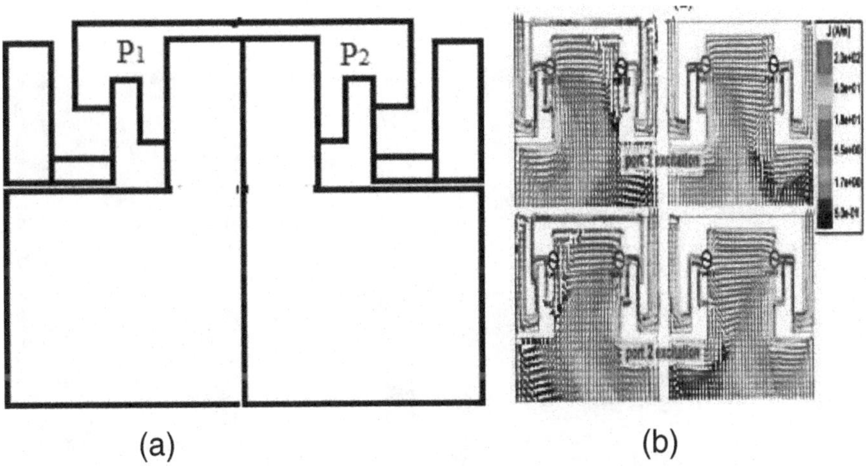

FIGURE 10.7 Dual-port antenna (a) with neutralization line and (b) surface current distribution at 2.4 GHz [19].

frequency band of GSM 1800 and 1900, UMTS, LTE 2300 and 2500, and 2.4-GHz WLAN bands. The NL elements with two symmetric antennas result in multiband operation with better impedance matching [21]. The optimum positioning of feed point by analyzing the flow of current in the antenna results in better isolation. The mutual coupling within the folded monopole antennas are minimized by connecting NL with the antenna elements [22].

10.2.4 Isolation through Metamaterial Techniques

Alternate method of minimizing mutual coupling in the MIMO antenna is by placing the human made metamaterial (MMT) structures between the antenna radiators. Optimization of the properties such as permittivity and permeability of a material controls the mutual coupling of MIMO antenna.

The most general structures of metamaterial are split ring resonator (SRR) and complementary split ring resonator (CSRR). The proper etching of the T-shaped structure on the ground plane improves the antenna bandwidth. Joining the three layers of MMT improves the coupling and results in 16 dB isolation [23]. Nowadays, the new structure of surface-integrated SRR is used for reducing mutual coupling and achieves an ECC of 0.02 [24–25]. The antenna miniaturization using CSRR technique also shifts the resonant frequency of operation from high to low. The surface current at the respective patches and the bottom layer depict the current flow at the resonant frequency of operation as shown in Figure 10.8.

The CSRR implementation with negative permittivity using MMT achieves a better mutual coupling reduction. The use of CSRR obtains an isolation of 15 dB [26]. The shape of ring trap between the antenna elements controls the mutual coupling and achieves an isolation of 14 dB and ECC less than 0.01. The implementation of electromagnetic bandgap (EBG) structures results in better mutual coupling [27].

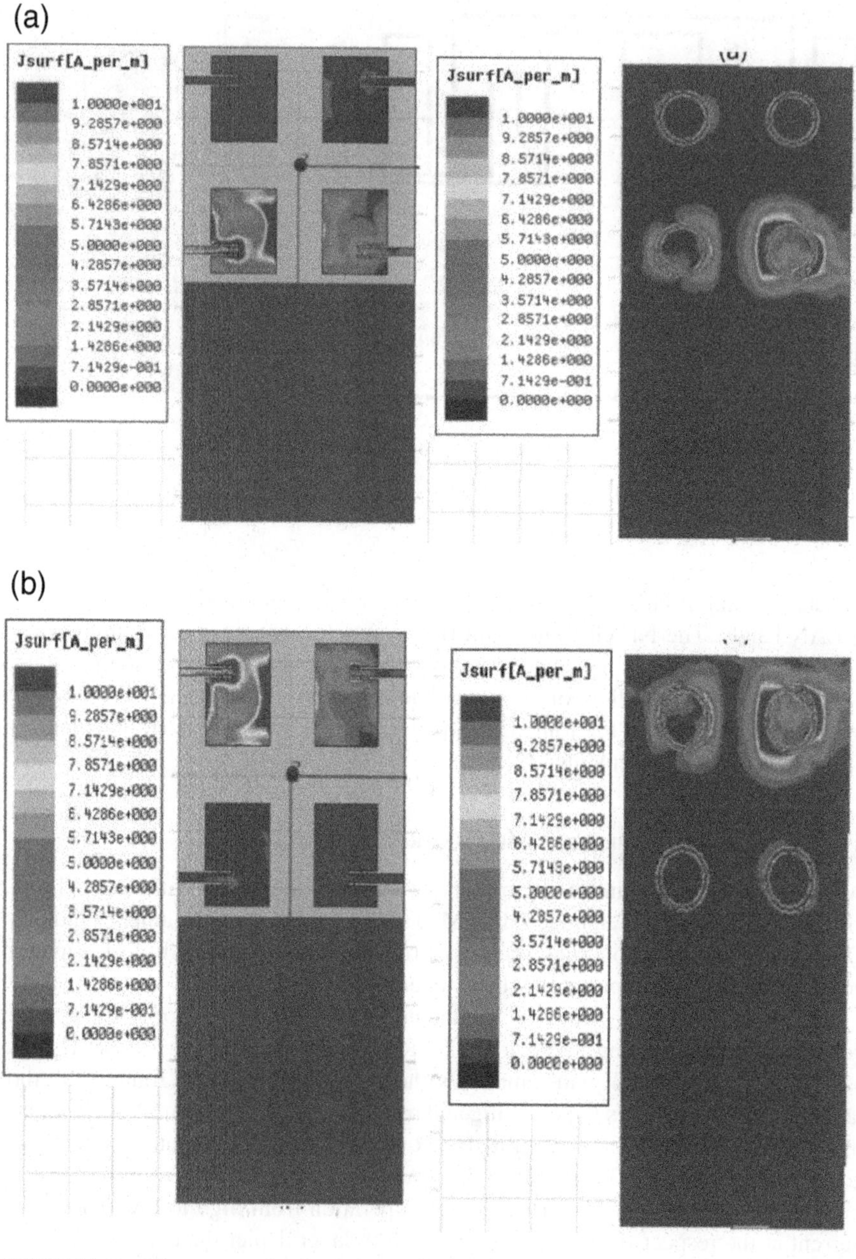

FIGURE 10.8 The current distribution at 2.48 GHz for active (a) patch 1 upper and bottom layer and (b) patch 3 upper and bottom layer [25].

The different types of mutual coupling reduction techniques are reported to improve the performance of MIMO antenna parameters such as isolation, correlation, gain, and directivity. A variety of isolation enhancement methods are discussed in MIMO antenna design, yet there were plenty of new novelty required to minimize the mutual coupling. Various techniques such as decoupling network (DN) technique, parasitic element (PE) technique, neutralization line (NL) technique, and isolation through metamaterial techniques are discussed and compared. The study on recent isolation techniques in recent research methods provides a novel idea towards the future technologies like mm-wave and 5G. The major challenges in the design of MIMO antenna is the controlling aspects of mutual coupling and this might be the research focus for future MIMO antenna design technology.

10.3 CONCLUSION

Various methods of decreasing mutual coupling and improving isolation in a MIMO antenna such as decoupling network, parasitic element technique, implementation of NL, and different MMT geometries are discussed. The investigation of various isolation reduction techniques for the future generation 5G wireless communication is a research area. The antenna designs for the future smart devices require wideband and multiband operations where minimizing the mutual coupling is the major challenging task.

REFERENCES

[1] A. C. J. Malathi, and D. Thiripurasundari, "Review on isolation techniques in MIMO antenna systems", *Indian Journal of Science and Technology*, vol. 9, pp. 1–10, 2016.

[2] S. Chen, Y. Wang, and S. Chung, "A decoupling technique for increasing the port isolation between two strongly coupled antennas", *IEEE Transactions on Antenna and Propagation*, vol. 56, pp. 3650–3658, 2008.

[3] A. C. K. Mak, C. R. Rowell, and R. D. Murch, "Isolation enhancement between two closely packed antennas", *IEEE Transactions on Antenna and Propagation*, vol. 56, pp. 3411–3419, 2008.

[4] J. Baek, and J. Choi, "The design of a LTE/MIMO antenna with high isolation using a decoupling network", *Microwave and Optical Technology Letters*, vol. 56, pp. 2187–2190, 2014.

[5] Y. K. Choukiker, S. K. Sharma, and S. K. Behera, "Hybrid fractal shape planar monopole antenna covering multiband wireless communications with MIMO implementation for handheld mobile devices", *IEEE Transactions on Antenna and Propagation*, vol. 62, pp. 1483–1488, 2014.

[6] Q. Kewei and G. Decheng, "Compact tunable network for closely spaced antennas with high isolation", *Microwave and Optical Technology Letters*, vol. 58, pp. 65–69, 2016.

[7] Z. Li, M. Han, and J. Choi, "Compact dual-band MIMO antenna for 4G USB dongle applications", *Microwave and Optical Technology Letters*, vol. 54, pp. 744–748, 2012.

[8] W. Lee and B. Jang, "A small 4 by 4 MIMO antenna system for LTE smart phones", *Microwave and Optical Technology Letters*, vol. 58, pp. 2668–2672, 2016.

[9] Y. Liu, L. Yang, Y. Liu, J. Ren, J. Wang, and X. Li, "Dual-band planar MIMO antenna for WLAN application", *Microwave and Optical Technology Letters*, vol. 57, pp. 2257–2262, 2015.

[10] Y. Yu, J. Ji, W. Seong, and J. Choi, "A compact MIMO antenna with improved isolation bandwidth for mobile applications", *Microwave and Optical Technology Letters*, vol. 53, pp. 2314–2317, 2011.

[11] B. Park, J. Choi, S. Park, T. Yang, and J. Byun, "Design of a decoupled MIMO antenna for LTE applications", *Microwave and Optical Technology Letters*, vol. 53, pp. 582–586, 2011.

[12] M. Bilal, R. Saleem, M. F. Shafique, and H. A. Khan, "MIMO application UWB antenna doublet incorporating a sinusoidal decoupling structure", *Microwave and Optical Technology Letters*, vol. 56, pp. 1547–1553, 2014.

[13] Z. Li, Z. Du, M. Takahashi, K. Saito, and K. Ito, "Reducing mutual coupling of MIMO antennas with parasitic elements for mobile terminals", *IEEE Transactions on Antenna and Propagation*, vol. 60, pp. 473–481, 2012.

[14] M. Ayatollahi, Q. Rao, and D. Wang, "A compact high isolation and wide bandwidth antenna array for long term evolution wireless devices", *IEEE Transactions on Antenna and Propagation*, vol. 60, pp. 4960–4963, 2012.

[15] H. S. Singh, G. K. Pandey, P. K. Bharti, and M. K. Meshram, "A low profile triband diversity antenna for WLAN/Wimax/Hiperlan applications with high isolation, *Microwave and Optical Techn. Letters*, vol. 57, pp. 452–457, 2015.

[16] S. Soltani, and R. D. Murch, "A compact planar printed MIMO antenna design", *IEEE Transactions on Antenna and Propagation*, vol. 63, pp. 1140–1149, 2015.

[17] M. S. Khan, A. D. Capobianco, M. F. Shafique, B. Ijaz, A. Naqvi, and B. D. Braaten, "Isolation enhancement of a wideband MIMO antenna using floating parasitic elements", *Microwave and Optical Technology Letters*, vol. 57, pp. 1677–1682, 2015.

[18] H. Huang, Y. Liu, and S. Gong, "Broadband omnidirectional dual-polarized antenna with high isolation", *Microwave and Optical Technology Letters*, vol. 57, pp. 1848–1852, 2015.

[19] S. Su, C. Lee and F. Chang, "Printed MIMO-antenna system using neutralization-line technique for wireless USB-dongle applications," *IEEE Transactions on Antennas and Propagation*, vol. 60, no. 2, pp. 456–463, Feb. 2012, doi: 10.1109/TAP.2011.2173450.

[20] H. Bae, Frances J. H. Park, T. Kim, N. Kim, D. Kim, and B. Lee, "Compact mobile handset MIMO antenna for LTE 700 applications", *Microwave and Optical Technology Letters*, vol. 52, pp. 2419–2422, 2010.

[21] Y. Wang, and Z. Du, "A wideband printed dual-antenna system with a novel neutralization line for mobile terminal", *IEEE Antennas and Wireless Propagation Letters*, vol. 12, pp. 1428–1431, 2013.

[22] W. H. Shin, S. Kibria, and M. T. Islam, "Hexa band MIMO antenna with neutralization line for LTE mobile device application", *Microwave and Optical Tech. Letters*, vol. 58, pp. 1198–1204. 2016.

[23] C. Hsu, K. Lin, and H. Su, "Implementation of broadband isolator using metamaterial-inspired resonators and a T-shaped branch for MIMO antennas", *IEEE Transactions on Antenna and Propagation*, vol. 59, pp. 3936–3939, 2011.

[24] Z. Liu, S. Qu, J. Wang, J. Zhang, M. Hua, Z. Xu, and A. Zhang, "Isolation enhancement of patch antenna array via metamaterial integration", *Microwave and Optical Technology Letters*, vol. 58, pp. 2321–2325, 2016.

[25] M. S. Sharawi, M. U. Khan, A. B. Numan, and D. N. Aloi, "A CSRR loaded MIMO antenna system for ISM band operation", *IEEE Transactions on Antenna and Propagation*, vol. 61, pp. 4265–4274, 2013.
[26] M. F. Shafique, Z. Qamar, L. Riaz, R. Saleem, and S. A. Khan, "Coupling suppression in densely packed microstrip arrays using metamaterial structure", *Microwave and Optical Technology Letters*, vol. 57, pp. 759–763, 2015.
[27] J. Kumar, "Compact MIMO antenna", *Microwave and Optical Technology Letters*, vol. 58, pp. 1294–1298, 2016.

11 Design and Enhancement of MIMO Antennas for Smart 5G Devices

Neeraja S. Jawali, Vidyadhar S. Melkeri and Gauri Kalnoor

11.1 INTRODUCTION

This chapter describes the design of 5G multiple-inputs and multiple-Outputs (MIMO) antenna structures. 5G is the fifth generation of mobile networks, a significant evolution from today's 4G LTE networks. 5G means a world where a massive number of machines are connected – also called an Internet of Things (IoT) – consisting of enhanced mobile broadband and providing ultra-low latency with high reliability and strong security connections to the automobiles. This new version of wireless technology provides a lot of advantages that include an increasing speed, reducing reluctance, and providing a more reliable connection for a wide range of users. MIMO is a means for improving the capacity of the radio link. Here, an inset feeding technique is used. The proposed antenna operates at millimetre-wave (mm-wave) frequency. The patch antenna is designed to resonate at 29 GHz frequency. The FR4 epoxy, a dielectric substrate, is used in the proposed design which has dielectric constant (ε_r) of 4.4 and the thickness (h) of the substrate is 1.6 mm. Ansys HFSS software is used to design and simulate the MIMO antenna system. At a desired frequency of 29 GHz, multiple antennas are resonated with better S11, VSWR, and bandwidth. An antenna is a transducer that changes over radio frequency (RF) fields into pivoting current or the other way round. There are both reception and transmission gathering devices for sending and receiving radio transmissions. Antennas play a huge part in the movement of all radio gadgets. They are used in remote neighbourhood, cell phone, and satellite communications.

DOI: 10.1201/9781003093558-14

The microstrip patch is a kind of antenna which is made up of radiating patch of specific shape on one side of dielectric substrate and ground plane on the other side. The microstrip antennas have a number of advantages that make mobile communication industries prefer them to the preservationist antennas for mobile to satellite communications, cell, GPS, satellite, remote LAN for PCs, Wi-Fi, Bluetooth innovation, RF ID gadgets, WiMax, and so on [7]. Future requirements in the field of remote portable correspondence, for example, high data transfer capacity, radio wire clusters for upgraded productivity, moveable versatile specialized gadgets, and data and checking amassing gadgets, have interested the considerations of numerous scientists in the field of correspondence building to grasp the plan and to examine the endless states of microstrip antennas (MSA) for upgrading the transmission capacity and other reception apparatus parameters to suit their needs [8]. Moderate microstrip patch antenna involves a couple of equal leading sheets separated by dielectric medium, which is expressed as substrate. In this design, the upper leading sheet or the "patch" is the reason for radiation where electromagnetic vitality borders off the edges of the patch and into the substrate. The lower directing sheet goes as a faultlessly reflecting ground plane, ricocheting vitality back through the substrate and into the free space. Generously, the patch is a slight conductor that is an impressive portion of frequency in degrees [3].

5G is the fifth-generation moveable framework. It is another overall remote standard following the 1G, 2G, 3G, and 4G frameworks. 5G will revolutionize the world more than 3G or 4G. 5G means a world where not only the people but all things are also connected. 5G technology will be a lot quicker than any other existing networks and will have better and improved video quality in terms of download rate and transfer speed. 5G can achieve an enhanced transmission speed at a minimum of 20 GB/sec. 5G uses better spectrum, has significantly lower latency, has increased bandwidth, and also beam steering and beamforming capacity than 4G [4]. With high speeds, prevalent dependability, and unimportant inertness, 5G will extend the portable biological system into novel fields. 5G will affect every industry, making more secured transport, remote social insurance, exactness in farming, and digitized coordinations a reality – and the sky is the limit from there [1].

Multiple-input multiple-output (MIMO) is a wireless technology that enhances the data capacity of an RF radio by means of various transmitting and receiving antennas.

In a MIMO framework, the same information is conveyed through numerous receiving wires over a similar transfer speed. Because of this characteristics, individual signals reach the receiving antenna through diverse paths, resulting in more consistent data. The data rate also surges by a factor determined by the number of transmitting and receiving antennas.

Advantages of MIMO system:

1. With the use of bounced and reflected RF transmissions, MIMO system offers improved signal quality via unobstructed path.
2. The increased throughput permits improved quality and amount of video sent over the system.

3 Multiple data streams decrease the number of lost data packets, which results in improved video or audio quality.

In a MIMO system, two or more antennas are used on both the receiving and the transmitting sides. As shown in Figure 11.1, Antenna 1, Antenna 3, Antenna 5, Antenna 7 are used as transmitting antennas and Antenna 2, Antenna 4, Antenna 6, Antenna 8 are used as receiving antennas. The mutual coupling or isolation between adjacent antennas is improved.

Indeed, the ongoing extraordinary advancements in the design of 5G base station antenna may be seen from writing that the introduced architecture does not satisfy all the 5G framework necessities. For example, the proposed antenna design in Ref. [2] has difficulty in fabrication and the design in Ref. [14-15] does not support the MIMO technology. In this chapter, we propose a mm-wave 5G antenna using MIMO technology as inset feeding technique resonating at 29 GHz with better S-parameter, VSWR, higher efficiency, and high bandwidth, which makes it a better choice for future 5G applications [6].

The software used is HFSS (High-Frequency Structure Simulator), a viable Finite Element Method (FEM) solver for electromagnetic systems from ANSYS. HFSS

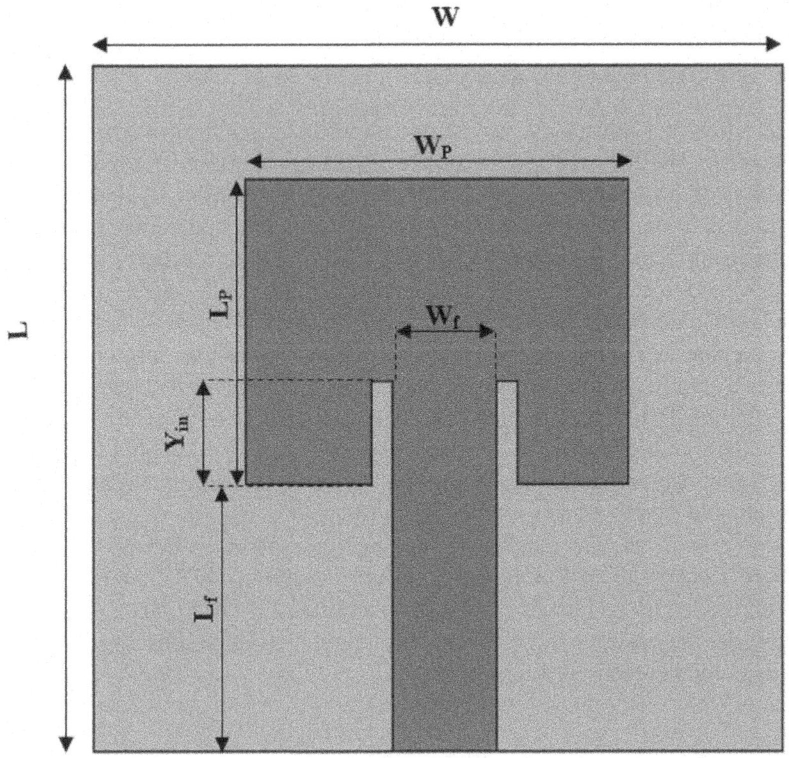

FIGURE 11.1 Inset feed antenna.

is used to integrate solid modelling, simulation, and visualization in a communal environment. It divides the complete structure into tetrahedral mesh, after which the boundary conditions are applied.

11.1.1 Typical Applications

1 Package modelling
2 PCB board modelling
3 Antenna mobile communication
4 Connectors
5 Waveguide
6 Filters

11.2 INSET FEEDING

An antenna feed line is a type of wire used to associate the antenna with the radio; it is also utilized to energize the transmission by indirect or direct contact. Various methods of feeding techniques are available. In this chapter, we will use the inset feeding technique for designing the MIMO antenna. It is an advancement made to the microstrip line feed. An inset feed is fed closer to the patch to reduce the input impedance [9].

11.3 DESIGN CALCULATIONS

The selection of a substrate is very important when designing a microstrip patch. An antenna cannot be designed without the appropriate choice of substrate parameters such as width, length, dielectric constant, and height. There are different kinds of materials that are accessible and are utilized as substrate depending upon the prerequisites and performance. Here, FR4 epoxy of thickness 1.6 mm with a relative permittivity of 4.4 is utilized as the substrate. The dimensions of the patch are designed using the formulas given below. The antenna feeding is structured cautiously to give proper or exact impedance matching. At high-signal frequencies, taking care of the feed line assumes a significant job for radio-wire execution. For better outcomes, taking care of the line ought to have an impedance equivalent to qualities of the impedance of fix. For appropriate feeding of antennas, the inset feeding method is utilized in this chapter. With a point inside the patch where the input impedance is equal to 50 Ω, the patch antenna is fed with a microstrip line [10].

Using the rectangular patch antenna equations [5] given below, the calculated length (L_p) and width (W_p) of the patch antenna is 1.8 mm and 3.26 mm, respectively, using the following parameters: frequency resonance (f_r) = 29 GHz, dielectric constant of antenna substrate ($\mathcal{E}_r$) = 4.4, and the height of antenna substrate (h) = 1.6 mm.

Designing the width of patch (W_p):

$$W_p = \frac{C}{2f_r}\sqrt{\frac{2}{\varepsilon_r + 1}} \tag{11.1}$$

Designing the effective dielectric (ε_{eff}):

$$\varepsilon_{eff} = \left(\frac{\varepsilon_r + 1}{2}\right) + \left(\frac{\varepsilon_r - 1}{2}\right)\left[1 + 12\left(\frac{h}{W_p}\right)\right]^{-1} \tag{11.2}$$

Designing the length extension (ΔL):

$$\Delta L = (0.412)h\left[\frac{\left(\varepsilon_{eff} + 0.3\right)\left(\frac{W_P}{h} + 0.264\right)}{\left(\varepsilon_{eff} - 0.258\right)\left(\frac{W_P}{h} + 0.8\right)}\right] \tag{11.3}$$

Designing the effective length of patch (L_{eff}):

$$L_{eff} = \frac{C}{2fr\sqrt{\varepsilon_{eff}}} \tag{11.4}$$

Designing the actual length of patch (L_P):

$$L_P = L_{eff} - 2\Delta L \tag{11.5}$$

Designing the patch width (W_g):

$$W_g = (6 \times h) + W_P \tag{11.6}$$

Designing the patch length (L_g):

$$L_g = (6 \times h) + W_P \tag{11.7}$$

Designing the inset depth:

$$y_0 = \frac{L_P}{\pi}\cos^{-1}\left(\sqrt{\frac{Z_{in}}{R_{in}}}\right) \tag{11.8}$$

OR

$$y_0 = 10^{-4}\begin{Bmatrix} 0.001699\varepsilon_r^7 + 0.13761\varepsilon_r^6 - 6.1783\varepsilon_r^5 + 93.187\varepsilon_r^4 \\ -682.69\varepsilon_r^3 + 2561.9\varepsilon_r^2 - 4043\varepsilon_r^1 + 6697 \end{Bmatrix}\frac{L_P}{2} \tag{11.9}$$

where Z_{in} = resonant input impedance and
R_{in} = resonant input resistance.

11.4 DESIGN AND SIMULATION OF MIMO ANTENNAS

After calculating the dimensions of patch by using the above equations, slots are placed at the ground plane to improve the S11 of the designed antenna (Table 11.1).

11.4.1 Design of MIMO Antenna Structures

See Figures 11.2 and 11.3.

11.4.2 Simulation Results

S-parameters or scattering-parameters are the elements of the S-matrix, which is a square matrix that characterizes the input–output connection between the ports in an electrical framework [11]. Supposing that the power is transmitted from the first port (port-1) to the second port (port-2) of a provided network. If there is no proper impedance matching of the network with the load terminating the port-2, then it is likely that most of the signal will get reflected towards port-1. This results in the loss which is known as "Return Loss", measured in dB. For the system to have lesser losses, there should be proper impedance matching. Figures 11.4 and 11.5 show the simulated S11 at 29 GHz is −33.2 dB, −35.8 dB, and −32.6 dB, respectively. As we can see, the designed system is working properly with proper impedance matching and having a bandwidth of nearly 4 GHz.

The voltage standing wave ratio (VSWR) is the measure of how thoroughly the source impedance and the load impedance are matched. In other words, it is the quantity of how thoroughly the antenna is to a perfect 50 Ohms. VSWR is the utility of reflection coefficient, which defines the power reflected from the antenna [13]. If the VSWR esteem is high, it implies that there is higher impedance mismatch. The VSWR value must range between 1 and 2. Figure 11.6 shows the VSWR of various designed MIMO antennas. The obtained VSWR values are 1.04, which means that there is better impedance matching and hence better transmission. Thus, the power reflected is only 0.002 dB.

Radiation pattern is the portrayal of the energy emanated by an antenna. These are diagrammatical portrayals of the dispersion of emitted energy into space, as a purpose of direction [12]. The simulation results for radiation patterns have the highest gain of 3 dB (Figure 11.7).

Another parameter of MIMO system is Envelope Correlation Coefficient (ECC). ECC are presented to evaluate some of the diversity capabilities of MIMO system. For a two-element MIMO system, ECC can be calculated using the following formula:

$$\rho = \frac{\left|S_{11}S_{12} + S_{21}S_{22}\right|^2}{\left(1-\left(\left|S11\right|^2 + \left|S_{21}\right|^2\right)\right)\left(1-\left(\left|S_{22}\right|^2 + \left|S_{12}\right|^2\right)\right)}$$

The lower the correlation, the higher data throughput is supported by MIMO system. For MIMO system, we need $\rho < 0.5$. From Figure 11.8 ECC obtained is 0.05, so it satisfies the condition.

TABLE 11.1
Dimensions of the proposed antenna

Parameters	Width of substrate (W)	Length of substrate (L)	Width of patch (W_p)	Length of patch (L_p)	Height of substrate (h)	Width of feed line (Wf)	Length of feed line (Lf)	Inset depth (Y_{in})	Inset gap
1 × 1 antenna dimensions (mm)	13.7	13.2	3.26	1.8	1.6	1	3.35	0.8	0.15
2 × 2 antenna dimensions (mm)	13.7	22	3.26	1.8	1.6	1	3.35	0.8	0.15
4 × 4 antenna dimensions (mm)	13.7	38.04	3.26	1.8	1.6	1	3.35	0.8	0.15

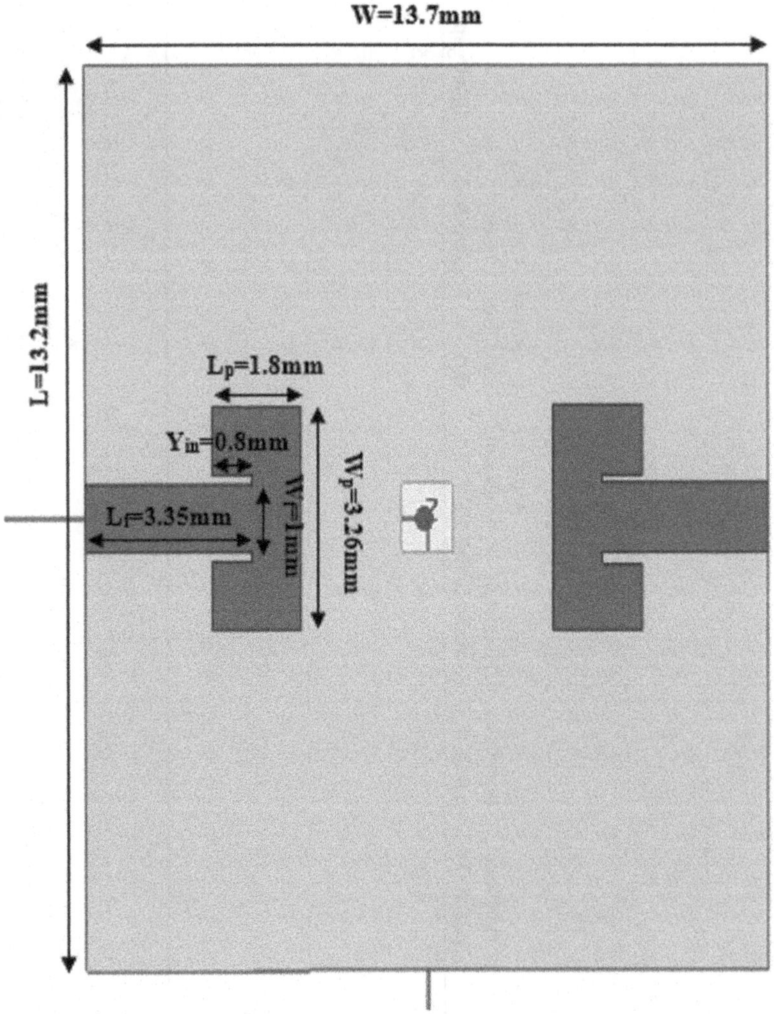

FIGURE 11.2 1 × 1 MIMO antenna structure.

11.5 CONCLUSION

Inset feed rectangular MSA functioning at 29 GHz is designed and its implementation analysis is presented in this chapter. The projected antenna is proficient of receiving and transmitting signals at its operational range. ECC and mutual coupling of MIMO system are improved. The projected antenna with better impedance matching has accomplished radiating signals with 75% efficiency. The return loss is −33.20 dB with a bandwidth of nearly 4 GHz. The future scope of this chapter is to improvize the gain so that it can produce strong signals that can radiate in the desired direction and enhance the radiation pattern of the designed antennas, while also fabricating the prototype antenna.

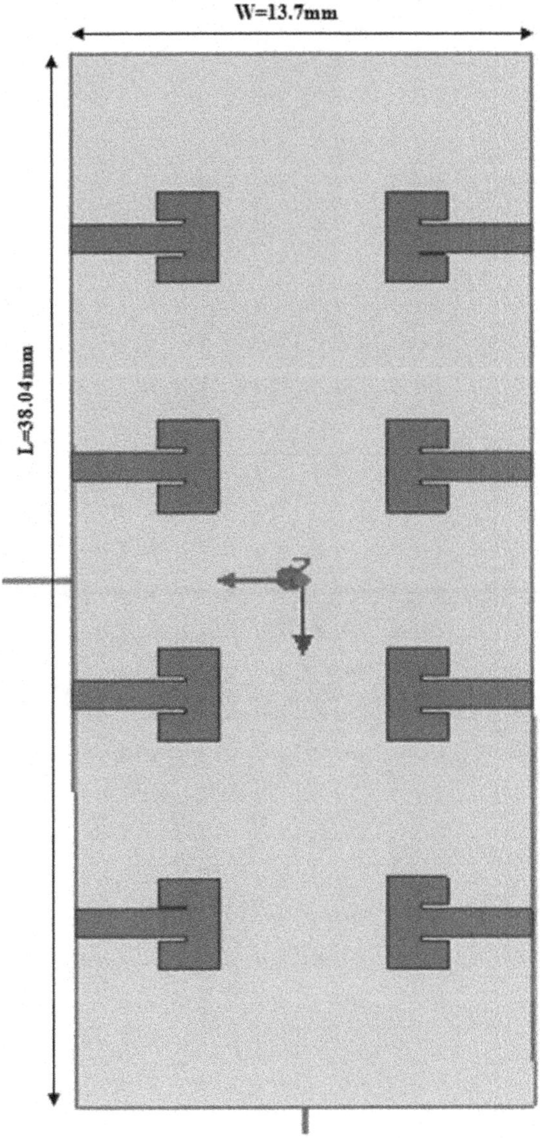

FIGURE 11.3 4 × 4 MIMO antenna.

REFERENCES

[1] A. Zhao and Z. Ren, "Size reduction of self-isolated MIMO antenna system for 5G mobile phone applications". *IEEE Antennas and Wireless Propagation Letters*, 18(1), 2019, pp. 152–156.

[2] P. M. Sunthari and R. Veeramani, "Multiband microstrip patch antenna for 5G wireless applications using MIMO techniques", *2017 First International Conference*

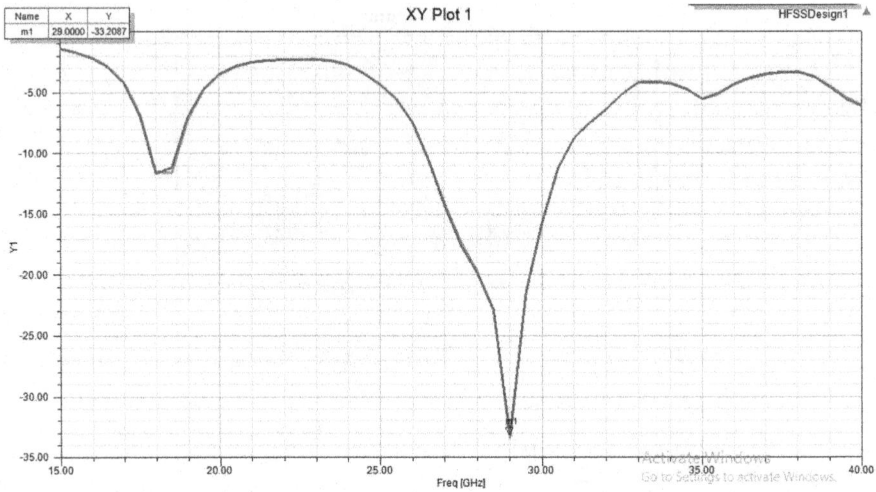

FIGURE 11.4 S11 of the 1 × 1 MIMO antenna.

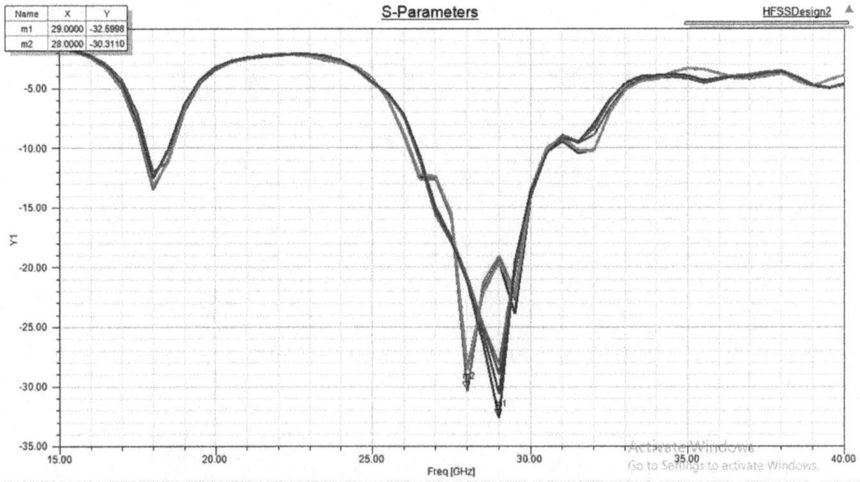

FIGURE 11.5 S11 of 4 × 4 MIMO antenna.

on Recent Advances in Aerospace Engineering (ICRAAE), Coimbatore, 2017, pp. 1–5, doi: 10.1109/ICRAAE.2017.8297241.

[3] S. B. Patil, R. D. Kanphade, and V. V. Ratnaparkhi, "Design and performance analysis of inset feed microstrip square patch antenna for 2.4 GHz wireless applications", *2015 2nd International Conference on Electronics and Communication Systems (ICECS), Coimbatore*, 2015, pp. 1194–1200, doi: 10.1109/ECS.2015.7124773.

[4] J. Singh, "Insert feed microstrip patch antenna", *International Journal of Computer Science and Mobile Computing (IJCSMC 2016)*, 5(2), February 2016, pp 324–329.

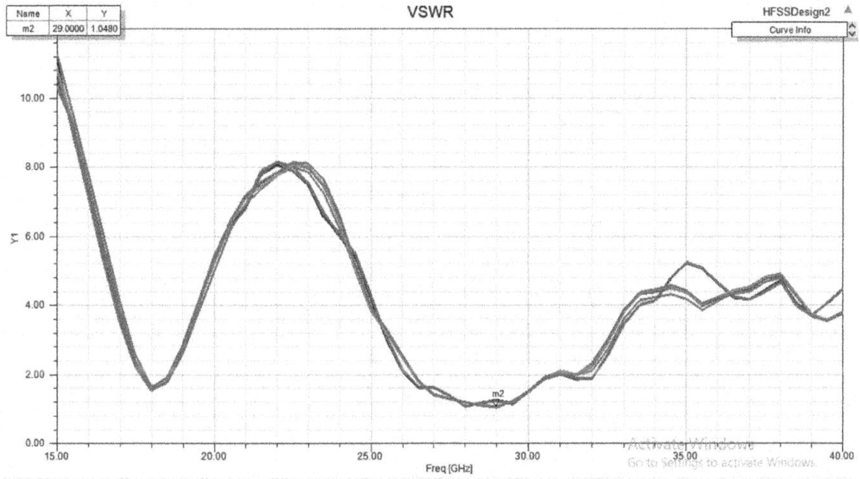

FIGURE 11.6 VSWR of 4 × 4 MIMO antenna.

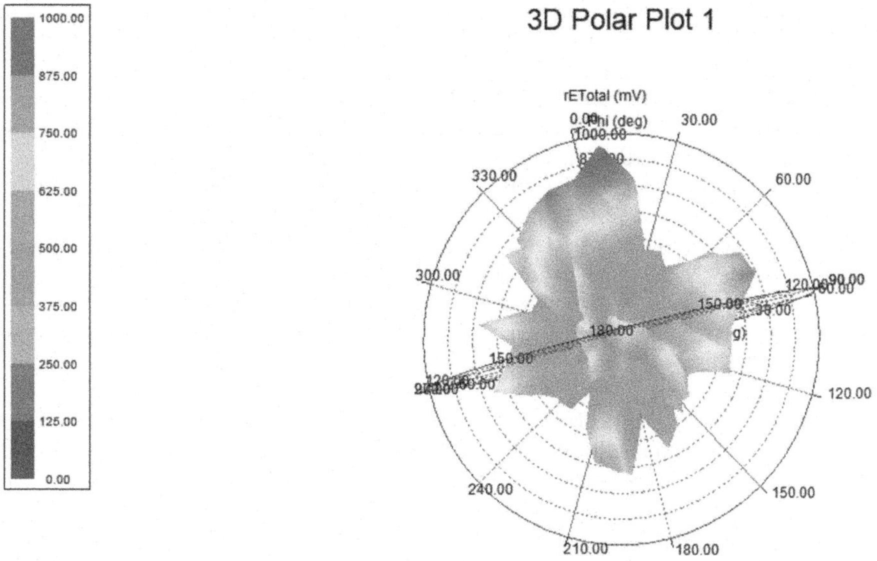

FIGURE 11.7 Radiation pattern of 4 × 4 MIMO antenna.

[5] H. T. Chattha, "4-port 2-element MIMO antenna for 5G portable applications," *IEEE Access*, 7, 2019, pp. 96516–96520, doi: 10.1109/ACCESS.2019.2925351.

[6] J. Salai Thillai Thilagam and P. K Jawahar, "Design of microstrip patch antenna for wireless communication". *International Journal of Applied Engineering Research*, 6(18), 2011.

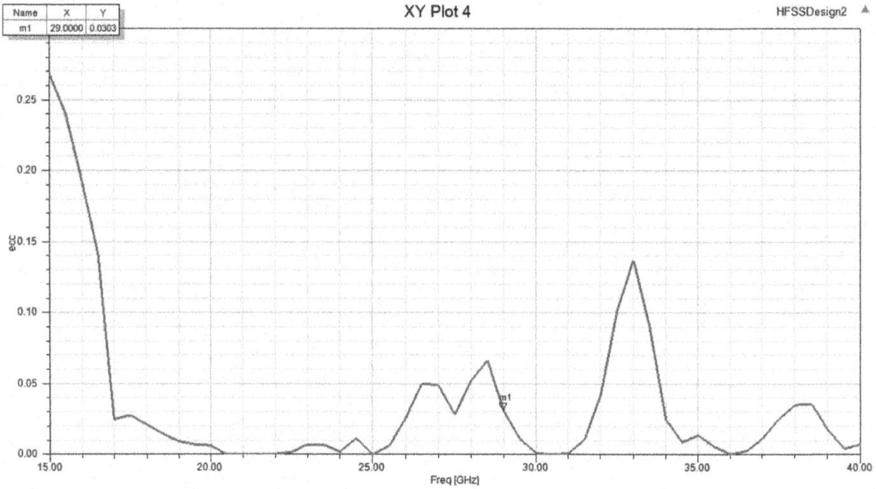

FIGURE 11.8 ECC of 4 × 4 MIMO system.

[7] Y. Bhomia, V. Gupta, A. Bhardwaj, and N. Akhtar, "To design and simulate microstrip patch antenna by line feed," *2017 International Conference on Electrical and Computing Technologies and Applications (ICECTA), Ras Al Khaimah*, 2017, pp. 1–5, doi: 10.1109/ICECTA.2017.8251923.

[8] L. C. Paul, M. Ur Rashid, M. Mowla, M. Morshed, and A. K. Sarkar, "Performance enhancement of an inset feed rectangular microstrip patch antenna by adjusting feeding position for wireless applications", *3rd International Conference on Electrical, Electronics, Engineering Trends, Communication, Optimization and Sciences (EEECOS 2016), Tadepalligudem*, 2016, pp. 1–5, doi: 10.1049/cp.2016.1549.

[9] K. R. Mahmoud and A. M. Montaser, "Performance of tri-band multi-polarized array antenna for 5G mobile base station adopting polarization and directivity control", *IEEE Access*, 6, 2018, pp. 8682–8694, doi: 10.1109/ACCESS.2018.2805802.

[10] Y. Liu, Y. Lu, Y. Zhang, and S. Gong, "MIMO antenna array for 5G smartphone applications", *2019 13th European Conference on Antennas and Propagation (EuCAP), Krakow, Poland*, 2019, pp. 1–3.

[11] V. S Melkeri, D. B Ganure, S. N. Mulgi, R. M. Vani, and P. V. Hunagund, "Size reduction with multiband operating square microstrip patch antenna embedded with defected ground structure". *International Journal for Research in Applied Science and Engineering Technology (IJRASET)*, 6(3), March 2018, pp. 1446–1454.

[12] V. S. Melkeri, S. L. Mallikarjun, and P. V. Hunagund, "H-shaped-defected ground structure (DGS) embedded square patch antenna", *International Journal of Advance Research in Engineering and Technology (IJARET)*, 6(1) January 2015, pp. 73–79.

[13] V. S. Melkeri, M. Lakshetty, and P. V. Hunagund, "Microstrip antenna with defected ground structure: A review". *International Journal of Electrical and Electronics and Telecommunication Engineering (IJEETE)*, 46(1), February 2015, pp. 1492–1496.

[14] V. S. Melkeri, R. M. Vani, and P. V. Hunagund, "Chess board shape defected ground structure for multi band operation of microstrip patch antenna" *Journal of Computational Information Systems (JCIS)*, 14(5), October 2018, pp. 73–78.

[15] V.S. Melkeri and P. V. Hunagund, "Concentric ring defect on ground plane for enhancement of parameters and physical size reduction for 2.4 GHz antenna" *Journal of Computational Information Systems (JCIS)*, 15(1), January 2019, pp. 17–22.

[16] V. S. Melkeri and P. V. Hunagund, "Study and analysis of SQMSA with square shaped Defected Ground Structure: Performance enhancement and virtual size reduction," *2017 IEEE International Conference on Antenna Innovations & Modern Technologies for Ground, Aircraft and Satellite Applications (iAIM), Bangalore*, 2017, pp. 1–5, doi: 10.1109/IAIM.2017.8402516.

[17] *M. B. Yadav, B. Singh, and V. S. Melkeri, "Design of rectangular microstrip patch antenna with DGS at 2.45 GHz", 2017 International Conference of Electronics, Communication and Aerospace Technology (ICECA), Coimbatore, 2017, pp. 367–370, doi: 10.1109/ICECA.2017.8203706.*

Part IV

Fractal and Defected Ground Structure Microstrip Antennas

12 Compact Rhombus Rectangular Microstrip Fractal Antennas with Vertically Periodic Defected Ground Structure for Wireless Applications

T. Poornima, E. L. Dhivya Priya and K. R. Gokul Anand

12.1 INTRODUCTION

A novel compact rhombus-shaped rectangular microstrip antenna using fractal geometry with a Vertically Periodic Defected Ground Structure (VPDGS) is presented in this chapter. Incorporating fractal geometry on the ground plane enhances the bandwidth. The Defected Ground Structure (DGS) of the microstrip antenna is used to suppress higher-order harmonics and mutual coupling between adjacent elements. Good performance in terms of return loss and cross-polarization for improving radiation characteristics are obtained using the operation bands. The proposed antenna

DOI: 10.1201/9781003093558-16

is more efficient and analyzed at the center frequency of 9.75 GHz and 14.2 GHz. The designed parameters are optimized and its characteristics are investigated using Ansys HFSS software. The obtained antenna operates in X-band and Wi-Max applications. Microstrip antenna turns out to be well-known and pulled in incredible consideration in the wireless world for their eminent potential benefits including easy to analyze and fabricate, low cost, excellent radiation characteristics, and capability of single, dual, triple, and various frequency operation with reduced hardware complexity [1]. This type of antenna is widely used in radar applications, satellite communications, biomedical, and aerospace applications. Feed line is the transmission line, which connects antenna and radio. Generally, antenna feed is the location of the antenna used to connect or attach the receiver or antenna [2]. The main purpose of feed line is to attain the compatibility with Integrated Circuits (IC) and more adaptable in terms of operating or center or resonant frequency, impedance matching, pattern, and polarization [3, 4]. Coaxial feed is opted, as they can be placed at any desired position and also supports in maintaining the impedance matching. Though microstrip antenna has several advantages, low gain, narrow bandwidth, and large size are its major disadvantages [5]. This in turn limits the applications of microstrip antenna in other fields. Enormous techniques are evolved to overcome the limitations [6, 7]. The recently used prominent techniques are increasing the length and width of the dielectric substrate, using various shapes of patch, and decreasing the dielectric constant. Rectangular, circular, and square-shaped patches are widely used in recent years [8]. The main purpose of this chapter is to incorporate the fractal geometry with VPDGS for modern wireless applications [9–11]. Though there are several shapes, the microstrip antenna is designed as a rhombus slot in order to provide better isolation between the microstrip elements and ensure mutual coupling. Apart from the conventional periodic structure, DGS have attracted a wide interest on telecommunication due to its extensive applicability in antennas and microwave circuits [12]. It is introduced in the ground plane of microstrip patch antenna for improving low gain, cross-polarization, and narrow bandwidth [13]. VPDGS is mainly adopted to attain the compact size of microstrip antenna. Fractal geometry is employed in VPDGS to resonate antenna in various frequencies and its performance mainly depends on enhancing bandwidth, return loss, gain, and cross-polarization. The designed antenna are simulated for results using Ansys High Frequency Structure Simulator (HFSS) – 3D Electromagnetic (EM) simulation software and analysis of the S parameters and return loss in dB.

Antenna design and configuration are proposed in Section 12.2 with the detailed parameters especially for substrate and patch. In Section 12.3, simulation results are analyzed for the proposed antenna. Finally, the conclusion is presented in Section 12.4.

12.2 ANTENNA DESIGN AND CONFIGURATION

The representation of coaxial feed in conventional antenna is shown in Figure 12.1. Figure 12.2 shows the design and analysis of the proposed antenna. The radiating patch and feeding mechanism are printed on the top side of the substrate, whereas the

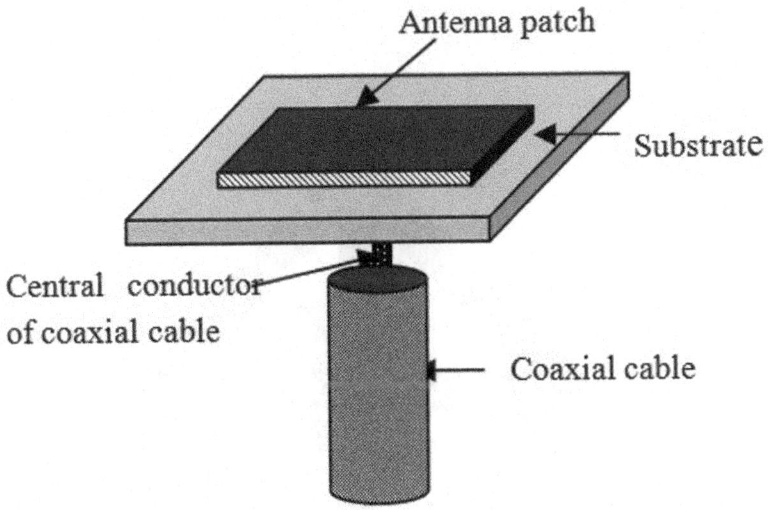

FIGURE 12.1 Representation of coaxial feed in conventional antenna (Google source [14]).

ground plane is printed on the bottom side. Coaxial feed technique is used to achieve 50 Ω characteristics impedance.

A rhombus rectangular patch microstrip antenna is designed and simulated in this chapter. The design parameters are optimized and investigated in detail. The proposed microstrip antenna is printed with the dimension of 70 × 80 mm^2, Rogers RT/Duroid 5870™ with the dielectric constant 2.33 and substrate thickness of 0.8 mm. The width of the microstrip feed line is fixed at 4 mm to attain 50 Ω characteristics impedance. The main objective of overall work is to enhance bandwidth, return loss, and cross-polarization for improving radiation characteristics in the operation bands. The proposed antenna is designed with HFSS software. The patch is operated at different frequencies between 9.75 GHz and 14.2 GHz. Figures 12.1 and 12.3 show the representation of coaxial feed to the conventional antenna and proposed antenna. The basic schematic configuration of proposed antenna is shown in Figure 12.2.

The performance of antenna parameters is analyzed to improve gain, bandwidth, and return loss. The parameters are calculated as in Ref. [15].

12.2.1 Directivity

Directivity is defined as the ratio of the radiation intensity in the given direction from the antenna to the radiation intensity averaged in overall directions [15].

$$D = \frac{4\Pi U}{Prad} \tag{12.1}$$

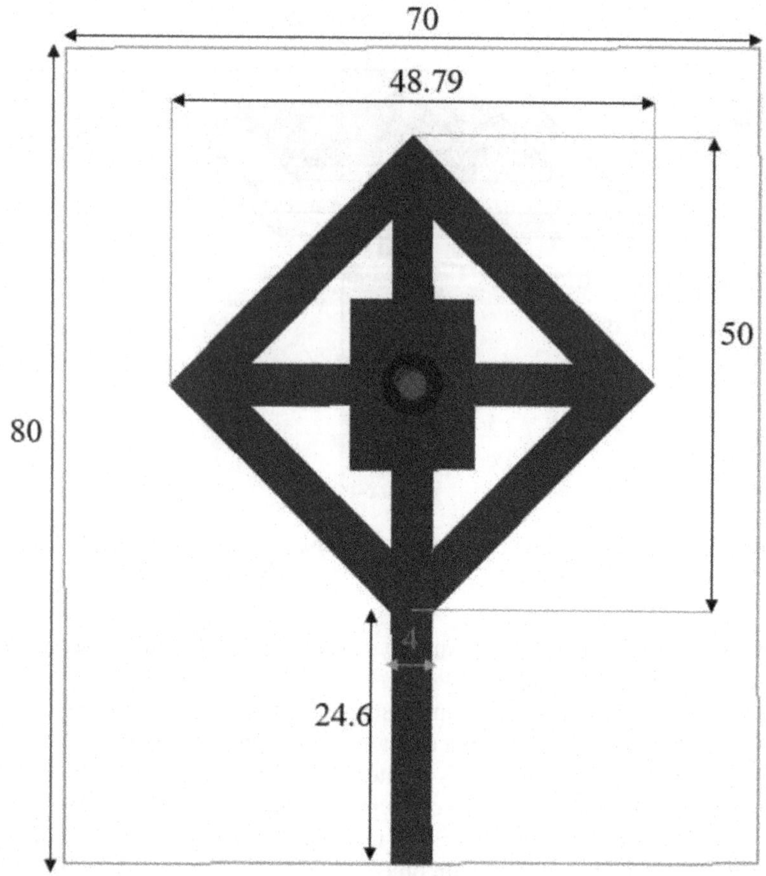

FIGURE 12.2 Geometry and configuration of the proposed antenna (unit: mm) (simulated design using Ansys HFSS).

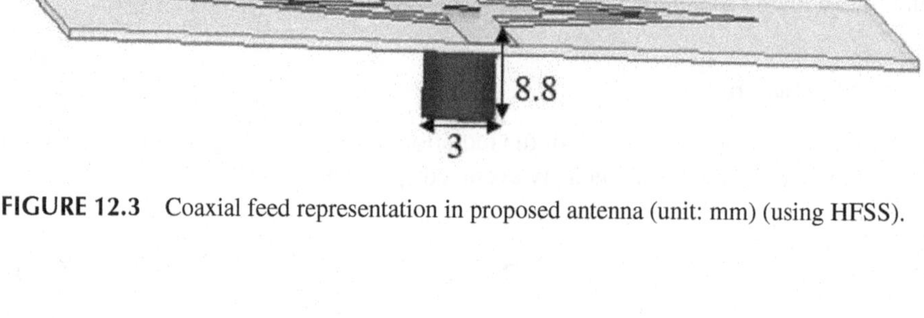

FIGURE 12.3 Coaxial feed representation in proposed antenna (unit: mm) (using HFSS).

12.2.2 Gain

Gain of an antenna is termed as the ratio of intensity in a given direction to the radiation intensity obtained when the power accepted by the antenna is radiated isotopically [15].

$$\text{Gain} = 4\Pi \frac{\text{Radiation Intensity}}{\text{Total Input}(\text{accepted})\text{power}} \tag{12.2}$$

12.2.3 Bandwidth

In general, antenna bandwidth is defined as the range of frequency within the performance of the antenna [15]. The bandwidth of broadband and narrowband is defined by the following relation:

$$\text{Bandwidth (BW)} = F_{ii} - F_{i} \tag{12.3}$$

where F_{ii} is the higher frequency and F_i is the lower frequency.

12.2.4 Return Loss

Return loss is otherwise termed as reflection loss from the load to source. It is defined as the reflection of the signal power given from source to load in a transmission line [8]. It is usually expressed as a ratio in dB relative to the transmitted signal power. The expression for return loss is given by:

$$\text{RL} = 10\log\frac{\text{Pr}}{\text{Pi}} \tag{12.4}$$

where Pr is the reflected power and Pi is the inserted power or power given to the load in the transmission line.

12.2.5 Cross-Polarization

Cross-polarization is defined as X-pol in antenna slang and is the ability to maintain the radiated or received polarization purity between horizontally and vertically polarized signals in the transmission line [15, 16].

12.2.6 Antenna with VPDGS

Defects or slots incorporated and etched on the ground plane of the microstrip antenna are merely termed defected ground structure (DGS) [17]. Apart from the conventional periodic structures, DGS has attracted great attention to improve various parameters such as gain, return loss, cross-polarization, and bandwidth [18]. It is mainly adopted to overcome the limitations of microstrip antennas in terms of single resonant frequency, low bandwidth with perfect impedance matching, and polarization problems [19]. Also DGS helps to attain band stop characteristics and to suppress higher mode harmonics and mutual coupling [20]. The factors that affect the periodic DGS are the shape of DGS units, distance between two DGS units, and the distribution of the different DGS. Two types of periodic structures, such as horizontally and vertically periodic DGS, have been used recently in microwave devices and circuits [21]. The most prominent one chosen as VPDGS, which helps in achieving the compact size of the antenna.

12.2.7 Antenna with Fractal Geometry

The proposed antenna with VPDGS is subjected to the fractal design. Fractal antennas are self-similar designs to increase the effective length or maximize the perimeter of material, which can transmit and receive electromagnetic waves within a given total surface area or volume [22, 23]. It is also defined as the irregular segments that are the set of complex geometry. Fractal antenna resonate at different frequencies influenced by the size of the ground plane [24–27]. The performance of the fractal antenna mainly depends on directivity, gain, return loss, VSWR, bandwidth, and so forth.

12.3 SIMULATION AND EXPERIMENTAL RESULTS

In this section, the rhombus rectangular patch antenna is constructed with VPDGS and fractal geometries. The proposed antenna is designed using Ansys HFSS. This type of simulator is based on the Finite Element Method (FEM) and is used to calculate return loss, bandwidth, and gain. The major advantage of this simulator is to reduce the fabrication cost because the best antenna is obtained by internal optimization and is best for further fabrication processes. The simulation results are presented and discussed. Figure 12.4 shows the simulated results of return loss of proposed antenna. It can be seen that the proposed antenna is varied from 9.8 GHz to 112.69 GHz. The return loss obtained is –27.88 dB with the gain of 5.27 dB and bandwidth of 670 MHz. Return loss with negative sign shows that the proposed antenna is not affected by huge loss while transmitting signals. The rhombus patch antenna also radiates at the frequency of 112.69 GHz with the return loss of –26.39 GHz, gain of 9.78 dB, and bandwidth of 780 MHz. These results are obtained by the function of w (w = 4 mm) and the corresponding results are presented in Figure 12.4.

The proposed antenna with VPDGS is presented in Figure 12.5. Simulation results are also discussed with the return loss, gain, and bandwidth. The main advantage of introducing VPDGS in rhombus rectangular patch antenna is to attain a high return loss, which reduces the transmission losses of signals. Also DGS has its own characteristics and creates a good performance of devices with its perfect geometry and size [28, 29]. The main purpose of DGS is to achieve a compact antenna in turn enhancing the operating bandwidth and gain. Mutual coupling has been reduced between two networks, while suppressing the higher order harmonics and unwanted cross-polarization. From the literature survey, VPDGS has an advantage of 44.4% size reduction when compared to horizontally periodic defected ground structure

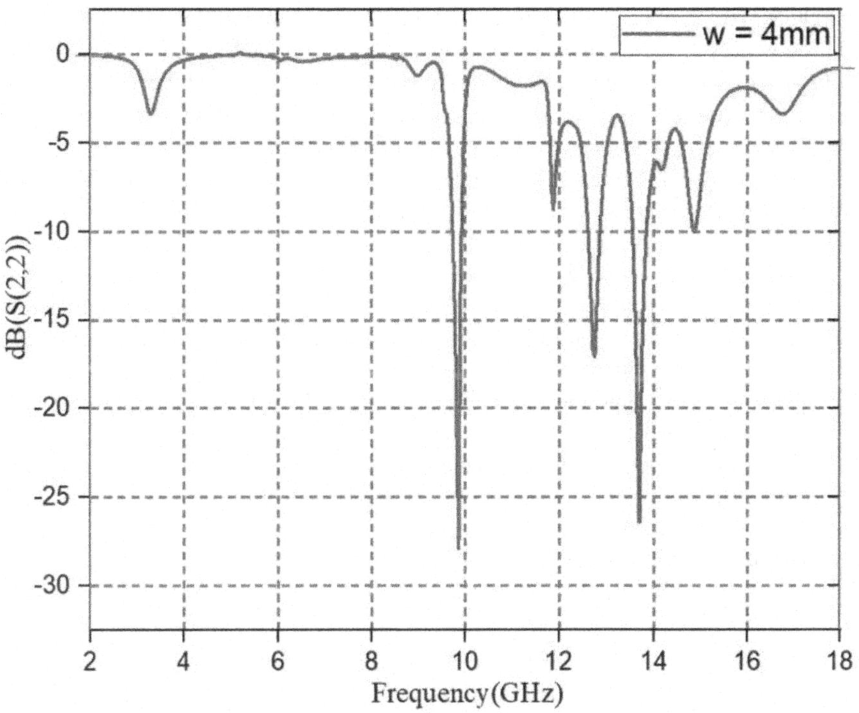

FIGURE 12.4 Simulated return loss of the proposed antenna (using HFSS).

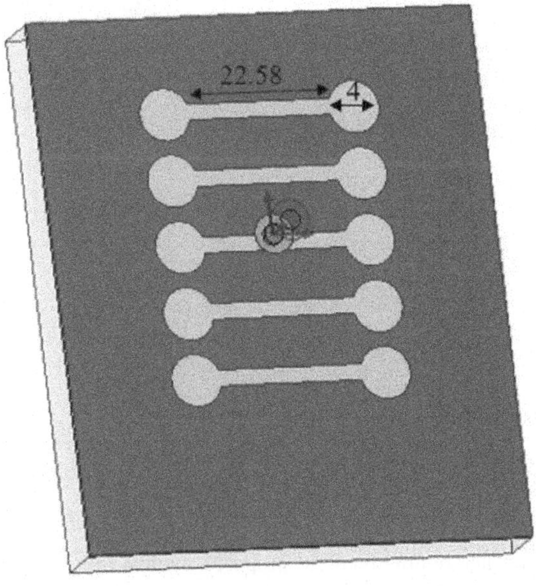

FIGURE 12.5 Geometry and configuration of proposed antenna with VPDGS (using HFSS).

(HPDGS). HPDGS with size reduction of 38.5% is reported in Ref. [13]. VPDGS can be used widely in matching network of an amplifier.

Figure 12.5 shows the proposed rhombus antenna with vertically periodic defected ground structure (VPDGS). The geometry and its corresponding configuration with unit as mm is also presented.

The simulated results of return loss of the proposed antenna with VPDGS is presented in Figure 12.6. The proposed antenna is varied from 9.75 GHz to 14.26 GHz. The return loss obtained is −17.95 dB with the gain of 12.32 dB and bandwidth of 734 MHz at 9.75 GHz. Return loss with negative sign shows that the proposed antenna is not affected by huge loss while transmitting the signals. The rhombus patch antenna also radiates at the frequency of 14.26 GHz with the return loss of −31.20 GHz, gain of 14.86 dB, and bandwidth of 870 MHz. These results are obtained by the function of D (D = 4 mm) and the corresponding results are presented in Figure 12.8.

Incorporating Fractal geometry to the proposed antenna with VPDGS is shown in Figure 12.7. The irregular segments, which are the set of complex geometry, are in turn called fractal designs [30–33]. Fractal antennas resonate at different frequencies and are influenced by the size of the ground plane [34–36].

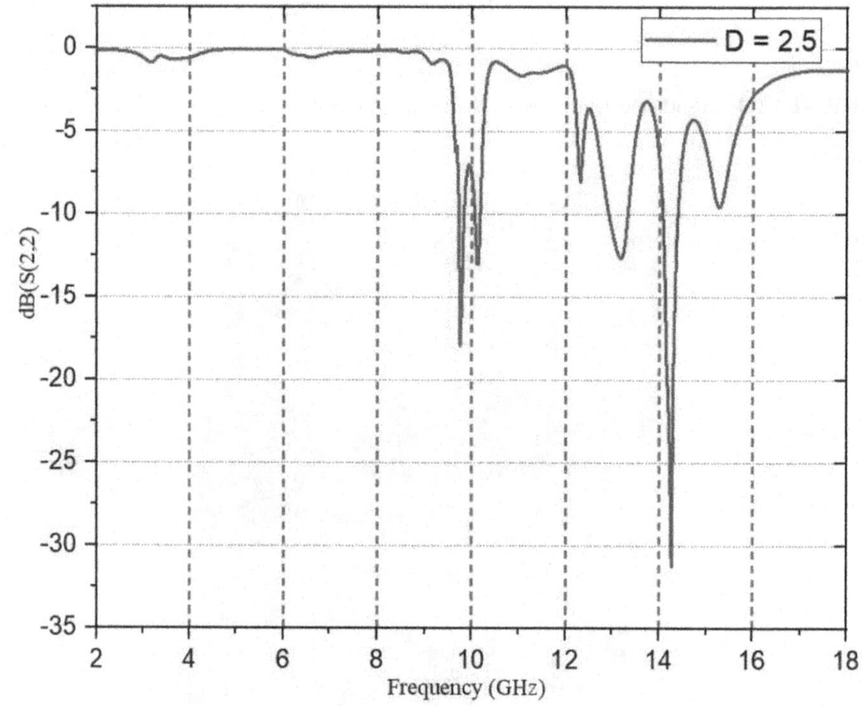

FIGURE 12.6 Simulated return loss of the proposed antenna with VPDGS (using HFSS).

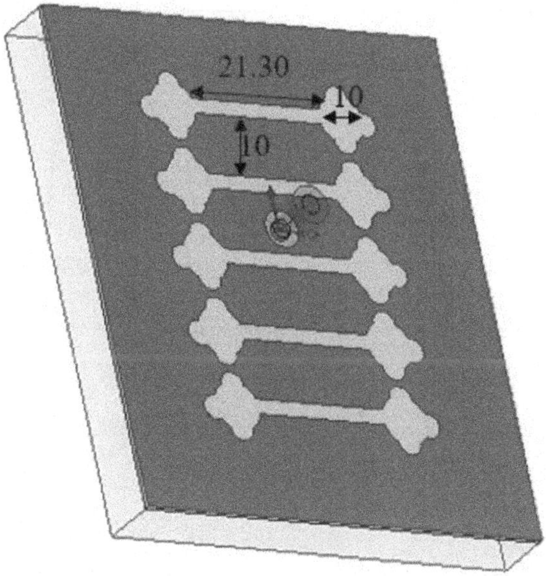

FIGURE 12.7 Geometry and configuration of proposed antenna with fractal geometry and VPDGS (unit: mm) (using HFSS).

Figure 12.8 shows the simulated results of return loss of the proposed antenna with VPDGS and fractal geometry. The proposed antenna is varied from 9.75 GHz to 14.19 GHz. The return loss obtained is −12.28 dB with the gain of 112.72 dB, and bandwidth of 736 MHz at 9.75 GHz. The rhombus patch antenna with fractal design also radiates at the frequency of 14.19 GHz with the return loss of −28.65 GHz, gain of 14.86 dB, and bandwidth of 975 MHz. These results are obtained by the function of D (D = 4 mm) and the corresponding gain results are presented in Figure 12.9.

12.4 CONCLUSION

The compact rhombus rectangular patch antenna with good performance is presented and investigated. We have discussed the antenna by etching the dumb-bell-shaped VPDGS and fractal geometry on the ground plane. The proposed antenna radiates at high frequencies and well suited for X-band and Wi-Max applications. The reflection co-efficient of the proposed antenna simulated using HFSS meets the design requirements. Good performance in terms of return loss and cross-polarization for improving radiation characteristics are obtained in the operation bands. The coaxial feed is introduced to avoid the surface wave radiation and also supports for attaining good impedance matching. Further optimization can make the antenna work on other applications too. Consequently, the proposed antennas are expected to be a good candidate in various wireless systems.

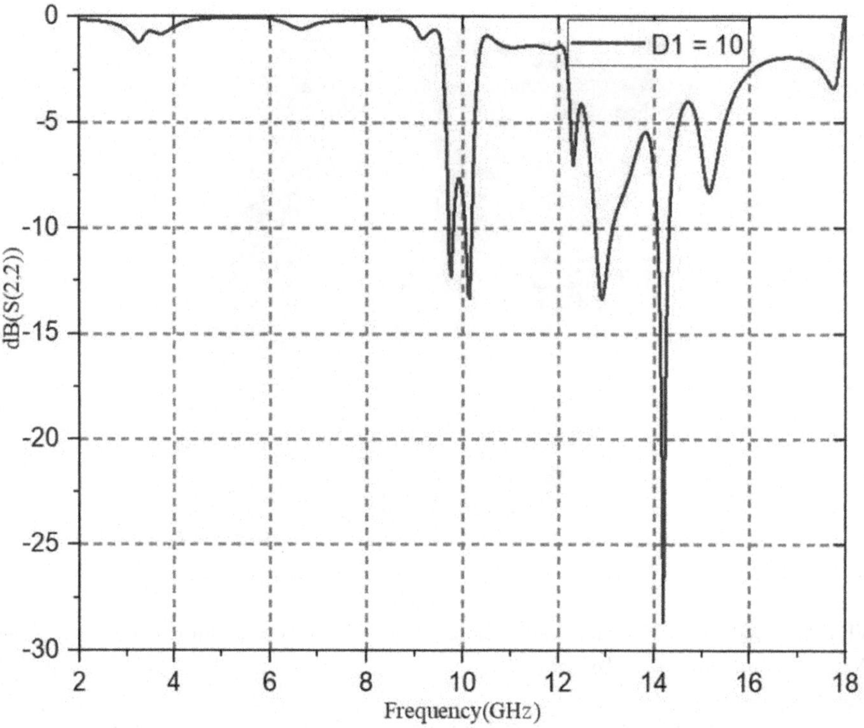

FIGURE 12.8 Simulated return loss of the proposed antenna with VPDGS and fractal geometry (using HFSS).

REFERENCES

[1] Saidulu, V., Kumara Swamy, K., Srinivasa Rao, K., and Somasekhar Rao, P. V. D. (2013), "Performance analysis of circular microstrip patch antenna with dielectric superstrates", *IOSR Journal of Engineering*, Volume 3, Issue 9, pp. 39–51, 2012.

[2] Sharma, V., Jangid, K. G., Bhatnagar, D., Yadav, S. and Sharma, M. M. (2014), "A compact CPW fed modified circular patch antenna with stub for UWB applications", *International Conference on Signal Propagation and Computer Technology*, pp. 214–217, IEEE-2014.

[3] Nayna, T. F. A., Baki, A. K. M. and Ahmed, F. (2014), "Comparative study of rectangular and circular microstrip patch antennas in X band", *International Conference on Electrical Engineering and Information & Communication Technology*, pp. 1–5, IEEE-2014.

[4] Ayyappan, M., Manoj, B., and Rodrigues, S. (2016), "Low return loss circular microstrip patch antenna at 5.8 GHz for wide-band applications", *International Conference on Electrical, Electronics, and Optimization Techniques*, pp. 3023–3026, IEEE-2016.

[5] Liu, X., Li, Y., Liang, Z., Zheng, S., Liu, J. and Long, Y. (2016), "A method of designing a dual-band sector ring microstrip antenna and its application", *IEEE Transactions on Antennas and Propagation*, pp. 1–6, IEEE-2016.

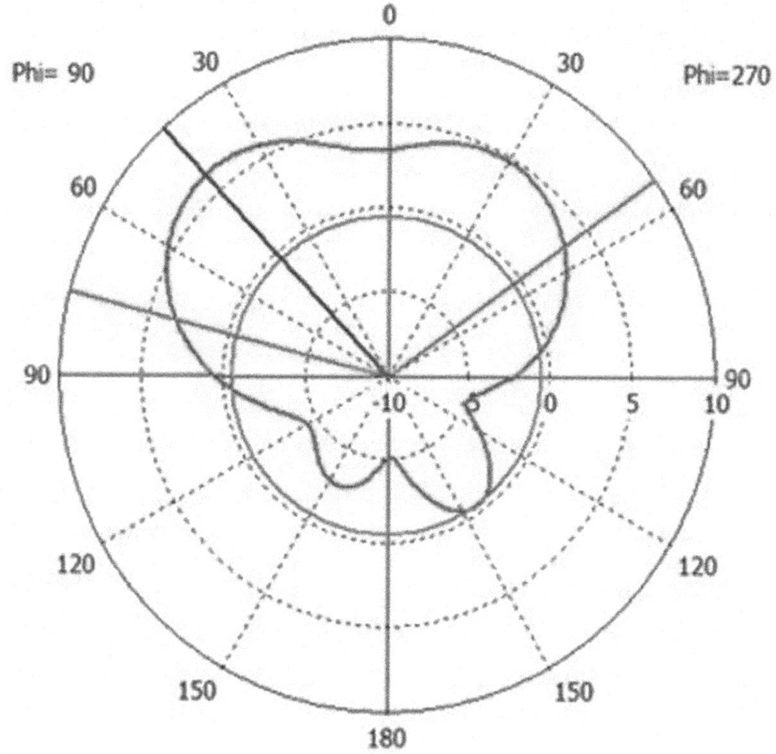

FIGURE 12.9 Gain of the proposed antenna (using HFSS).

[6] Sibi Chakravarthy, S., Sarveshwaran, N., Sriharini, S., and Shanmugapriya, M. (2016), "Comparative study on different feeding techniques of rectangular patch antenna", pp. 1–6, IEEE-2016.

[7] Saddam Hossain, Md., Masud Rana, Md., Shamim Anower, Md. and Kumar Paul, A. (2016), "Enhancing the performance of rectangular patch antenna using E shaped slot technique for 5.75 GHz ISM band applications", *5th International Conference on Informatics, Electronics and Vision*, pp. 449–453, IEEE-2016.

[8] Balanis, C. A. (2005), Antenna Theory Analysis and Design, 3rd ed. John Wiley & Sons Publications.

[9] Kushwahal, N. and Kumar, R. (2014), "Study of different shape electromagnetic band gap (EBG) structures for single and dual band applications", *Journal of Microwaves, Optoelectronics and Electromagnetic Applications*, Volume 13, Issue 1, pp. 16–30.

[10] Shaka, S. M., Prashant, R. T, Vani, R. M., and Hunagund, P. V. (2014), "Study of microstrip antenna using various types of EBG cells", *International Journal of Advanced Research in Electrical, Electronics and Instrumentation Engineering*, Volume 3, Issue 5, pp. 9680–9686.

[11] Saroha, N. and Goyat, M. (2015), "Design and simulation rectangular and circular patch antenna with EBG substrates", *International Journal of Scientific Research and Management*, Volume 3, Issue 7, pp. 3407–3410.

[12] Tumsare, K. V. and Zade, P. L. (2016), "Microstrip antenna using EBG substrate", *International Journal on Recent and Innovation Trends in Computing and Communication*, Volume 4, Issue 12, pp. 73–76.

[13] Khandelwal, M. K., Kanaujia, B. K. and Kumar, S. (2017). "Defected ground structure: fundamentals, analysis, and applications in modern wireless trends", *International Journal of Antennas and Propagation*, Volume 2017, Article ID 2018527, p. 22.. https://doi.org/10.1155/2017/2018527.

[14] Ndujiuba, C. U. and Oloyede A. O. (2015), "Selecting best feeding technique of a rectangular patch antenna for an Application", *International Journal of Electromagnetics and Applications*, Volume 5, Issue 3, pp. 99–107.

[15] Gopal, E. and Singhal, A. (2018), "Design and simulation of circular rectangular microstrip patch antenna for wireless applications", *Engineering and Technology Journal*, Volume 3, Issue 1, pp. 361–365.

[16] Mahatthanajatuphat, C., Saleekaw, S. and Akkaraekthalin, P. "A rhombic patch monopole antenna with modified Minkowski fractal geometry for UMTS, WLAN and mobile WiMAX applications," *Progress in Electromagnetics Research*, Volume 89, pp. 57–74, 2009.

[17] Kumar, C. and Guha, D. (2012), "Nature of cross-polarized radiations from probe-fed circular microstrip antennas and their suppression using different geometries of defected ground structure (DGS)", *IEEE Transactions on Antennas and Propagation*, Volume 60, Issue 1, pp. 92–101.

[18] Khandelwal, M. K., Kanaujia, B. K., Dwari, S., Kumar, S. and Gautam, A. K. (2014), "Analysis and design of wide band microstrip-line-fed antenna with defected ground structure for Ku band applications", *AEU—International Journal of Electronics and Communications*, Volume 68, Issue 10, pp. 951–957.

[19] Khandelwal, M. K., Kanaujia, B. K., Dwari, S., Kumar, S., and Gautam, A. K. (2014), "Bandwidth enhancement and cross-polarization suppression in ultrawideband microstrip antenna with defected ground plane", *Microwave and Optical Technology Letters*, Volume 56, Issue 9, pp. 2141–2146.

[20] Chiang, K. H. and Tam, K. W. (2008), "Microstrip monopole antenna with enhanced bandwidth using defected ground structure", *IEEE Antennas and Wireless Propagation Letters*, Volume 7, pp. 532–535.

[21] Khandelwal, M. K., Kanaujia, B. K., Dwari, S., Kumar, S. and Gautam, A. K. (2015), "Triple band circularly polarized compact microstrip antenna with defected ground structure for wireless applications", *International Journal of Microwave and Wireless Technologies*, Volume 8, Issue 6, pp. 943–953.

[22] Khanna, G. and Sharma, N. (2016), "Fractal antenna geometries: A review", *International Journal of Computer Applications*, Volume 153, Issue 7, pp. 29–32.

[23] Wagh, K. H. (2015), "A review on fractal antennas for wireless communication", *International Journal of Review in Electronics and Communication Engineering*, Volume 32, Issue 2.

[24] Kharat, K., Dhoot, S., and Vajpai, J. (2015), "Design of compact multiband fractal antenna for WLAN and WiMAX applications", *International Conference on Pervasive Computing (ICPC)*.

[25] Singh, M., and Sharma, N. (2016), "A design of star shaped fractal antenna for wireless applications", *International Journal of Computer Applications*, Volume 134, Issue 4, pp. 41–43.

[26] Poonam, S., Jadhav, M. M. (2016), "Review on fractal antennas for wireless communication", *International Journal of Latest Trends in Engineering and Technology (IJLTET)*, Volume 6, Issue 3, pp. 451–454.

[27] Ranjan, A., Singh, M., Kumar Sharma, M. and Singh, N. (2014), "Analysis and simulation of fractal antenna for mobile Wimax", *International Journal of Future Generation Communication and Networking*, Volume 7, Issue 2, pp. 57–64.

[28] Chakraborty, S., Dey, S., Guha, R., Pramanik, S., Gangopadhyaya, M. and Chakraborty, M. (2015), "Design of frequency tuned circular microstrip antenna with angular unconnected DGSǁ", *International Journal of Advanced Research in Electronics and Communication Engineering (IJARECE)*, Volume 4, Issue 3, pp. 426–431.

[29] Thatere, A., and Zade, P. L. (2020), "Defected ground structure microstrip antenna for WiMAX", In *Optical and Wireless Technologies* (pp. 333–346). Springer, Singapore.

[30] Vidyashree, D. and Mathpati, M. S. (2015), "Overview on recent development of fractal antenna in communication", *International Journal of Computer Application.*

[31] Sharma, N., Kaur, A., and Sharma, V., (2016), "A novel design of circular fractal antenna using inset line feed for multiband applications", *2016 IEEE 1st International Conference on Power Electronics, Intelligent Control and Energy Systems*, ICPEICES-2016, pp. 1–4.

[32] Srivastava, P. and Singh, O. P. "A review paper on fractal antenna and their geometries", *Advances in Electrical & Information Communication Technology*, AEICT-2015, Volume 153, pp. 29–32, 2016.

[33] Nair, R. G., (2016), "Comparative study of multiband antennas with fractal geometry", *International Journal of Innovative Research In Electrical, Electronics, Instrumentation and Control Engineering*, Volume 4, Issue 2, pp. 16–18,.

[34] Saini, R. and Prakash, D. (2013), "CPW fed rectangular shape microstrip patch antenna with DGS for WLAN/WiMAX application", *International Journal of Advanced Research in Computer Science,* Volume 4, Issue 11, pp. 199–203.

[35] Kumar, S. B., and Singhal, P. K. (2018), "Investigation of circularly polarized fractal antenna at 2.1 GHz for LTE application", *International Journal of Engineering Research and Development*, Volume 14, pp. 44–50.

[36] Shambavi, A. A. and Alex, Z. C. (2012), "Design and analysis of fractal antenna for UWB applications", *2012 IEEE Students' Conference on Electrical Electronics and Computer Science.*

13 Design and Analysis of Heptagon-Shaped Multiband Antennas with Multiple Notching for Wireless Applications

K. Sumathi, S. Thenmozhi and V. Seethalakshmi

13.1 INTRODUCTION

The heptagon-shaped multiband antenna with notching at different frequencies is proposed for several wireless applications. The antenna is designed with less-loss FR4 substrate of 1.6 mm thickness. In the first design, a simple heptagon-shape is constructed and analysis is done. In the second design, in order to have notch frequencies and multiband operations, over the heptagon, seven small heptagons are formed in each face and simulated. A heptagon is introduced in the patch in the third design to have multiband operation with five notched frequencies. The simulation results of all the three antennas are compared and analyzed. The proposed antenna is suitable for various wireless applications like WLAN and WiMax. The antenna is designed using ADS software. With the advent of new developments in wireless industry, there is a lot more demand for multiband antennas. The antenna that resonates over multiple frequencies is called a multiband antenna, as proposed in this chapter.

An antenna with three main parameters namely ultra-wideband, formation of composite antenna, and high frequency of operations have been explained in Ref. [1]. The advantages and disadvantages of microstrip antennas are discussed in this chapter. The authors prove that microstrip antenna still appears to be a promising device in wireless applications. In Ref. [2], the authors have designed an inverted-F multiband antenna that can be applied as a dual-band antenna or a tri-band antenna based on the applications just by initially tuning the inverted-F part. The branch element and

DOI: 10.1201/9781003093558-17

coupled element are adjusted to obtain the other resonant frequencies such as 2.4 GHz and 5 GHz that can be used for wireless LAN and cellular applications. In Ref. [3], an antenna that has triple narrowband notched features have been proposed. The developed structure works in the frequency band between 3.1 GHz and 14 GHz, along with three notched bands from 4.2 GHz to 6.2 GHz, 6.6 GHz to 7.0 GHz, and 12.2 GHz to 14 GHz. In Ref. [4], the authors have demonstrated a multiband antenna along with the SAR model. The folded loop and a meandered planar inverted-F antenna (MPIFA) used along with a supportive branch line aids in hepta-band LTE/GSM/ UMTS applications.

An octahedron-shaped compact dual-band-notched ultra-wideband antenna that is used for WiMax and wireless-LAN applications has been developed in Ref. [5]. This is achieved by fractal geometry, changing the front loading plane along with the ground plane with a microstrip feed which develops high impedance bandwidth in the range of 2.4 GHz to 19.5 GHz. In Ref. [6], an urn-shaped antenna with three different notch bands for WiMax, wireless-LAN, and X-band downlink satellite system in Ku band has been designed. The coaxial-fed patch antennas for various shapes like rectangle, triangle, and circle have been examined in Ref. [7]. Each shape has specific advantages in any of the antenna parameters for the X-band application.

In Ref. [8], a rectangular patch antenna has been designed for WLAN applications. The design of the antenna is very simple, which can be used for wide applications in large scale. The authors have presented a novel microstrip-fed ultra-wideband antenna that is capable enough to reject four different bands of interference in Ref. [9]. The authors have used analytical hierarchy process (AHP) of preferring an antenna from the earlier antenna designs or giving importance to them based on many important factors like bandwidth, substrate, and total number of notches. The authors describe how a square-patch antenna with the slots can be designed to obtain various resonant frequencies like 3.3 GHz, 3.9 GHz, 4.4 GHz, and 8.8 GHz in Ref. [10]. The slots are placed in such a way that the antenna becomes a lightweight structure, which is an added advantage. In Ref. [11], the two main objectives in designing an antenna like the choice of substrate material among FR-4, RO4003, GML1000, and RT/Duroid 5880 and the appropriate shape selection among H-shape, E-shape, S-shape, and U-shape are described. From the detailed performance comparison, it was concluded that the substrate with lower dielectric constant improves the bandwidth and the one with higher dielectric constant enhances the gain and directivity. The hexagonal-shaped 2 × 2 MIMO microstrip patch multiband antenna is developed for various wireless-based applications [12]. Different configurations of MIMO antenna are simulated on FR4 substrate with dielectric constant of 1.6 and the analysis are carried out using ADS software. In this chapter [13], the author proposes a square-patch antenna that resonates at 5.4 GHz. The antenna is developed with the FR4 substrate and is observed to produce a minimal return loss of –40 dB making it useful for wireless LAN applications.

13.2 PROPOSED ANTENNA DESIGN WITH SIMULATION RESULTS

The step-by-step antenna design is illustrated in design (1), (2), and (3) respectively. The proposed antenna is shown after three design modifications.

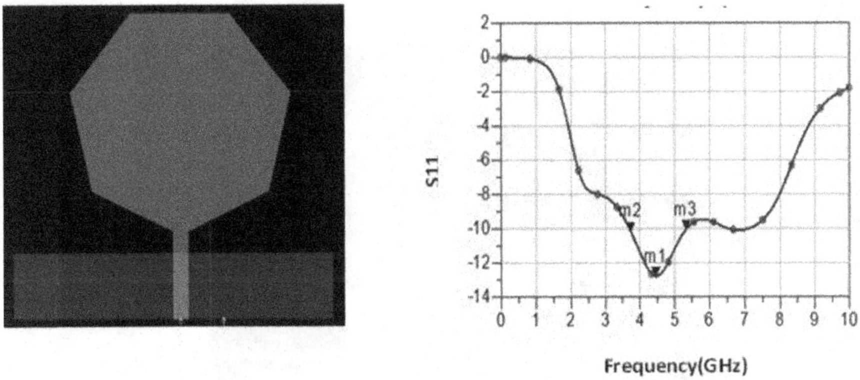

FIGURE 13.1 Design 1 and return loss graph.

TABLE 13.1
Dimensions of antennas

Parameters	Dimensions (mm)	Parameters	Dimensions (mm)
Lp	30	Wf	1.7
Wp	35	Wg	35
L	35	Lg	7
W	37	Oh	2
Lf	9	Ih	1

13.2.1 Design 1

In the rectangular-patch antenna, the heptagon shape is formed and is shown in Figure 13.1. This is the shape with seven faces. It shows the basic antenna design and the corresponding return loss graph. The return loss is found to be less for this design. The radiation pattern is good and obeys the omnidirectional pattern. The overall dimension of the proposed antenna is 37 × 35 × 1.6 mm^3.

The dimension of the antenna is shown in Table 13.1 and the corresponding antenna design with the specifications are provided in Figure 13.2.

13.2.2 Design 2

To increase the return loss and notching, seven smaller heptagons are created in each face of the main heptagon, which is shown in Figure 13.3. The dimensions of the smaller heptagons are shown in Table 13.1. The current distribution can also be observed from Figure 13.3.

This design yields dual-band operation with triple notch frequencies at 4 GHz, 9.18 GHz, and 9.96 GHz with return loss of −3.2 dB, −2.2 dB, and −1.6 dB, respectively (Figure 13.4).

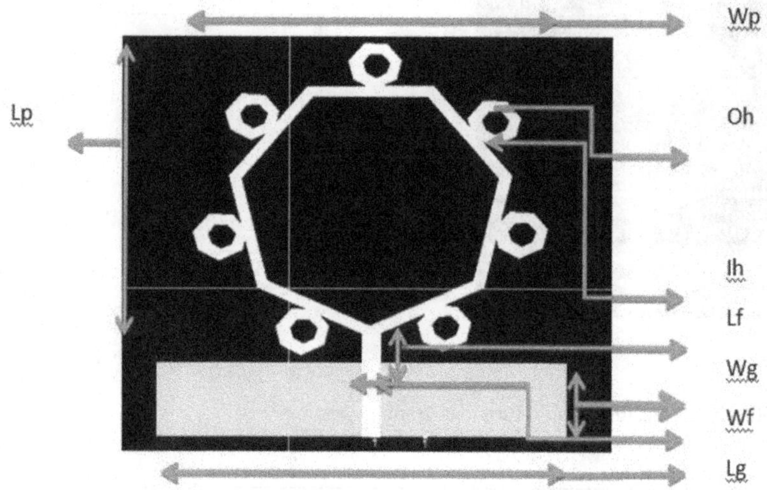

FIGURE 13.2 Antenna design with dimensions.

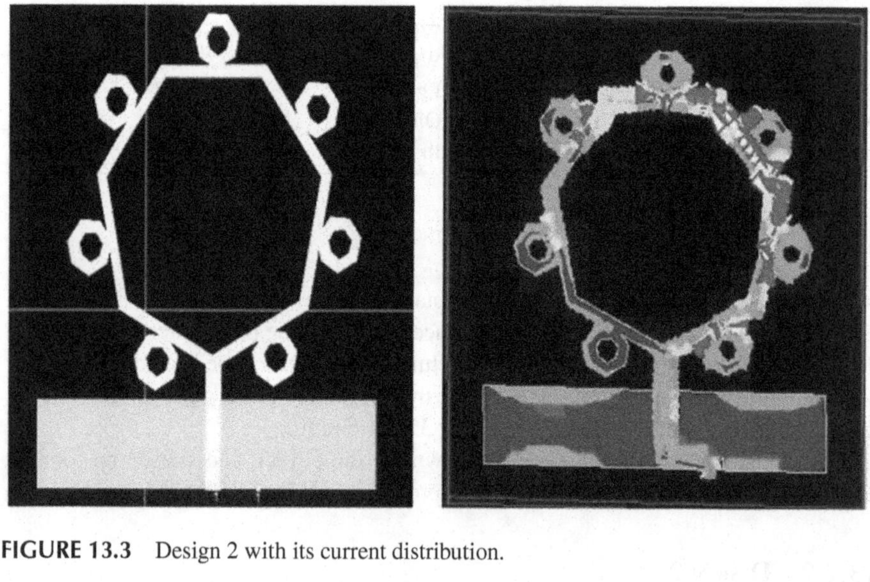

FIGURE 13.3 Design 2 with its current distribution.

13.2.3 Design 3

In Design 3, a heptagon which is smaller than the basic heptagon shape is introduced in the middle of the patch to get multiband operation with multiple notching. The proposed design resonates at five different frequencies at 1.5 GHz, 3.3 GHz, 4.5 GHz, 5.9 GHz, and 6.5 GHz. Notching occurs at four different frequencies namely at 1.6 GHz, 3.9 GHz, 9 GHz, and 11 GHz.

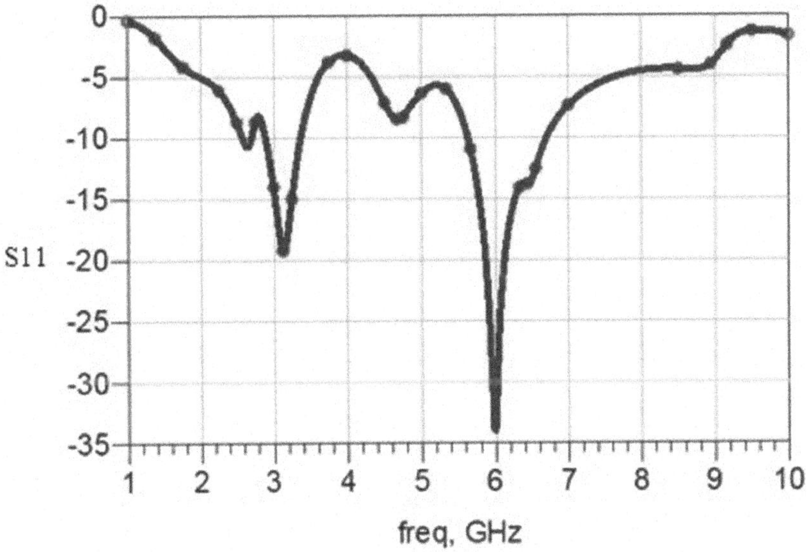

FIGURE 13.4 Design 2 with its return loss.

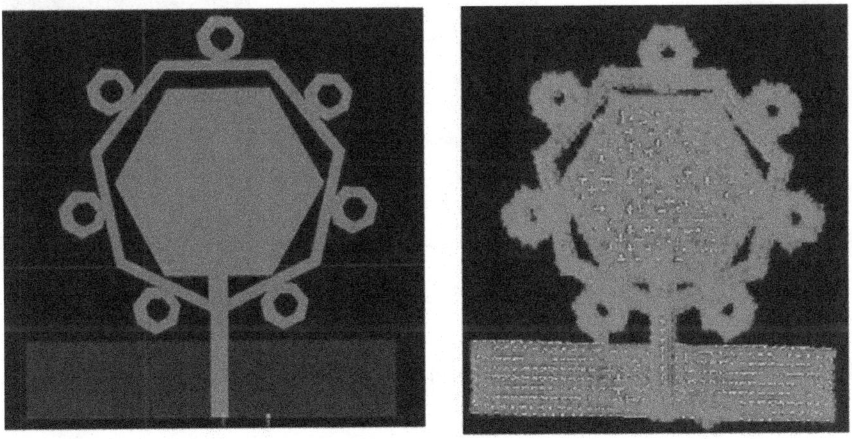

FIGURE 13.5 Design 3 with its current distribution.

The proposed design with the corresponding current distribution and return loss characteristics are shown in Figures 13.5 and 13.6, respectively. Radiation pattern and far-field radiation pattern of Design 3 are illustrated in Figure 13.7.

For all the resonant frequencies, the VSWR value lies in the range of 1–2. The gain is plotted with respect to frequency in Figure 13.8. The gain of the antenna differs from 0.5 to 4.53. The directivity of the heptagon-shaped proposed antenna ranges from 2.4 to 7.3 with the attena efficiency varying between 30% and 78%.

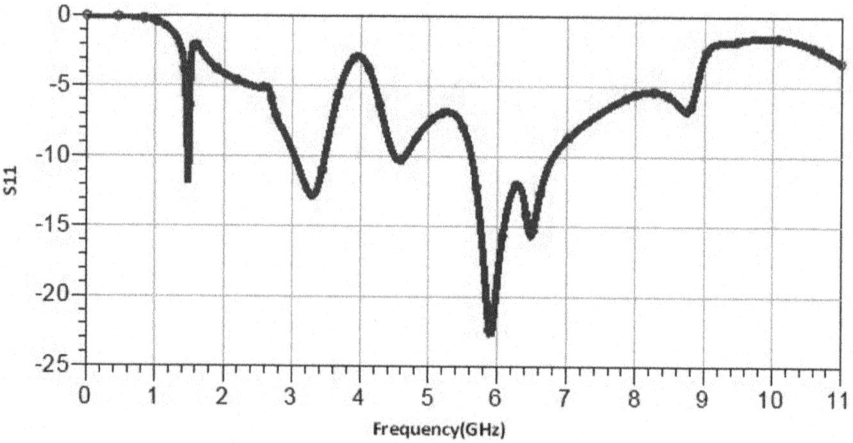

FIGURE 13.6 Design 3 with its return loss.

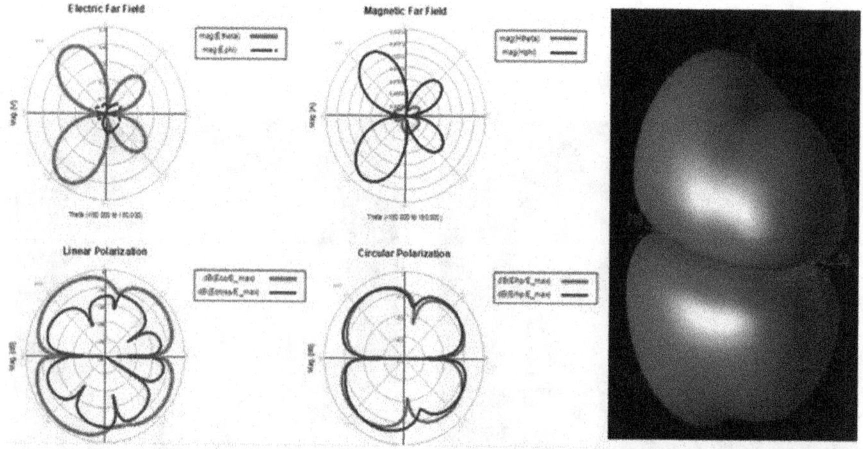

FIGURE 13.7 Radiation pattern and far-field radiation pattern of Design 3.

13.3 CONCLUSION

The heptagon-shaped multiband antenna is recommended to operate over five different frequencies with four notch frequencies at 1.6 GHz, 3.9 GHz, 9 GHz, and 11 GHz. Three variants of heptagon antenna design are suggested in this chapter. The first design results are not significant in terms of return loss. Second and third designs are proposed to have better return loss along with notching at multiple frequencies. The proposed antenna is suitable for different wireless applications like WLAN and WiMax. The simulation reveals that the proposed antenna has different frequency tuning, better gain, and directivity.

REFERENCES

[1] James, J., Hall, P., Wood, C., and Henderson, A., "Some recent developments in microstrip antenna design". *IEEE Transactions on Antennas and Propagation*, 29(1), pp. 124–128, 1981.

[2] Liu, D. and Gaucher, B., "A new multiband antenna for WLAN/cellular applications," *IEEE 60th Vehicular Technology Conference, VTC2004-Fall.2004, Los Angeles, CA*, Vol. 1, pp. 243–246, 2004.

[3] Li, Y., Li, W., and Ye, Q., "A reconfigurable triple-notch-band antenna integrated with defected microstrip structure band-stop filter for ultra-wideband cognitive radio applications", *International Journal of Antennas and Propagation*, pp. 1–13, 2013.

[4] Hong, Y., Tak, J., Baek, J., Myeong, B., and Choi, J., "Design of a multiband antenna for LTE/GSM/UMTS Band Operation". *International Journal of Antennas and Propagation*, pp. 1–9, 2014.

[5] Mishra, G. and Sahu, S., "Modified octahedron shaped antenna with parasitic loading plane having dual band notch characteristics for UWB applications", *International Journal of RF and Microwave Computer-Aided Engineering*, 26(5), pp. 426–434, 2016.

[6] Sharma, M., Awasthi, Y., Singh, H., Kumar, R., and Kumari, S., "Compact printed high rejection triple band-notch UWB antenna with multiple wireless applications", *Engineering Science and Technology, an International Journal*, 19(3), pp. 1626–1634, 2016.

[7] Kiruthika, R. and Shanmuganantham, T., "Comparison of different shapes in microstrip patch antenna for X-band applications", *International Conference on Emerging Technological Trends (ICETT), Kollam*, 2016.

[8] Singh, S., "Design and fabrication of microstrip patch antenna at 2.4 Ghz for WLAN application using HFSS", *IOSR Journal of Electronics and Communication Engineering*, 1(1), pp. 1–6, 2016.

[9] Zehforoosh, Y., Mohammadifar, M., Mohammadifar, A., and Ebadzadeh, S., "Designing four notched bands microstrip antenna for UWB applications, assessed by analytic hierarchy process method", *Journal of Microwaves, Optoelectronics and Electromagnetic Applications*, 16(3), pp. 765–776, 2017.

[10] Sneha Talari, A. Navya, C. H. Vamshi Krishna, and G. Bhanuprakash, "Design and analysis of rectangular microstrip patch antenna with different substrates", *International Journal of Advance Engineering and Research Development*, 4(04), 2017.

[11] Ghosal, A., Majumdar, A., Das, S. K. and Das, A., "A multiband microstrip antenna for communication system", *2018 Emerging Trends in Electronic Devices and Computational Techniques (EDCT), Kolkata*, pp. 1–4, 2018.

[12] Sumathi, K., and Abirami, M., "Hexagonal shaped fractal MIMO antenna for multiband wireless applications", *Analog Integrated Circuits and Signal Processing*, 103(3), pp 277–287, 2020.

[13] Thenmozhi, S., Maheswar, R., Nandakumar, E., Jayarajan, P., Dhivya Priya, E. L., and Ganeshprabhu, S., "A novel compact square patch antenna for wireless LAN applications", *6th International Conference on Advanced Computing and Communication Systems (ICACCS)*, 2020.

Part V

Microstrip Antennas in Vehicular Communication

14 Multifunctional Integrated Hybrid Rectangular Dielectric Resonator Antennas for High-Speed Communications

Sovan Mohanty and Baibaswata Mohapatra

14.1 INTRODUCTION

The design techniques and problem-solving approaches in microwave and millimeter-wave antennas are distinct from those of long-wave antennas. This chapter provides a systematic approach to design multifunctional and multiband-based integrated hybrid rectangular dielectric resonator antenna. The proposed antenna offers maximum impedance bandwidth of 22% while operating from 4.63 GHz to 10.84 GHz. The multifunctionality of the radiator can be achieved by controlling the energy coupling between the dielectric resonator relative to the wideband microstrip patch and the inclusions. It provides ultra-high reliability, low latency, and tenability suitable for modern cellular tunable Internet of Thing–based global tracking system and high-speed communication. The antenna as a front-end device for wireless applications has undergone a spectacular evolution from narrowband to wideband to the multiband operation. The multiband technology is playing a major role in high-speed communication and navigation systems. It integrates applications like ranging, localization, and communication in the near-field personal area and portable wireless sensor network, and far-field applications such as short pulse radar, time-hopping spread spectrum system. It enters into operations such as battlefield monitoring through precision

DOI: 10.1201/9781003093558-19

radar, congestion, and traffic control with the collision detection system, medical imaging in healthcare, environmental monitoring, and accurate localization in a positioning system. By using a wide spectrum at extremely low power, the multiband antenna provides robustness in dealing with multipath environment and interference from the narrowband technologies [1].

The channel capacity can be defined as the maximum mutual information between the transmitter and the receiver. It should be maximized to provide high data throughput. It is a function of signal power, noise levels, number of antennas, and channel that is affected by maximum antenna field interactions [2]. It is given by Equation (14.1) [3]:

$$C = BW \log_2 \left[\det \left(I_R + \frac{\rho}{N_T} HH^T \right) \right] \tag{14.1}$$

where BW = operating bandwidth, $N_R \times N_R$ matrix = arrangements of the receiver antenna, I_R = number of receiver antennas, ρ = signal to noise ratio, N_T= number of transmitting antennas, and HH^T = channel matrix. Therefore, the performance and the efficiency of the transmitting and receiving systems depend mainly upon its end devices. The traditional, fundamental figure of merits to realize the performance of radiating structures like impedance matching, return loss, gain, directivity, and so on are not sufficient enough to optimize the performance of a multiband system. A high-speed system deals with a pulse of extremely short duration, which is prone to ringing effect due to the presence of small ripples (tens of picoseconds) after each pulse [4–6]. It initiates dispersion or delays in the time domain to reduce the speed of operation. This inherent ringing effect can be neutralized by designing the radiating structure to have a constant phase center and stabilization of the group delay [7]. For a high-speed system, each frequency component must radiate at the same time, and the phase should vary linearly with frequency. Thus, system stability can be determined by taking the phase difference between the different components. Therefore, the focus of the modern antenna designers is to design an active, integral antenna with a miniaturized, conformal, electrically small structure for the wireless platform. The concentration is to achieve phase linearity, low latency, ultra-high reliability in the millimeter-wave band [8].

The microstrip patch antenna behaves as a narrowband device due to high surface wave loss and circuit-based capacitive arrangement, where the magnitude of reactive current density ($\omega\varepsilon''$) is much greater than the magnitude of dissipative current density $(\sigma + \omega\varepsilon'')$ resulting in high-resonance quality factor [9].

$$Q = \frac{\omega\varepsilon' |E|^2}{(\sigma + \omega\varepsilon'')|E|^2} = \omega \frac{\text{Peak density of electric energy}}{\text{Average density of power dissipated}} \tag{14.2}$$

Therefore, to radiate maximum power the radiator should have a low Q-factor. The bandwidth is inversely proportional to the Q-factor, so the bandwidth can be enhanced by lowering the Q-factor [10–11]. However, a low Q-factor leads to lower performance and stability. The efficiency of an antenna indicates power radiated into the space.

$$\eta_T = \eta_r \left(1 - |S_{11}|^2\right) \tag{14.3}$$

where η_T is the total antenna efficiency, η_r is the radiation efficiency, $|S_{11}|$ is the absolute reflection coefficient. High radiation efficiency can be achieved by reducing the losses and improving the antenna matching profile [12]. The real challenge occurs while matching the reactive impedance at any band. While designing, the tradeoff has to be made between relatively good bandwidth response and the system losses.

The narrowness in bandwidth performance of a single-layer microstrip antenna can be improved by altering the coupling mechanism and critically top-loading it with a dielectric resonator antenna. Therefore, there will be a shift in resonant frequency and modification of the Q-factor. However, the most sensitive parameters will be co- and cross-polarization.

The dielectric resonator antenna (DRA) offers many attractive features such as low surface wave loss, high radiation efficiency, high bandwidth, and different radiation patterns through modes control [13]. It provides a stable and uniform radiation pattern over the entire band of frequency where radiation bandwidth is equal to the impedance bandwidth. The time harmonic electromagnetic field produced within a dielectric is due to the displacement of charge by rotation or separation, relaxation of the bond under the influence of the applied electric field, and the variation in the electron cloud. This phenomenon leads to the occurrence of polarization within a finite, non-zero amount of time, and constant phase center. Therefore, the dielectric is responsible for the modification of the phase fronts. Rectangular DRA (RDRA) provides tremendous practical advantages over other structures in terms of bandwidth control, handling of evanescent modes, setting up aspect ratio, and so on [14, 15].

In this chapter, we report our investigations on multifunctional hybrid rectangular dielectric resonator antenna with a maximum of two inclusions. The proposed antenna offers maximum impedance bandwidth of 22% while operating from 4.63 GHz to 10.84 GHz. It can be achieved by changing the spatial parameters of the RDRA relative to the microstrip patch antenna and the inclusions. The surface current concentration can be enhanced through inclusions and make it viable for the creation of different mode patterns. Through proper matching, mode degeneracy can be avoided, and there will be no major shift in the output characteristics. The principle of operation is presented in Section 14.2; various methods to improve bandwidth are given in Section 14.3; the structure of the proposed antenna is described in Section 14.4; Section 14.5 provides detailed analysis of the results obtained; and finally, the conclusion and future scope are presented in Section 14.6.

14.2 PRINCIPLE OF OPERATION

For the high profile, electrically large antenna current distribution gets severely attenuated as fields are radiated. The fundamental principle to achieve multiband design is to activate different zones to generate different field patterns under a range of frequency of operation [16]. But all the frequency should be transmitted at the same time. Therefore, the conditions of lag and ringing are not acceptable.

Scattering and diffraction have to be given due importance while designing a microwave antenna. The radiated output field pattern is the superposition of the diffracted field components and the field due to the transmission line.

$$F(\xi,n) = A(\xi,n)e^{-j\varphi(\xi,n)} \tag{14.4}$$

where (ξ,n) is the coordinate point over the radiator, $A(\xi,n)$ is the amplitude distribution, and $\varphi(\xi,n)$ is the intrinsic phase difference arising from different path length from the origin to the various field points. The phase behavior of multiband system depends upon the group delay (GD) as presented in Equation (14.5) [7]:

$$GD(f) = -\frac{1}{2\pi}\frac{d\varnothing(f)}{df} \quad at \quad f = f_0 \tag{14.5}$$

The phase error over the radiator induces phase distortion and intensity distributions. It may result in the reduction of gain, broadening up the main lobe, and enhancement of the side lobes. The resultant field will be maximized and in a direction normal to the antenna when contributions made by all the field components are in phase and uniform. The path length phase vector $\mathrm{ksin}\theta(\xi\cos\varnothing + n\sin\varnothing)$ is a linear function of the coordinate on the radiator.

The overall transfer function of the antenna depends upon the variation of matching profiles, gain, polarization w.r.t. frequency, and should be stable over the entire operating band [17]. An important figure of merit to measure the performance of the multiband antenna is the pattern stability factor (PSF). It is defined as the average frequency domain correlation pattern $C(\theta_0,\varnothing_0)$ over all possible directions. $C(\theta_0,\varnothing_0)$ is the average value of spatial correlation factor over the whole range of possible operating direction for the antenna within the solid angle (Ω) and $(\theta_0,\varnothing_0)$is the reference direction [7].

$$PSF = \frac{\int C(\theta_0,\varnothing_0)d\Omega}{\int d\Omega} \tag{14.6}$$

In practice, for the best performance of the multiband antenna, the value of PSF should be greater than 0.95. However, for an arbitrary solid angle (Ω), PSF decreases as antenna bandwidth increases, which affects the performance of the system.

RDRA radiates like a short magnetic dipole. The mode within the rectangular DRA is of two types – confined and non-confined modes [18]. The resonator satisfies two magnetic wall conditions given by Equation (14.7a) and (14.7b) [19–20].

$$E.n = 0 \tag{14.7a}$$

$$n \times H = 0 \tag{14.7b}$$

where E and H denote electric and magnetic field intensity and n is the unit vector normal to the surface of the resonator. The confined mode satisfies both Equation (14.7a) and (14.7b) whereas non-confined mode satisfies only Equation (14.7b). The rectangular DRA satisfies only Equation (14.7b). The dimension of the RDRA can be calculated from the dielectric waveguide model [21–22]. It supports TE_x, TE_y, and TE_z mode in the x-, y-, z-directions, and the resonant frequency is a function of the dimensions. The lowest order mode will be $TE_{x\delta 11}$, $TE_{y1\delta 1}$, and $TE_{z11\delta}$. The resonant frequency of $TE_{x\delta 11}$ can be obtained by solving the transcendental equation.

$$k_x \tan\left(\frac{k_x d}{2}\right) = \sqrt{(\varepsilon_r - 1)k_0^2 - k_x^2} \tag{14.8}$$

where $k_0 = \frac{2\pi}{\lambda_0} = \frac{2\pi f_0}{c}, k_y = \frac{\pi}{w}, k_z = \frac{\pi}{b}$

$$k_x^2 + k_y^2 + k_z^2 = \varepsilon_r k_0^2 \tag{14.9}$$

The transcendental equation can be solved for selected ratio of w/b as a function of d/b w.r.t normalized frequency F, which can be defined as

$$F = \frac{2\pi w f_0 \sqrt{\varepsilon_r}}{c} \tag{14.10}$$

Here, f_0 is the resonant frequency. The value of the normalized frequency F can be determined from the F vs. (d/b) plot for different values of (w/b). Equation (14.10) can be effectively written as

$$f_{GH_z} = \frac{15F}{w_{cm}\pi\sqrt{\varepsilon_r}} \tag{14.11}$$

where resonant frequency is represented in GHz scale and 'w' is in cm. The radiation Q-factor can be obtained from the total stored energy and the total power radiated by the RDRA [22].

There is an introduction of distributed capacitance and inductance into the structure through the inclusions that will act as a microwave bandstop filter. The bandstop filter provides excellent frequency selectivity and tuning around the resonant frequency. The induced resonating current flow inside the broken loops forms the behavior similar to a magnetic dipole. The square split ring geometry of the resonator provides numerous possibilities to tune the distributed elements and to provide more magnetic coupling with the transmission line [23]. The distributed capacitance per unit length can be determined by the following equation [24]:

$$C_{pul} = \frac{\beta}{\omega Z_0} \tag{14.12}$$

where β is the propagation constant and Z_0 is the characteristics impedance of two coupled metallic strip as shown in Figure 14.1. However, the value of distributed inductance is difficult to compute.

14.3 BANDWIDTH IMPROVEMENT TECHNIQUES

The major factors that contribute to the characterization of bandwidth are (i) impedance bandwidth, (ii) radiation efficiency, (iii) power gain, (iv) beam width, (v) beam directions, (iv) polarization, and (v) side lobe level [25–27]. The impedance bandwidth determines input impedance, radiation resistance, and antenna efficiency; whereas, pattern bandwidth determines gain, side lobe level, beam direction, and polarization. With suitable structural modifications, the narrowband nature of the linear resonant monopole antenna can be up-converted into a multiband antenna. The lower operating limit can be decided by defining the longest electrical path of the current, and the upper limit can be decided by the aspect ratio and the architecture of the ground plane. The geometry of the antenna has to be modified in such a way that the electrical path length of the current flowing on the external edge should be increased. It requires an experiment with different Euclidean shapes such as triangular, square, bow-tie, trapezoidal, circular, and elliptical. Out of these shapes, circular and elliptical provide the highest bandwidth. However, there is always a tradeoff between the radiation stability and the impedance bandwidth [28]. The current distribution can be altered by adopting the following methods:

i. *Use of parasitic element*: The parasitic elements are those that are not connected directly to the source. These passive elements add the effect of mutual coupling to accomplish partial overlapping between nearby bands. The monopoles placed close to each other should have a different shape to avoid filtration out of the frequency.
ii. *Use of short pin*: The short circuit pin is to be used in the zone of high current density. It causes the creation of additional modes to enhance the impedance bandwidth. It can reduce the physical size of the antenna.
iii. *Asymmetric feed*: Through asymmetric feed, the current symmetry within the antenna is lost. It greatly affects the output radiation pattern but the bandwidth in the upper limit increases.
iv. *Double feed*: The two feed points are symmetrically placed w.r.t. the center of the antenna. It reduces the horizontal current and there is a reduction in cross-polarization.
v. *Broadband matching techniques*: The term broadband matching deals with real loads and not for matching arbitrary impedance that varies with frequency. The broadband matching technique comes under the scope of the transmission line model. Effective elimination of the reactive components can be achieved by analyzing the current amplitude w.r.t. horizontal and

vertical components for a particular frequency. It is essential to ensure even distribution of the current. In a planar monopole antenna over a ground plane, the horizontal components are present in the side and around the feed. As the separation between the ground plane and antenna is electrically small, the field due to the horizontal current will remain in between the antenna and ground plane. Therefore, the horizontal current contributes less to the overall radiation pattern.

vi. *Setting up the lower limit of the frequency band is important*: The lower limit can be reduced by enhancing the current path along the perimeter. To achieve high bandwidth, graded change is preferable than that of abrupt cut.

14.4 STRUCTURE OF THE PROPOSED ANTENNA

Figures 14.1 and 14.2 show the side, top, and rear view of multifunctional hybrid rectangular dielectric resonator antenna with two inclusions. The wideband antenna is printed onto a piece of printed circuit board (PCB) (FR4 epoxy, $\varepsilon_r = 4.4$, $\tan\delta = 0.02$) with a thickness of 1.6 mm. As shown in Figure 14.1, two inclusions of external dimensions 4 × 4 mm^2 each are cut onto the upper portion of the radiator at the upper half of the centerline of the substrate. These inclusions are responsible for reducing the effect of the ground plane on the output characteristics of the radiator. The inclusion contains an open-ended stub in series with a short-ended stub at the center. The resonant frequency of the inclusions can be approximated by the following equations [23]:

$$\omega_0 = \frac{1}{\sqrt{LC}} \tag{14.13}$$

$$\omega_0 = \frac{\omega}{\sqrt{\tan(\beta l \text{ short ended})\tan(\beta l \text{ open ended})}} \tag{14.14}$$

$$\beta = \frac{\omega}{c}\sqrt{\varepsilon_{eff}} \tag{14.15}$$

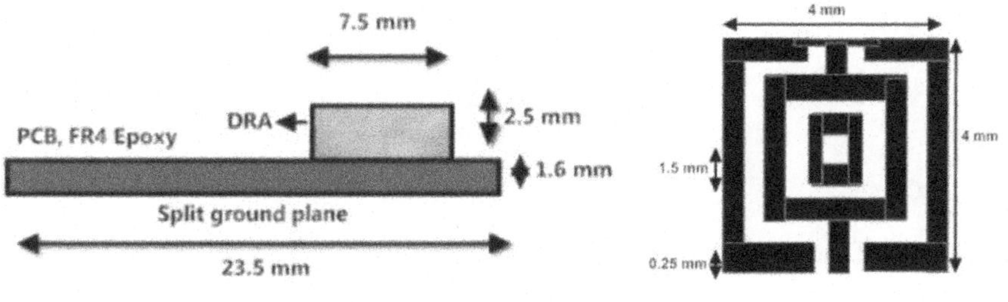

FIGURE 14.1 Side view of the proposed antenna and structure of one of its inclusions.

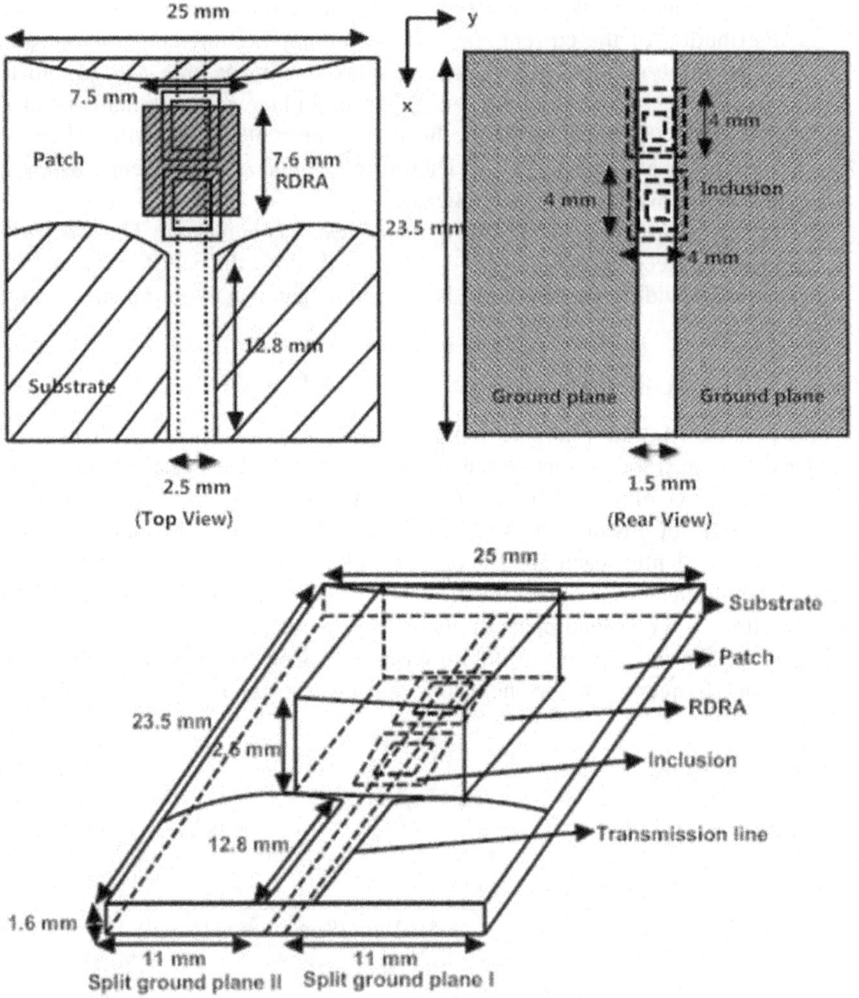

FIGURE 14.2 Top view, rear view, and structure of the total hybrid DRA.

If open ended = short ended = 1, then

$$\frac{f_0}{f} = \frac{1}{\tan\left(\frac{\omega}{c}\sqrt{\varepsilon_{eff}}\, l\right)} = \frac{1}{\tan\left(\frac{2\pi}{\lambda_g}\right) l} \tag{14.16}$$

The length of the inner stub will be

$$l = \frac{\lambda_g}{8} \tag{14.17}$$

where λ_g is the guided wavelength and ε_{eff} is the effective permittivity of the open- and short-ended center stub.

A rectangular DRA (Zirconia, ECCOSTOCK, Emerson & Cuming Microwave Products N.V., $\varepsilon_r = 4.4$, $\tan\delta = 0.02$) of dimension d = 7.6 mm, w = 7.5 mm, and width 2.5 mm is connected from the source. Two finite ground planes (split ground plane I and split ground plane II) each with dimensions of 23.5 × 11 mm^2 is etched at the bottom side of the substrate. As shown in the figure, a patch with a top curve shape is placed over the substrate. This shape increases the electrical length of the current path to decide the lower edge of the frequency of operation. It helps in miniaturizing the antenna size by enhancing the length of the effective current path. Figures 14.3 to 14.6 show vector components of electric field intensity 'E' in V/m, magnetic field intensity 'H' in A/m, and 2D surface current density 'J' in A/m^2. The introduction of

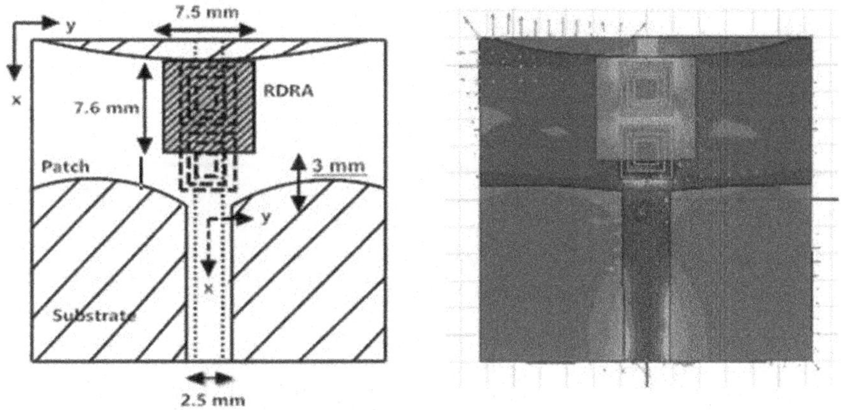

FIGURE 14.3 RDRA is positioned 3 mm away along –x direction and its E (V/m), H (A/m), J (A/m^2).

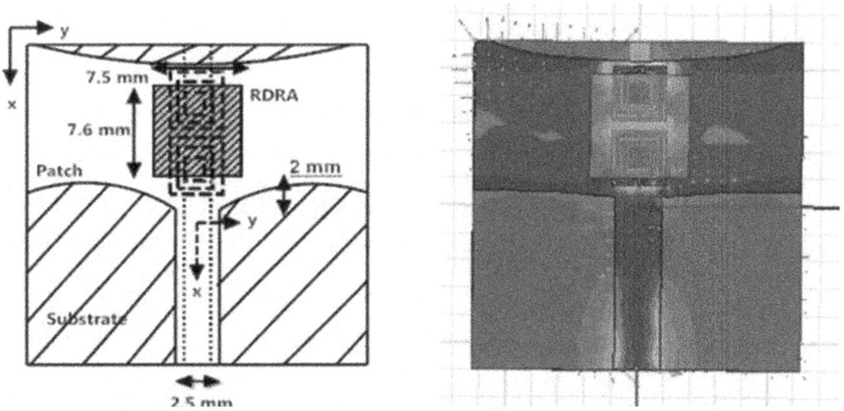

FIGURE 14.4 RDRA is positioned 2 mm away along –x direction and its E (V/m), H (A/m), J (A/m^2).

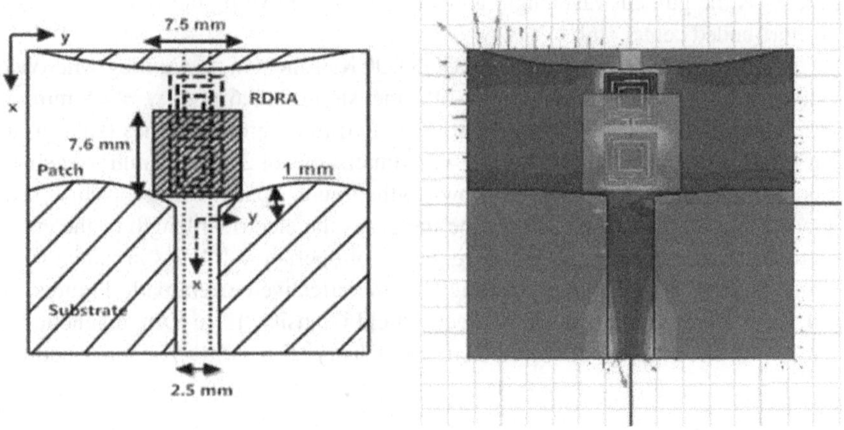

FIGURE 14.5 RDRA is positioned 1 mm away along –x direction and its E (V/m), H (A/m), J (A/m^2).

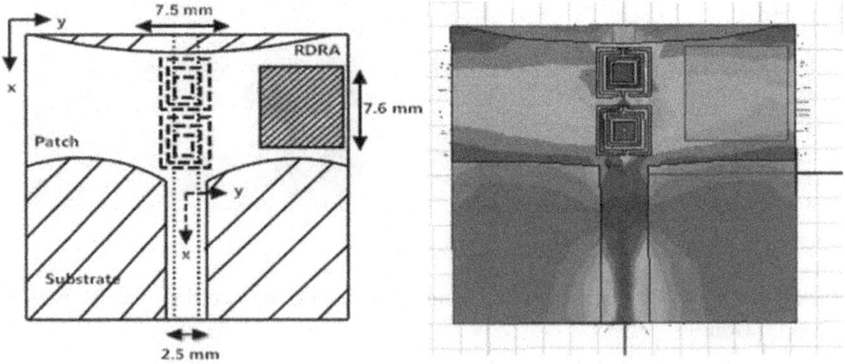

FIGURE 14.6 RDRA is positioned at the corner of the patch and its E (V/m), H (A/m), J (A/m^2).

an etched strip on the ground plane improves the matching profile and provides isolation at the lower operating frequency. The bandwidth enhancement of the patch and its impedance tuning can be achieved by loading the antenna on the surface of the patch by a dielectric resonator. The dielectric resonator is pasted by using a thin layer of conducting epoxy. In Figure 14.6, DR is placed in the middle of the non-radiating edge experimentally to improve the overall matching profile.

14.5 RESULTS AND DISCUSSIONS

For the distributed analysis to extract circuit information from a 3D model like DRA by using Maxwell's equation is too critical. To analyze the performance of the device at any frequencies, scattering parameter (S-parameter) is used. The S-parameter

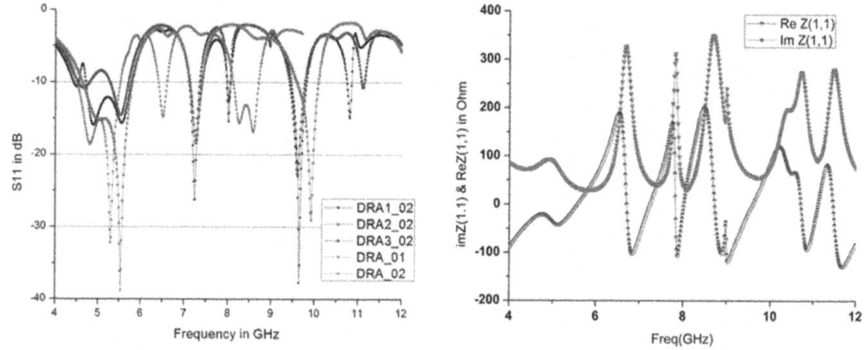

FIGURE 14.7 Plot of return loss $|S_{11}|$ versus frequency and rectangular plot of impedance versus frequency.

TABLE 14.1
Multiband response of the radiator in different configurations

	Band-I (GHz)	Band-II (GHz)	Band-III (GHz)	Band-IV (GHz)	Band-V (GHz)	Max. f_r (GHz)	$\lvert S_{11}\rvert_{max}$ (dB)
DRA1_02	4.71–5.78	7.13–7.41	7.97–8.05	9.45–9.81	10.74–10.84	9.64	37.73
DRA2_02	4.52–4.80	5.28–5.71	7.10–7.35	7.96–8.04	9.41–9.77	7.23	26.24
DRA3_02	4.52–4.81	5.28–5.71	7.10–7.35	7.96–8.04	9.41–9.90	7.24	26.18
DRA_01	4.70–6.12	7.37–7.64	9.76–10.59	–	–	5.80	32.16
DRA_02	4.63–5.80	7.12–7.40	9.58–10.09	–	–	5.53	38.71

employs the concept of wave propagation for describing the device. It is related to the transmission and returns loss relative to a given characteristics impedance. Impedance bandwidth is the range of frequency over which antenna can be matched to its transmission feed line. It can be described in terms of VSWR ≤ 2. Figure 14.7 compares the $|S_{11}|$ response against the frequency of the antenna at various positions of the RDRA with respect to the inclusions and the wideband patch. Table 14.1 shows the multiband response of the radiator in different configurations.

DRA1_02 indicates position 1 of the RDRA as shown in Figure 14.3, DRA1_02 indicates position 2 of the RDRA as shown in Figure 14.4, DRA1_03 indicates position 3 of the RDRA as shown in Figure 14.5, DRA_01 indicates position 2 of the RDRA with one inclusion, DRA_02 indicates right-edge position of the RDRA with two inclusions as given in Figure 14.6. It is observed that this antenna is resonating from a minimum of 4.63 GHz to 10.84 GHz. The lower and upper portion of the frequency band is almost constant and impedance matching is very sensitive to the position of the RDRA, inclusions, and the finite ground plane. This antenna shows multiband and multi frequency response within this frequency range of operation. Figure 14.7 shows the rectangular plot of impedance versus frequency of the structure as shown in Figure 14.4. It is observed that even if impedance bandwidth varies, the

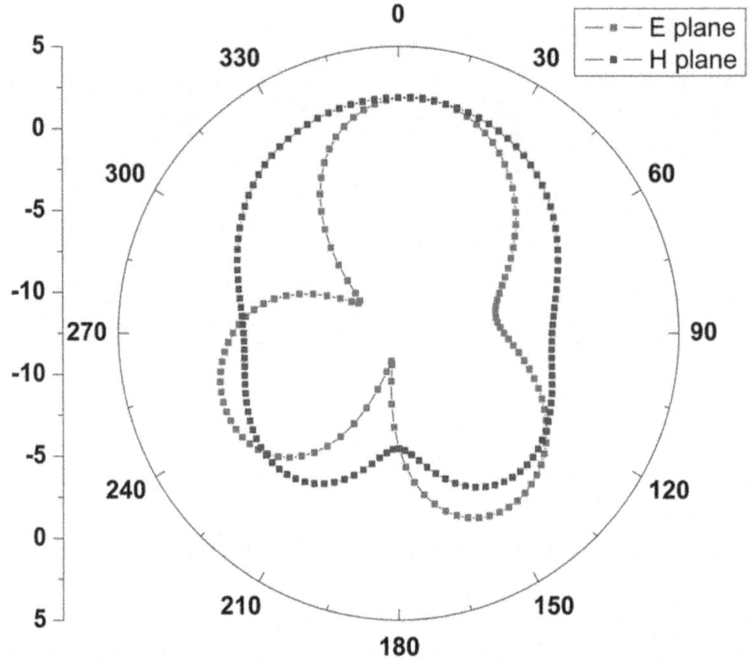

FIGURE 14.8 Radiation pattern in E- and H-planes (10 dB/div).

output characteristics remain constant. Therefore, output characteristics at the center position as shown in Figure 14.4 are studied in detail.

Gain is the primary design consideration of an antenna. Factors affecting gain are: (i) back lobe interference effect, (ii) consideration of phase error, and (iii) design of the feed pattern [29, 30]. The efficiency of the antenna is the product of a fraction of total power anticipated by the antenna and the efficiency of the antenna in concentrating the available energy into the peak of the main lobe. Maximum gain can be realized with uniform current distribution. It modifies the phase front to obtain maximum directive. Therefore, at high speed, wideband application phase distribution should be uniform. In addition to gain, all the side lobes should be suppressed. It has been reported that maximum gain and maximum side lobe levels are generally incompatible. Necessary compromise is being made in antenna design by adjusting the amplitude distribution over the radiator. The portion of the output pattern possessing maxima and contained within the angular region bounded by the adjacent minima is called the main lobe. The subsidiary maxima are called side lobes. The symmetry of the elements is generally planes of symmetry and referred as principal plane of the pattern, as shown in Figure 14.8.

The output field at any point has both E_θ and $E_\varnothing$ components. Polarization of the field is usually expressed with reference to x- and y-axes rather than with a spherical coordinate system. E_x is the principal polarization component and E_y is the cross-polarization component. Symmetry properties of the antenna w.r.t. the principal plane leads to zero cross-polarization in those planes. The co- and cross-polarization levels

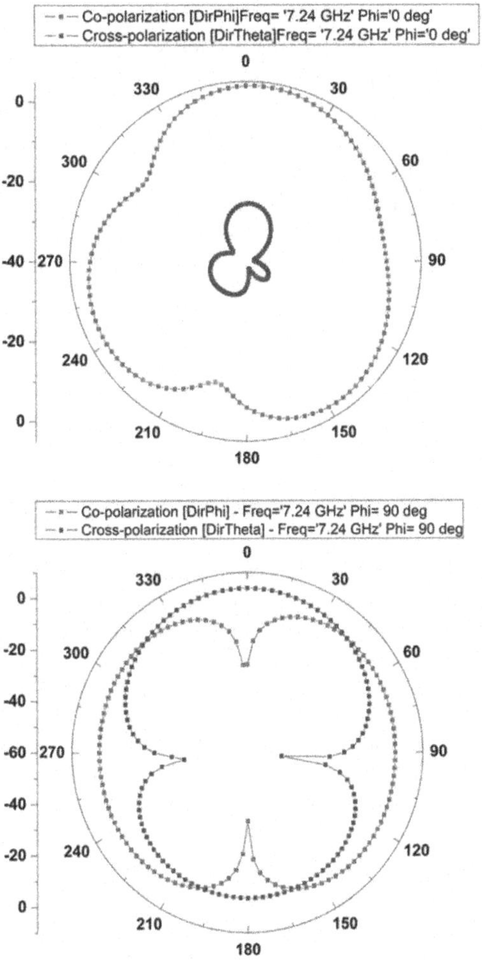

FIGURE 14.9 Radiation pattern indicating H- and E-planes co- and cross-polarization (10 dB/div).

of E-and H-planes are shown in Figure 14.9. Cross-polarization studies should be made on all antennas in which the side lobe specifications are very stringent [31]. The difference between E-plane co- and cross-polarization is about −20 dB at an angle of 90° and 270°; whereas, in H-plane cross-polarization level is around −28 dB. Here cross-polarization in E-plane pattern is very sensitive to the variation of the RDRA. Therefore, received signal sometimes become noisy as SNR decreases.

14.6 CONCLUSIONS

Narrow bandwidth provides serious limitations on a wider range of applications of the modern wireless system. Multifunctional integrated hybrid rectangular dielectric resonator antenna provides practical innovative extension and coupling of wideband microstrip patch antenna to increase impedance bandwidth while operating over 4.63 GHz to 10.84 GHz covering C, L, and Ku band of operations. This novel compact design provides impeccable phase linearity, high radiation efficiency with exceptional bandwidth response. Apart from this, it radiates an omnidirectional pattern over the operating frequency range. This embedded antenna is suitable to be used in cellular tunable IoT-based global tracking module for operations such as tracking, smart metering, a sensor-based smart home which concentrates mainly on 0.5 GHz to 10.4 GHz. For stacked configuration, with an appropriate switching mechanism, this antenna can function as a reconfigurable antenna for cellular and cognitive radio platforms.

REFERENCES

[1] D. Guha and Y. M. M. Antar, *Microstrip and Printed Antennas New Trends, Techniques and Applications*, 1st edition, John Wiley and Sons Ltd, United Kingdom, 2011, pp. 305–340.

[2] S. Ramo, J. R. Whinnery, and T. V. Duzer, *Fields and Waves in Communication Electronics*, 3rd edition, Wiley.

[3] Simon Haykin, *Communication System*, 5th edition, John Wiley and Sons Ltd, United Kingdom, 2009.

[4] D. M. Pozar, *Microstrip Patch Antenna*, 1st edition, IEEE Press, USA and Canada, 1992, pp. 203–259.

[5] K. L. Wong, *Compact and Broadband Micro-Strip Antenna*, John Wiley and Sons Ltd, New York, 2002.

[6] Y. Huang, K. Boyle, *Antenna from Theory to Practice*, John Wiley and Sons Ltd, USA, 2008.

[7] D. Valderas, J. I. Sancho, D. Pente, C. Ling, and X. Chen, *Ultra Wide Band Antenna Design and Application*, Imperial College Press, United Kingdom, 2011.

[8] R. B. Waterhouse, *Microstrip Patch Antennas: A Designer's Guide*, Khuwer Academic Publishers, Massachusetts, USA, 2003.

[9] Roger F. Harrington, *Time Harmonic Electromagnetic Fields, Series on Electromagnetic Wave Theory*, IEEE Press, USA and Canada, 2001.

[10] Y. Coulibaly, and T. A. Denidni, "Broadband microstrip fed dielectric resonator antenna for X-band applications", *IEEE Antenna and Wireless Propagation Letters*, Vol. 7, pp. 341–345, 2008.

[11] K. M. Luk, and K. W. Leung, *Dielectric Resonator Antennas*, Research Studies Press, Baldock, Herfordshire, UK, 2003.

[12] L. V. Blake, *Antennas*, Artech House, John Wiley & Sons, USA, 1966.

[13] A. Petosa, A. Ittipiboon, Y. M. M. Anter, D. Roscoe, and M. Cuhahi, "Recent advances in Dielectric Resonator Antenna Technology", *IEEE Antenna and Propagation Magazine*, Vol. 40, No. 3, pp. 35–48, June 1998.

[14] R. K. Mongia, and P. Bhartia, "Dielectric resonator antennas – a review and general design relations for resonant frequency and bandwidth", *International Journal of Microwave and Millimeter-Wave Computer-Aided Engineering*, Vol. 4, No. 3, 1994, pp. 230–247.

[15] R. S. Yaduvanshi, and H. Parthasarathy, *Rectangular Dielectric Resonator Antennas, Theory and Design*, Springer, Germany, 2016.

[16] F. Wang, C. Zhang, H. Sun, and Y. Xiao, "Ultra-wideband dielectric resonator antenna design based on multilayer form", *Hindawi International Journal of Antennas and Propagation*, Vol. 2019, Article ID 4391474.
[17] X. Li, Y. Yang, F. Gao, H. Ma, and X. Shi, "A compact dielectric resonator antenna excited by a planar monopole patch for wideband applications", *Hindawi International Journal of Antennas and Propagation*, Vol. 2019, Article ID 97344781.
[18] R. K. Mongia, and A. Ittipiboon, "Theoretical and experimental investigations on rectangular dielectric resonator antennas", *IEEE Transaction on Antennas and Propagation*, Vol. 45. No. 9, pp. 1348–1356, Sept. 1997.
[19] R. D. Richtinger, "Dielectric resonator," *Journal of Applied Physics*, Vol. 10, June 1939, pp. 391–398.
[20] A. Okaya, and L. F. Barash, "The dielectric microwave resonator," *Proceedings of the IRE*, Vol. 50, Oct. 1962, pp. 2081–2092.
[21] M. H. Neshati, and Z. Wu, "The determination of the resonance frequency of the TE111 mode in a rectangular dielectric resonator antenna application", *11th International Conference on Antennas and Propagation, 17–20 Apr 2001*, Conference Publication No. 480 IEE 2001.
[22] A. Petosa, *Dielectric Resonator Antenna Handbook*, Artech House, Norwood, MA, 2007.
[23] O. A. Safia, L. Talbi, and K. Hettak, "A novel artificial magnetic material based on a CPW series connected resonator and its implementation in original applications", *European Conference on Antenna and Propagation*, pp. 2756–2760, 2013.
[24] R. Marques, F. Mesa, J. Martel, and F. Medina, "Comparative analysis of edge- and broadside-coupled split ring resonators for meta material design – theory and experiment", *IEEE Transactions on Antennas Propagation*, Vol. 51, No. 10, pp. 2572–2581, Oct. 2003.
[25] Y. Coulibaly, and T. A. Denidni, "Broadband microstrip fed dielectric resonator antenna for X-band applications", *IEEE Antenna and Wireless Propagation Letters*, Vol. 7, pp. 341–345, 2008.
[26] K. M. Luk, M. T. Lee, K. W. Leung and E. K. N. Yung", Technique for improving coupling between microstrip line and dielectric resonator antenna", *Electronics Letters*, Vol. 35, No. 5, March 4, 1999.
[27] S. Mohanty, A. Khan, and B. Mohapatra, "Embedded rectangular dielectric resonator antenna for Ku-band applications", *Elsevier SSRN Proceedings*, pp. 2214–7583, Feb 2020, https://doi.org/10.2139/ssrn.3549250.
[28] A. Petosa, "Design of microstrip-fed series array of dielectric resonator antennas", *IEE Electronics Letter*, Vol. 31, No. 16, pp. 1306–1307, Aug. 1995.
[29] N. A. Shalaby, and S. E. Sherbiny, "Mutual coupling reduction of DRA for MIMO applications," *Advanced Electromagnetics*, Vol. 8, No. 1, pp. 75–81, May 2019.
[30] S. Mani, and L. Edeswaran, "High gain multiband stacked DRA for WiMax and WLAN applications,"*American Journal of Applied Sciences*, pp. 779–785, 2017.
[31] S. Mohanty, and B. Mohapatra, "Leaky wave-guide based dielectric resonator antenna for millimeter-wave applications," *Transaction on Electrical and Electronic Materials*, 2020, https://doi.org/10.1007/s42341-020-00240-w.

Part VI

Importance and Use of Microstrip Antennas in IoT

15 Importance and Use of Microstrip Antennas in IoT

Opportunities and Challenges

D. Ganeshkumar and K. Jaikumar

15.1 INTRODUCTION

In this chapter, the importance and uses of a microstrip antenna is illustrated. The microstrip antenna plays a major role in wireless communication systems. The Internet of Things (IoT) is one of the important technologies of the twenty-first century. In general, the antennas are considered as the array of elements or conductors acting as an interface between the transmitter and the receiver. There are different types of antennas suitable for different applications. IoT is the network of things that are embedded with sensors or other devices where data exchange or communication within or with other systems and devices occurs over the internet. It allows monitoring and controlling of various processes efficiently. For monitoring those devices that communicate wirelessly, a good performing compact antenna is required for data transfer either once a day or twice a day or continuously. Thus, the advancement in antenna development plays a great role in influencing the wireless technologies. Since a microstrip antenna is known for its compatibility, it has a high use in IoT applications. The antenna is considered as the energy transducer through which electromagnetic waves are received or sent out. IEEE defines antenna as "a part of

DOI: 10.1201/9781003093558-21

transmitting or receiving system that is designed to radiate or receive electromagnetic waves" [1]. Following are the different types of antennas:

- Wire antenna
- Array antenna
- Aperture antenna
- Lens antenna
- Printed antenna

The microstrip antenna is the common version of a printed antenna. Photolithography technique is used in the fabrication of printed antennas. The microstrip antenna comprises of a thin metallic patch or a strip placed above the ground plane with a dielectric in-between. The size and shape of the patch determines the frequency of operation and performance of antenna. The microstrip antennas can be designed in different dimensions and are thus suitable for many applications, including IoT applications. As the name indicates, the size of a microstrip antenna is small. The fabrication does not require a separate component since it is designed on the PCB itself. The cost of the antenna is also less and easy to integrate with other components. Due to its numerous advantages, a microstrip antenna has a great impact in IoT applications.

IoT (Internet of Things) is nothing but extending the power of internet all over the world. Connecting things to internet brings the ability to transfer the data without human-to-human intervention. In connecting things, "things" can be a sensor in automobile, a person with heart monitor implant, or any other object which has the ability to transfer data. IoT can be applied for manufacturing, transportation, waste management, traffic control, and so on. Internet of Things uses processor, controller, and antenna for data transfer. The data is shared by IoT gateway or stored in a cloud [2].

IoT makes work smart and also helps in automation. Some of the benefits of IoT include:

- Monitoring and controlling overall processes
- Saving time
- Helps in making better decisions in business

Some of the real-world applications of IoT include industrial Internet of Things (IIoT), consumer IoT, smart wearables, healthcare, smart building, and smart farming system. IoT paves the way for "smart". The increased use of IoT application in modern communication system has also increased the demand for antenna, especially in microstrip antenna.

15.2 IMPORTANCE AND USE OF MICROSTRIP ANTENNAS IN IOT

15.2.1 Importance of Microstrip Antennas in IoT

A microstrip antenna (MPA) is primarily a two-dimensional structure that uses "patch" of one-half wavelength long, so that a metal surface acts as a resonator. The microstrip antenna uses conventional microstrip fabrication technique.

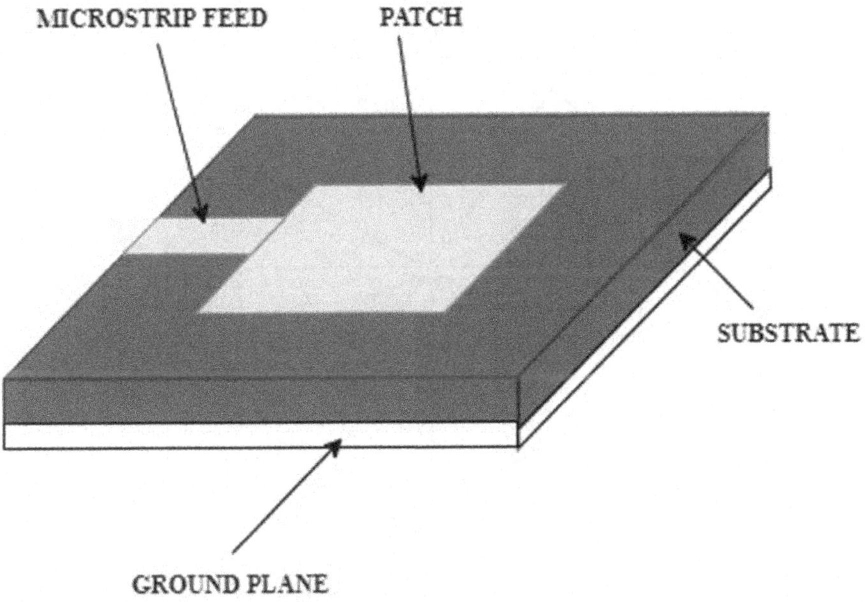

FIGURE 15.1 Microstrip antenna.

Most MPA uses FR4 as the dielectric substrate. One side consists of radiating patch and ground plane on the other side. Based on the shape and the size, different types of microstrip antennas are available.

Figure 15.1 shows the pictorial view of a microstrip antenna. The patch is made of copper or gold with common shapes such as rectangular, elliptical, triangular, or circular. On the substrate, the feed lines and patch are photo-etched. For higher efficiency and better radiation, a thick dielectric substrate with low dielectric constant is required, but such configuration increases the size of the antenna.

Antennas of small size may affect the common behavior. There must be a tradeoff between the antenna size and the antenna performance. The electromagnetic energy source guided to patch through excitation gives negative charges around one side of patch (feed point) and positive charges on the other side. The charge difference creates electric fields which are responsible for radiations [3]. The electromagnetic wave may radiate into space or diffracted and reflected back into space or remain trapped in the substrate.

Figure 15.2 represents the radiation pattern of a microstrip antenna. The polarization is determined by the direction of wave radiated, which is identical to the direction of electric field. It is said to be linearly polarized, when the path of electric field vector follows a line, and is said to be circularly polarized, when the vector rotates in a circle. Axial ratio is used to characterize the polarization.

An antenna may be small but requires large attention. Antennas with poor performance drain batteries, which affect connection to cloud-based systems. Antenna size, radiation pattern, gain need to be in consideration while designing. As mentioned,

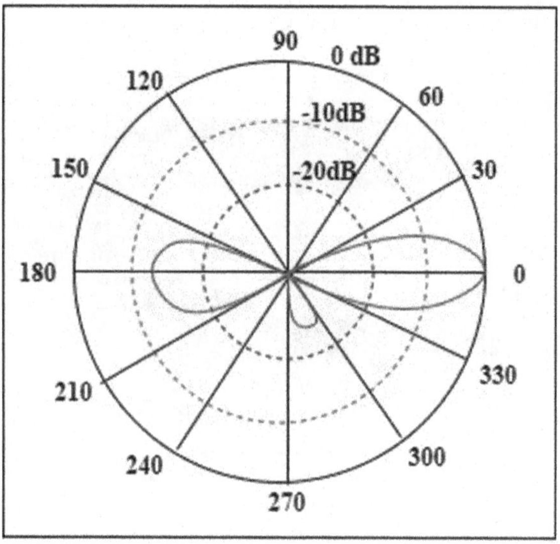

FIGURE 15.2 Radiation pattern.

due to easy fabrication and compatibility, microstrip antenna is used in IoT. One of the key factors is that microstrip antenna allows easy integration of IoT devices compared to conventional antenna. It also has improved gain and bandwidth. The radiation pattern of antenna is broad with narrow frequency bandwidth and low radiative power. The distinct characteristics of microstrip antenna are:

- Discrete separate components are not essential since antennas are fabricated as a part of Printed Circuit Board (PCB).
- Simple and inexpensive to design.
- Easy integration with IoT components.
- Good performance and maintain efficiency.
- Their geometry includes rectangular, circular, star, or spiral with complex combinations. These complex dimensions and shapes provide unique characteristics which are useful in many applications.
- Since shapes are adjustable, it is used for real-world consideration.
- Lightweight and robust.

It is possible to design microstrip antenna providing signal rejection and fading. There may be some drawbacks in microstrip antenna but they are suitable for many applications. One application of IoT includes monitoring the data or process; antenna is the essential element for a monitoring device that communicates wirelessly. Efficiency in monitoring is maintained by reducing noise and fading, hence performance of antenna plays a significant role [4]. Thus, a microstrip antenna becomes one of the essential elements in IoT and its applications.

TABLE 15.1
Comparison of various types of microstrip antennas

Characteristic features	Microstrip patch antenna	Printed dipole antenna	Microstrip travelling-wave antenna
Profile	Thin	Thin	Thin
Fabrication	Very easy	Easy	Easy
Shape	Any shape	Rectangular and triangular	Rectangular and circular
Polarization	Linear and circular	Linear	Linear and circular
Dual-frequency operation	Possible	Possible	Possible

15.2.2 Uses of Microstrip Antenna in IoT

The microstrip antenna generally involves many applications. It is used in the fields of healthcare, mobile communication, and RFID. Wearable antennas are also available. In telemedicine, satellite navigation, remote sensing, and mobile applications, microstrip antenna is utilized substantially.

Internet of Things (IoT)–based applications require integration with wireless communication technology to make the data readily available. The spectrum regulatory authorities are releasing new frequency bands at various ranges due to the growth in demand for IoT, the antenna for IoT application should match the spectrum range. Since microstrip antenna is suitable for wireless communication system, it resonates at more than one frequency and can be designed with different dimensions it becomes the suitable antenna for most IoT applications.

There are different types of microstrip antennas that are applicable for IoT applications and make the integration process effective. The antenna can be in different shapes and dimensions (Table 15.1). Some of them are:

- Microstrip patch antenna
- Microstrip dipoles
- Printed slot antenna
- Microstrip travelling wave antenna

15.2.3 Satellite Communication

In wireless communications, like satellite communication, circularly polarized radiation pattern is essential. In such a case, a circular or square microstrip antenna can be used. Different dimensions of antenna can be designed for different applications [5].

A microstrip patch antenna array has been applicable for IoT radio frequency-link application. Each patch is designed using Factor V antigen substrate and are of the same dimension. The performance of antenna depends on the substrate used along with its thickness.

The microstrip patch antennas are also designed with a function of dual frequency. One rectangular microstrip antenna and two slotted patches are used. The slots used are connected to the rectangular one. By etching multiple slots on patch helps in operating at dual frequency. The antenna operates at 2.39 GHz and 3.15 GHz and hence used in applications of IoT where large bandwidth is needed for operating dual frequency bands.

The planar microstrip patch antenna is designed for high frequencies using planar structure. The dimensions and feeding position need to be properly maintained and so radiation or reception of signals in appropriate band of frequency must be proper and hence preferred for IoT applications like GPS [6].

15.2.4 Smart Home Applications

One of the applications includes the concept of smart home allowing efficiency at different aspects. These antennas are also involved in weather monitoring, one of the applications of IoT. The rectangular patch antenna with different slots is used. The slots decrease the resonating frequency. In general, the resonating frequency in S-band range is shifted to L-band range thus reducing the size of antenna. In this modern era, devices such as mobiles and other wireless devices designed are smaller in size. Thus, this compact antenna is suitable to integrate with the small devices thereby minimizing the device size and are used in many applications.

The antennas within the handheld devices need to be much smaller and high performing ones. Scaling down of antenna without affecting the performance is done by introducing slots. A technology called Defected Ground Structure (DGS) is popularly used for bandwidth enhancement. It modifies gain, radiation pattern as per desired application requirements. Therefore, it allows proposed antenna to operate at dual or triple distinct frequencies which can be used in satellite applications too.

A meander line microstrip patch antenna has been designed for IoT applications by making inverted S-shape meander line connected with slotted rectangular box. A capacitive load and parasitic patch with shaped ground are attached. This antenna provides high gain compared to conventional one. It can be used with IoT sensors efficiently.

15.2.5 Real-Time Weather Monitoring Systems

The microstrip antenna also plays a vital role in aircraft real-time weather monitoring system using IoT. In this application, a compact rectangular microstrip patch antenna is used with different slots of miniature structure. Many numbers of slots are used such that resonating frequency can be reduced and to miniaturize the antenna. S-band resonating antenna is used for developing the microstrip antenna. To shift the resonating frequency from one slot to another, the frequency range has been extended to the L-band. By using such slots of different shapes and resonating frequencies, the antenna size is reduced. Real-time weather monitoring is used in aircraft to monitor the pressure variations inside the aircraft, turbulence, and the weather changes in the external environment. To detect the changes in the weather, a GPS antenna is used with the frequency of 1.2 GHz to 5 GHz. Inset feeding technique is also used [7].

For an IoT application, the normalized U-shaped structure microstrip antenna was used which has the gain from 1.3 dB to 2.54 dB, bandwidth from 134.1 MHz to 167.6 MHz, and return loss from –20.32 dB to –26.56 dB while compared with the conventional antenna [8].

15.2.6 Medical Applications

It has been investigated by the researcher that microstrip antenna is an ideal choice for medical applications as it is small in size and its electrical performance is modified. Therefore, microstrip antenna can be used in medical applications such as tissue detection, treatment of tumors, and so on.

15.3 MULTIBAND MICROSTRIP PATCH ANTENNAS IN IOT

A truncated icosidodecahedron, which is the largest Archimedean solid, can be used in the multiband microstrip antenna with planar projections. Such planar projected patch antennas have the largest peripheral length rather than the normal peripherals. This peripheral will increase the circulating current path and the multiband propagation can be maintained. Defected ground structure produces slots in the ground plane structure to deviate the distribution of current inside the shield. Thus, the transmission line properties such as line inductance and line capacitance changes. The rectangular-shaped radiating elements are also used to enhance the bandwidth [9].

Determining the resonance nature of the truncated icosidodecahedron is based upon certain methods. A truncated icosidodecahedron with planar projection consists of a combination of one decagon, five regular hexagons, and five squares. To reduce the complexity in the design process, the equal area concept can be used. The equal area concept defines a circular microstrip patch antenna which is similar to an icosidodecahedron: both have equal surface areas. The surface area of both the structure is directly proportional to the equivalent capacitance of the circuit and the frequency in which the antenna operates.

By assuming the radius of the circular patch antenna, the surface area of circular patch antenna can be calculated. The microstrip line fed truncated planar projection of icosidodecagon-shaped multiband microstrip patch antenna covers ultra high frequency, microwave, and high frequency. Thus, this type of antenna can be used for IoT applications, which operate at a frequency range of WLAN and Wi-Fi.

15.4 CONCLUSION

From the above discussion, it can be concluded that the microstrip antenna plays a vital role in IoT applications such as medical, weather monitoring, smart home applications, and so on. The performance parameters such as gain and bandwidth can be improved by various feed techniques in the microstrip antenna. Due to its robustness, some of the applications failed to use microstrip antenna; this can also be improved using various feed techniques. Thus, we can conclude that the microstrip rules the IoT in the real-time applications.

REFERENCES

[1] C. A. Balanis, *Antenna Theory: Analysis and Design*. Hoboken, NJ, USA: Wiley, 2005.

[2] R. P. Meenaakshi Sundhari, and K. Jaikumar, "IoT assisted hierarchical computation strategic making (HCSM) and dynamic stochastic optimization technique (DSOT) for energy optimization in wireless sensor networks for smart city monitoring", *J. Comm.*, https://doi.org/10.1016/j.comcom.2019.11.032.

[3] Y. T. Lo, and S. W. Lee, *Antenna Handbook: Antenna Theory*, Vol. 2, USA: International Thompson Publishing, Inc, 1993.

[4] S. Chamaani and A. Akbarpour, "Miniaturized dual-band omnidirectional antenna for body area network basestations", *IEEE Antennas Wireless Propag. Lett.*, vol. 14, pp. 17221725, 2015.

[5] Dioum, A. Diallo, S. M. Farssi, and C. Luxey, "A novel compact dual-band LTE antenna-system for MIMO operation", *IEEE Trans. Antennas Propag.*, vol. 62, no. 4, pp. 2291–2296, Apr. 2014.

[6] J. Pei, A.-G. Wang, S. Gao, and W. Leng, "Miniaturized triple-band antenna with a defected ground plane for WLAN/WiMAX applications", *IEEE Antennas Wireless Propag. Lett.*, vol. 10, pp. 298–301, 2011.

[7] Y. Dong, H. Toyao, and T. Itoh, "Design and characterization of miniaturized patch antennas loaded with complementary split-ring resonators", *IEEE Trans. Antennas Propag.*, vol. 60, no. 2, pp. 772–785, Feb. 2012.

[8] A. A. Elijah and M. Mokayef, "Miniature microstrip antenna for IoT application", June 2020. https://doi.org/10.1016/j.matpr.2020.05.678.

[9] M. Borhani, P. Rezaei, and A. Valizade, "Design of a reconfigurable miniaturized microstrip antenna for switchable multiband systems", *IEEE Antennas Wireless Propag. Lett.*, vol. 15, pp. 822–825, 2016, doi: 10.1109/LAWP.2015.2476363.

16 Importance and Use of Microstrip Antennas in IoT

Shivleela Mudda, K. M. Gayathri and Mallikarjun Mudda

16.1 INTRODUCTION TO MICROSTRIP PATCH ANTENNAS

Advancement in the antenna technology made conventional antennas to overcome their limits in many demanding applications such as radar, 5G, IoT, and so on. Patch antenna can be used in remote accessing of radar, missiles, and aircraft applications with high fidelity and compact structure. Concepts utilized by Internet of Things (IoT) provide opportunities for new services and modernization. The microstrip antennas are used in many IoT applications mainly due to their reasonable features. Tremendous changes in the area of communication have increased the need for multipurpose and multi-application devices. Due to limitations of electromagnetic spectrum, the antenna must be flexible and adaptable to various practical situations. Thus, antennas can be reconfigurable and are highly suitable for advancing e-utility applications. The IoT possesses high priority for next stage modification and implementations. This develops the need of a wireless system, which can provide a common functional

DOI: 10.1201/9781003093558-22

platform for multiple applications. A compact size antenna, with frequency reconfigurable property and circular polarization, deserves an adequate system essential. In this chapter, Section 16.1 provides an introduction to the microstrip patch antenna; Section 16.2 introduces IoT; Section 16.3 discusses IoT applications, antenna design for IoT, IoT antenna challenges, and current trends in the design of IoT antenna; and Section 16.4 concludes the chapter.

According to IEEE definition, an antenna is a transducer that transfers electric wave signals into electromagnetic waves and vice versa [1]. It transmits or receives electromagnetic waves. Wireless systems' motive to be compact and portable demands antennas that occupy less space and are of considerable size.

The evolution of wireless communication system needs to design compact, lightweight, modest antennas that are adequate at providing high efficiency over a broad frequency range of frequency in both government and commercial fields. This growth has focused technologically much concentration in the design of microstrip patch antennas. Compared to other antennas, patch antenna shows many advantages with simple structure. Advantages include very lightweight, simple design, low cost, low profile to construct, and also consistent to planar and non-planar structures [2]. The design of microstrip antennas varies and that may go beyond in relating to another antennas.

The microstrip patch antennas have some disadvantages too. Considerable are the lower gain, limited bandwidth, and spurious wave radiations that curtail radiation efficiency. Different methods have been developed to overcome drawback of narrow bandwidth [3]. The design of large height substrate whose dielectric permittivity is low or a ferrite configuration yields improved frequency band but the method leads high-profile designs. Bandwidth may also be enhanced by using the proximity and aperture coupled feeding techniques; but these make implementation difficult. The multi-resonator stack configuration is also used to improve bandwidth but it has a large thickness prototype [3]. Electromagnetic bandgap structures are used presently to reduce the surface waves of the antenna. An array of patch elements is used to achieve high gain antenna [2].

16.2 INTRODUCTION TO IOT

The Internet of Things (IoT) portray the network of physically interrelated devices embedded with sensors and electrical and digital machines with a unique identity to allow these devices to communicate over network [4]. The main objective of IoT is to spread or exchange internet connectivity from standard gadgets like laptops, computers, and mobiles to comparatively dumb devices like washing machines and toasters. It can make "smart" our daily life activities virtually with the power of data collection, AI algorithm, and networks. The entire process of IoT begins with the devices themselves like smart phones, smart watches, and electronic appliances that help to communicate with the IoT platform.

16.2.1 Importance of Antenna Design in a Wireless or IoT Product

Proper antenna design, testing, and integration are important and should be apparent. The Antennas that are designed and tested precisely contribute to the success of a product [5] by (i) providing gain (for fixed products) that increases the radiated signal without consuming additional battery power; (ii) assuring uniform signal propagation in multiple directions (for wearable, handheld, and IoT products) so that communication is reliable; and (iii) maximizing signal level while minimizing the reception of noise through optimum location (particularly to minimize noise reception from other components in the product, avoid signal shielding from the product or housing, and to ensure the antenna is not de-tuned).

16.2.2 Fields Where the Term IoT Is Used

The IoT is applicable in various fields, such as the human wearable, health, homes, smart city, industries, and environmental monitoring [6].

- IoT sensing capability and connectivity make feasible in tracing the health status and other needed particulars of a person. This application not only makes user's routine life better but also helps an individual to take care of their health by following precautionary actions.
- IoT is often used for remote sensing, local monitoring, and managing different home appliances such as electronics. It also helps to maintain plants in the garden using an automated sprinkler system.
- IoT is used effectively for collecting and processing information generated by numerous infrastructures such as for traffic monitoring, street lamps, surveillance of cameras, and the power grid [7]. Those systems help in avoiding the traffic through the city centers and energy efficiency enhancement of transport systems.
- In industries, the major objectives of IoT are to optimize operations, improve productivity, resources saving, and costs reduction. It is also used to keep footprints of business assets, boost surrounding safety, and to maintain high quality and stability in the production process.
- IoT can help individuals to use, understand, and manage the resources that they have in a better way. Sensors used by IoT can help to protect wildlife, track wastage of water flows, monitor local weather, monitor use of natural resources, or give warnings before and after natural disasters.

16.3 MICROSTRIP PATCH ANTENNAS AND IOT

16.3.1 Microstrip Patch Antenna Design for IoT Applications

Development in design of the antenna has highly modified wireless technology. In recent years, communication systems' demand for micro antennas are increasing due to the maximum utilization of the Internet of Things (IoT). IoT generally uses microstrip patch due to its compatibility [8]. Main benefit of the microstrip antenna is its easy integration into IoT devices. For many IoT applications, wireless connectivity is a pivotal enabler, especially mobile, or where wired data communication is not practical. Prominent wireless protocols such as Bluetooth®, ZigBee®, Wi-Fi® operating in the LTE and UWB frequency bands can defend high data rates over small distances. Wireless standards like LoRa and Sigfox support longer distance connections at lower transfer rates operating in sub-GHz bands. The use of any of these protocols for designing an RF (radio frequency) subsystem requires professional skills because yielding a working solution is not a plug and play operation.

The patch antennas used in IoT devices exhibit GPS functionalities as signals travelling from the satellite are usually either right-handed circularly polarized (RHCP) or left-handed circularly polarized (LHCP). The patch antennas have capabilities for dual polarization; which can be achieved by tuning a perturbation element such as RF-MEMS, PIN diode, or varactor switches, and there remain many patch antennas that operate for polarization of one type between linear, LHCP, or RHCP. Most of the time, it is highly difficult to choose the perfect type of polarization matching the transmission. The IoT devices in which antennas like chip and PCB have been embedded have the advantage that they can fit in a small area, diminishing a sensor node's geometry. The PCB antennas consist of conductive elements offering higher gains than their chip-based counterparts.

Figure 16.1 shows an IoT design architecture that has a basic three-layered design module composed of user-defined applications, network protocols and securities, and perception and control layers. The perception layer facilitates accessing and recording of all the data availability of the system; therefore, an efficient system is essentially required to perceive signal data with efficient data rate control. Many

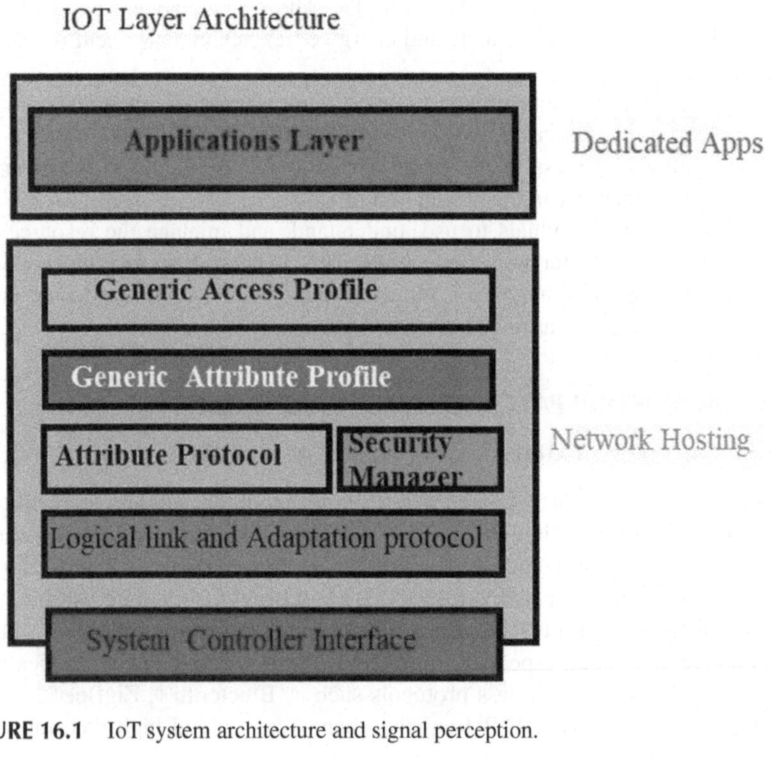

FIGURE 16.1 IoT system architecture and signal perception.

antenna topologies are available for the PCB antenna such as inverted-planar-F (PIFA), H, L, and folded-type monopoles [9]. Ground structure also has importance in the development of a patch antenna. In Figure 16.2, a smaller ground plate can limit the design significantly having a narrower bandwidth and improved radiation pattern. In practical, antenna radiator dimensions correspond to its gain; embedded chip antennas occupy more space on the board. However, sensor nodes have the characteristic of keeping a uniform shape that could permit easier enclosing and mounting in any environment having a PCB antenna. Directional antennas, such as Yagi, and sectoral antennas can be used to widen a radiation range or for base stations of IoT. The mechanical variation of high-gain directional antennas provides a larger communication distance than the omnidirectional antennas. The Yagi antennas are generally preferred by Supervisory Control and Data Acquisition system (SCADA) of IoT that allows improved data rate and also reliability. The sectoral antennas usually offer a wider beam width than Yagi-uda antennas.

From the basics of the antenna it is clear that the length of center-fed dipole antenna should be one-half of the signal wavelength in order to attain resonance. Smaller monopole antenna in combination with quarter wavelength connected as a return path close to a ground plane has a significant effect and it can be optimal for space-constrained IoT. Ground structure constitutes a form effectively of the unipolar conductor shown in Figure 16.2, so it must be sufficiently sized.

Inside the enclosure, if space constraints demand more flexibility in positioning the antenna, in such cases patch antennas can be placed at the side or along the longitudinal edge of the device. Compact coaxial cable or U-FL connector can be used

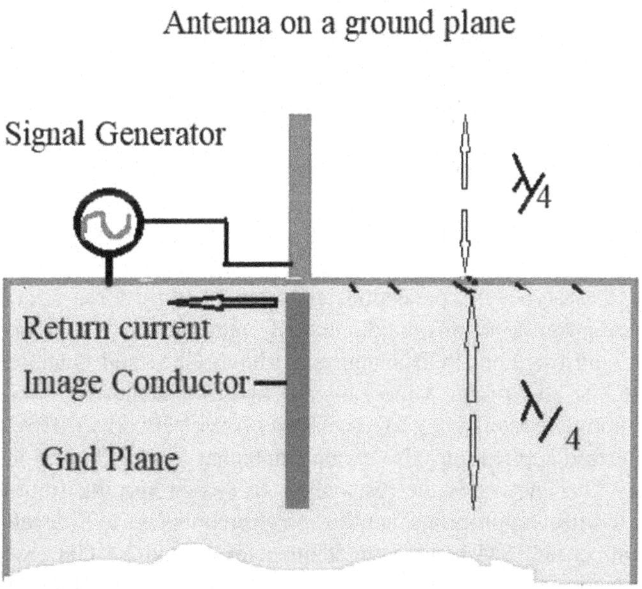

FIGURE 16.2 Effective use of ground plane.

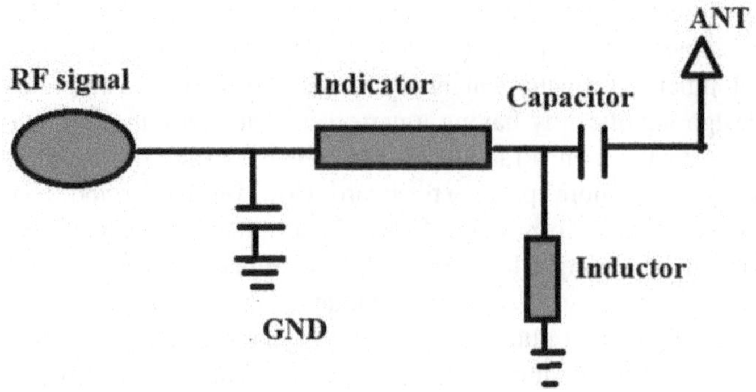

FIGURE 16.3 Impedance matching circuit to maximize power transfer.

to connect the PCB antenna to the main board [10]. In particular, if priority is high-volume and high-speed assembly, a suitable alternative is the stamped metal antenna. In case of extremely limited space, the technology of molded interconnect device (MID) can be considered for three-dimensional antenna design.

For antenna impedance matching with RF-amplifier output, loading the circuit board by adequate components or perfect impedance matching network needs to be placed. This requires a simple single-series capacitor and parallel inductor in case of a single-band antenna or a more complex network of dual-band antenna (Figure 16.3).

16.3.2 Design Challenges of Antennas for IoT Applications

The differing system of Internet of Things (IoT) supports numerous user-end devices that are with various RF parts, sensing elements, selectors, power unit, MEMS, energy harvesting concepts, and most of all, antennas. The selection and dimensions of the antenna are reliant on the application used and they differ according to frequency range, information transmission rate, and space availability [11]. Unlike 5G, IEEE 802.11ax and WI Gig are high data rate wireless networks utilizing immense continuing specter space. Network of IoT transfers and receives mini packets of data smartly using various network topologies such as star, mesh, and point-to-point. Many times, these types of connections are established by some frequent and relatively less composite kind of omnidirectional antenna, wire dipole antenna, whip, rubber duck, and fragment PCB antennas. Arduino GSM and Qualcomm-type IoT development kits also prefer analogous structures for achieving GPS, Wi-Fi, and Bluetooth communications [11]. Medical Body Area Networks (MBAN) and wearable computerized applications also demand antennas to be light or thin and comply with surface. This motivates the researchers to design specific front-end services customized to meet appropriate application prerequisites. IoT development platform uses unlicensed ISM bands with frequencies around 2.4 GHz; whereas, some Avionics services also incorporate aerials for licensed bands. Table 16.1 highlights widely used technologies and frequency bands for IoT applications.

TABLE 16.1
Desired frequency bands for IoT

Technologies	Frequencies	IoT Applications
Zigbee	915 MHz and 2.4 GHz	General (Smart home and commercial buildings)
Z-Wave	2.4 GHz	
Bluetooth	2.4 GHz	
Wi-Fi	2.4 GHz, 3.6 GHz, 16.9 GHz, 5 GHz, and 5.9 GHz	
Wireless HART	2.4 GHz	IIoT
ISA 100.11a 2.4GHz		
WBAN	2.4 GHz	Medical
MBAN	2,360–2,400 MHz	
WAIC	4,200–4,400 MHz	Avionics

TABLE 16.2
IoT system attributes

IoT primitives	Aspect	Aristocratic	Reliability	Security
Sensors	Physical	Yes	Yes	Yes
Aggregator	Virtual	Yes	Yes	Yes
Communication channel	Physical/virtual	Yes	Yes	Yes
e-Utility	Virtual/physical	Yes	Yes	Yes
Decision trigger	Virtual	Yes	Yes	Yes

Table 16.2 shows the basic attributes and risk analysis of IoT systems, which indicates that precise data rate and system adherence are the key aspects of the design analysis. This elaborates the need for fast switching devices to function among the sequenced IoT primitives. Moreover, effective transmission and control require risk analysis in IoT system on the basis of design primitives. Sequentially, perception layer elicits physical layer sensor attributes with high probability of false triggering risk in all three (aristocratic, reliability, and security) categories. On virtual or physical integrated development environment (IDE), communication channels as well as remote accessing experience high threat equally. The decision to trigger is end user interface (EUI), which leads all the RF stimuli and elicits challenging attributes to deal with RF access risk attributes. This develops the eventual need of effective RF perception and secure sensor attributes in IoT systems in both physical and virtual layers. The adequate re-configurability of an RF perception device (antenna) can reduce the threat of unwanted interference and thus enhance the system reliability in IoT devices.

Leaving space for antennas in IoT system is becoming a significantly challenging task as it comprises many wireless technologies. Hence, antenna design of IoT module faces constraints of narrowing footprints while upholding permissible performance under severe circumstances, namely interference, fading. Developed techniques for interference reduction, multiplexing operation, scheduling, and allocation of radio resource allow workspace for the implementation of effective IoT antennas [12]. Antenna system for UWB and RFID tags in IoT empowered environment are given particular emphasis on techniques of MIMO, massive MIMO, position tolerance fabrication technique, and position variance effect in antenna arrays.

Among many conflicts that must be attended, choosing and incorporating appropriate aerial have a serious effect on the RF subsystem, its transmission range, and power requirements [13]. Antenna is generally needed to fit within device enclosure with low profile and careful positioning for assured performance. Anyway, a relevant antenna should offer high efficiency and sufficient bandwidth over required range of frequencies.

16.3.3 Current Trends in the Design of Antennas for IoT Applications

IoT is a heterogeneous network that may be connecting around 7 billion devices by 2025 using distinct wireless technologies and standards such as 2G, 3G, 4G, Bluetooth, and Wi-Fi as predicted in 2015. Recently available 4G LTE network will not be adequate and efficient to encounter the demands of multiple device connectivity namely high data stream, large bandwidth, low-latency quality of service (QoS), and low interference.

The internet is one of the most important and efficient communicating platforms in human history. It is excessively used in business, biomedical science, communication, education, government, health monitoring systems, industrial applications, and many more applications. For IoT applications, various sensors are placed at different nodes that are communicated through the internet and enable data transmission for data exchange and data storage.

Different sensors have different sizes and different physical dimensions. Thus, there is a need for miniaturized antennas to be integrated along with the sensor and without loading the device. Hence application-specific antennas are designed for individual sensors or devices. The antenna designs have challenges like miniaturization, compactness, and low power.

As a greater number of sensors and circuit devices are connected using IoT, it also requires power management for various sensors; in most industrial, environmental, and logistical applications, various sensor nodes are placed at remotely large distances and powering them using batteries or other sources is difficult and impractical. Most of the sensors operate on low power. Low power supply to sensors can be achieved using RF energy harvesting because sensors are surrounded by many RF sources such as GSM/LTE, Wi-Fi, WI-Max, WLAN, and other sources. In an RF harvesting system, the RF power is extracted using antennas then given to rectifying circuits followed by power combiners and delivered to load; rectification is not possible.

16.3.4 RF Energy Harvesting

The concept of energy harvesting is to harvest energy from radio frequency waves to power various wireless sensor networks at distant, remote locations. When sensors are placed at remote conditions, the various existing RF sources are unpredictable. Thus, a single RF tone energy harvesting is not suitable [14]. Consequently, a countless number of sub-frequency bands harvester is preferred, which can increase the output voltages. For example, a 4-RF band rectenna harvests energy from two GSM bands (900/1800), Wi-Fi (2.4 Ghz), and Universal Mobile Telecommunications Systems (UMTS) bands concurrently. It improves efficiency of RF to DC conversion which is twice compared to the single RF source. The same applies to energy harvesting system from ambient sources like solar and RF. This device can power sensors in communication system, which works at WLAN 2.4 GHz ISM band, powering a sensor through autonomous working of a low power operated circuit for a sensor operated wirelessly, while it encapsulates a good "cold start" capability [15]. The designed harvester unit consists of a two-port rectangular microstrip patch antenna with slots, a 3D printed package, an RF transceiver, a solar cell, RF-DC converter, a circuit to manage power (PMU), and a microcontroller unit. One more trend of RF harvesting is to drive a battery-less sensor. The rectifying antenna is designed to harvest energy from a Zigbee device that operates at 2.45 GHz microwave power. The RF energy harvesting has several applications such as wireless charging technology, RF identification (RFID), and so on.

16.3.5 Fractal Antennas

The quick development of the wireless communication system leads to the design of wideband and multiband miniaturized antennas. To obtain these small antennas, fractal antennas are a very attractive design technique. Fractal antennas have the following structural properties:

a. Self-similarity property: The shape of the antenna will be repeated a number of times with a successive decrease in dimensions.
b. Space-filling property: The antennas are miniaturized using Hilbert curves having smaller elements with their space-filling property. This property provides an electrically lengthy structure which fits in a tiny space and lends itself to miniaturization.
c. Fractals are mostly frequency-independent antennas.

The various popular fractals are the Koch Curve, the Koch snowflake structure, Koch fractal bow tie, and Sierpinski Carpet. Fractal antennas have several advantages such as being tiny in size, high impedance matching, multiband/wideband operation, reduced use of coupling, and diminished material coupling in fractal array antennas. Fractal antennas have limitations such as low gain, complex design, and more number of iterations.

16.3.6 Compact Antenna for IoT Applications

Rapid growth in the wireless technology demands reduced operating frequency without altering the antenna profile [16–17]. Real-time weather monitoring systems

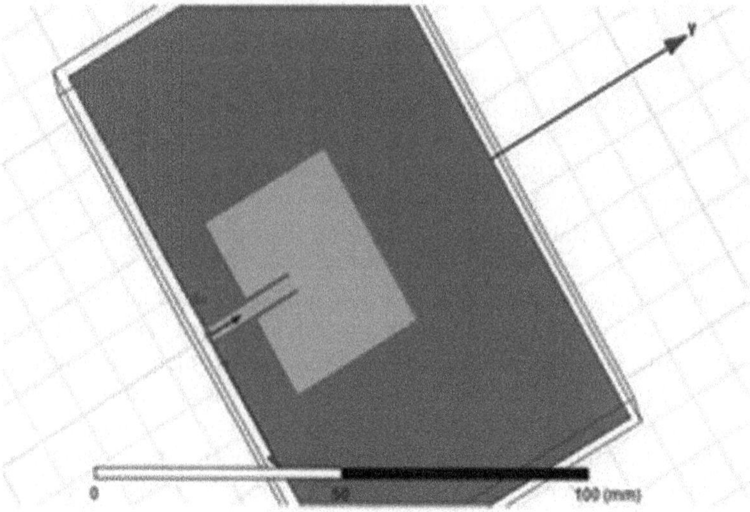

FIGURE 16.4 Rectangular patch microstrip antenna.

present in IoT spacecrafts/aircrafts are used for tracking turbulence, surrounding weather pressure deviations, and to conduct immediate action accordingly. For tracking the weather changes, GPS antenna tracker whose operating frequency is in LTE band is mostly used.

A compact microstrip antenna with rectangular shape patch and different shape slots are shown in Figures 16.4 and 16.5. The design of it uses etched slots on patch to lower the resonance frequency and for antenna miniaturization. The patch antenna frequency is in the range of S band. Distinct shape slots are tried on patch to switch frequency from S band to L band. They also result in the reduction of antenna dimensions. The design with distinct slot structures etched in the ground plane are shown in Figure 16.5.

Slots designed on the patch determine resonating frequency of antenna. Slots create some sort of discontinuity which has effects on length of current path causing an impact on the input impedance. As a result, multiple operating bands are obtained within the antenna system.

16.3.7 Multiband Microstrip Patch Antennas for IoT Applications

Many IoT applications demand tapered bandwidth. Practically bandwidth deviates from 10 MHz to 100 MHz. Some of the requisites of Internet of Things are omnidirectional radiation and flawless connectivity over a short period of time. The spectrum governing authorities have been issuing worldwide latest frequency bands at spectrum range from few sub GHz to several hundreds of GHz. Configured antenna for IoT would have to meet span of allocated spectrum. The microstrip patch antenna shown in Figure 16.6 utilizes the commonly available theory of antenna design with

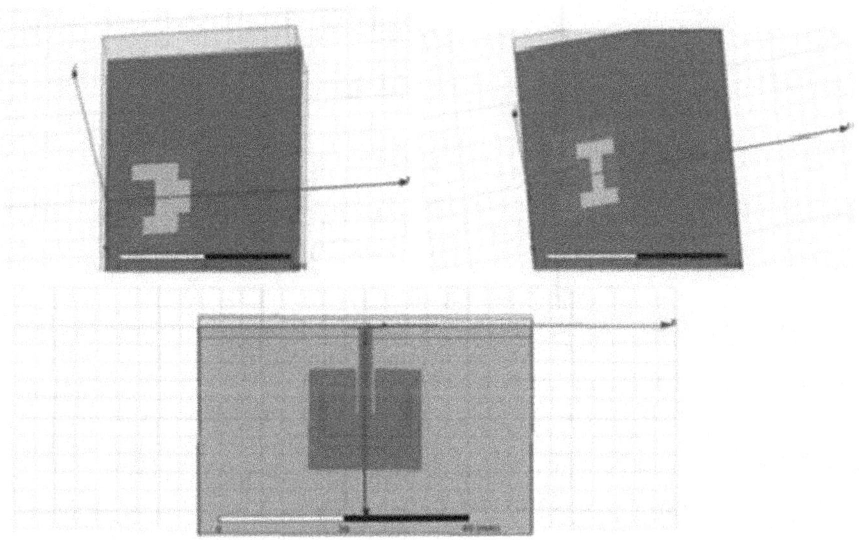

FIGURE 16.5 Structures with different-shaped slots.

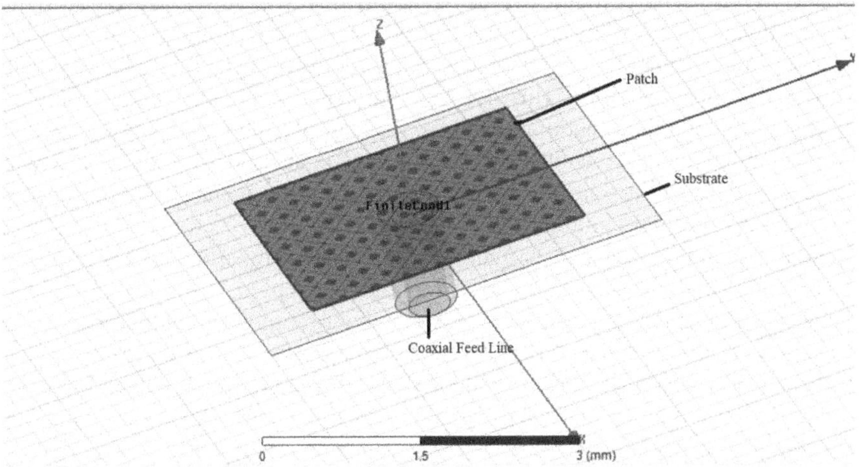

FIGURE 16.6 Simple planar microstrip patch antenna.

easily available simple co-planar composition [18–19]. It is suitable to propagate or catch the signals in the desired bands.

While designing the antenna, the important specification on which focus should be maintained is the desired narrow bandwidth. Omnidirectional radiation demand is accomplished by setting far-field radiations except at θ = 0° and θ = 180°.

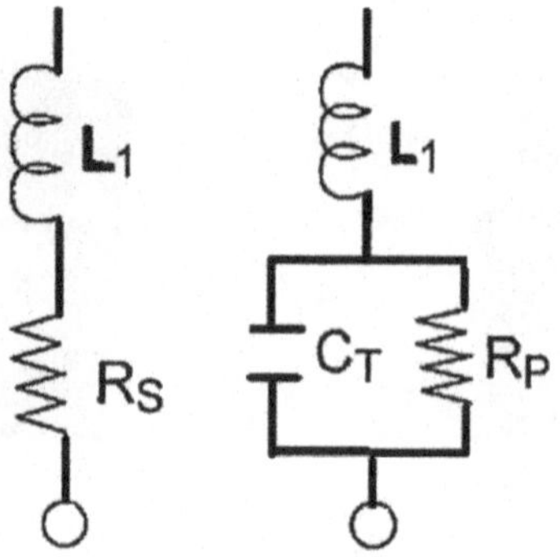

FIGURE 16.7 PIN diode equivalent circuit diagrams in ON and OFF states.

16.3.8 Multiband Frequency Reconfigurable Antennas (WLAN, Wi-Max, and C-Band IoT Applications)

The patch antennas are applied in remote accessing of radar, missiles, and aircraft with high fidelity and compact structure. The tremendous changes in the area of communication have increased the need for multipurpose and multi-application devices. As the electromagnetic spectrum is a limited source, radiator must be flexible as well as easily compliant to different practical conditions. Hence, reconfigurable antennas are highly applicable for advancing utility applications. IoT and Industrial Internet of Things (IIoT) are both top priorities for future stage modification and implementations. This develops the need for a wireless system, which can provide a common functional platform for multiple applications. An embedded system with frequency reconfiguration properties may lead to design analysis to a step ahead in utility enhancement. A compact size antenna, with frequency-reconfigurable property and circular polarization, deserves as an adequate system essential. Reconfiguration may be achieved in any of the properties such as frequency of operation, polarization, and directivity or radiation pattern. Reconfiguration in the frequency of antennas can reduce the requirement of large bandwidth and vast frequency spectrum allotment. Frequency reconfiguration can be obtained by varying the distribution of surface current of the radiating structure [20–23]. The frequency of the antenna can be reconfigured by using PIN diodes, RF-MEMS switches, and varactors (Figure 16.7).

The utility of the antenna becomes novel with reusable frequency applications and offers to server more applications by a single unit of antenna [24]. Among various available resources, PIN diode switching exhibits more stable response and better transient frequency shift. Technically, PIN diode is compatible for a large range of microwave frequency and satisfies the radiating possibilities of the structure.

16.4 CONCLUSION

The demand for microstrip patch antenna is rapidly increasing due to their many advantages. IoT is the buzzword used for short-range communication. Narrow bandwidth, low power, omnidirectional, and flawless connectivity over short time duration are key specifications of IoT. For IoT modules that use many wireless technologies, making space for antennas is becoming an increasingly notable challenge. Choosing the right antenna for an application imposes a key design challenge. There are various impressive new engineering techniques such as patch antenna using slots, fractal antennas, multiband antennas, and frequency reconfigurable antennas. These promise to alter antenna technology dramatically in the forthcoming years. These will resolve many issues and enable new antenna shapes and applications, which are not even possible with traditional antenna technology. The unique design approaches will address many of the challenges faced by the IoT applications.

REFERENCES

[1] IEEE Standard for Definitions of Terms for Antennas: IEEE Std 145-2013 (Revision of IEEE Std 145-1993), IEEE Standard, 20116.

[2] K. F. Lee and T. Kin-Fai, "Microstrip patch antennas – basic characteristics and some recent advances", *Proceedings of the IEEE*, vol. 100, no. 7, pp. 2169–2180, 2012.

[3] C. A. Balanis, *Antenna Theory: Analysis and Design*. Wiley-Interscience, UDA, 2005.

[4] C. A. Balanis, *Antenna Theory: Analysis and Design*. New York: John Wiley & Sons, Inc., 2016.

[5] S. Andreev, S. Balandin, and Y. Koucheryavy, *Internet of Things, Smart Spaces, and Next Generation Networks and Systems*. Springer, Germany 2015.

[6] S. C. Mukhopadhyay and N. K. Suryadevara, "Internet of Things: Challenges and opportunities", *Smart Sensors, Measurement and Instrumentation*, vol. 9, pp. 1–17, 2014.

[7] K. Yelamarthi, M. S. Aman, and A. Abdelgawad, "An application-driven modular IoT architecture", *Wireless Communications and Mobile Computing*, vol. 2017, Article ID 1350929, 16 pages, 2017.

[8] V. P. Kafle, Y. Fukushima, and H. Harai. "Internet of Things standardization in ITU and prospective networking technologies", *IEEE Communications Magazine*, vol. 54, pp. 43–49, 2016.

[9] V. K. Allam, B. T. P. Madhav, T. Anilkumar, and S. Maloji, "A novel reconfigurable band-pass filtering antenna for IoT communication applications", *Progress in Electromagnetics Research C*, vol. 96, pp. 13–26, 2019.

[10] A. Boukarkar, X. Lin, Y. Jiang, and X. Yang, "A compact frequency-reconfigurable 36-states patch antenna for wireless applications", *IEEE Antennas and Wireless Propagation Letters*, vol. 17, no. 7, pp. 1349–1353, July 2018.

[11] H. Tariq Chattha, N. Aftab, M. Akram, N. Sheriff, Y. Huang, and Q. Abbasi, "Frequency reconfigurable patch antenna with bias tee for wireless LAN applications", *IET Microwaves, Antennas Propagation*, vol. 12, no. 14, pp. 2248–2254, Nov. 2018.

[12] P. K. Bharti, H. Singh, G. Pandey, and M. Meshram, "Slot loaded microstrip antenna for GPS, Wi-Fi, and WiMAX applications survey", *International Journal of Microwaves Applications*, vol. 2, pp. 45–50, 2013.

[13] D. Wang, H. Liu, X. Ma, J. Wang, Y. Peng, and Y. Wu, "Energy harvesting for internet of things with heterogeneous users", *Wireless Communications and Mobile Computing*, vol. 2017, Article ID 1858532, 15 pages, 2017.

[14] K. Shafique, B. A. Khawaja, M. D. Khurram, S. M. Sibtain, Y. Siddiqui, M. Mustaqim, H. T. Chattha, and X. Yang. Energy harvesting using a low-cost rectenna for Internet of Things (IoT) applications. *IEEE Access* vol. 6, pp. 30932–30941, 2018.

[15] A. Azari, “A new super wideband fractal microstrip antenna”, *IEE Transactions on Antennas and Propagation*, vol. 59, pp. 1724–1727, 2011.
[16] A. Satheesh, R. Chandrababu, and S. Rao, “A compact antenna for IoT applications”, *IEEE Access*, 978-1-5090-3294-5/17/@2017.
[17] Y. Cai, K. Li, Y. Yin, S Gao., H. Wei, and L. Zhao, “A low-profile frequency reconfigurable grid-slotted patch antenna”, *IEEE Access*, vol. 6, pp. 36305–36312, Jun. 2018.
[18] Y. Zehforoosh and M. Rezvani, “A small quad-band monopole antenna with folded strip lines for Wi-MAX/WLAN and ITU applications”, *Journal of Microwaves, Optoelectronics and Electromagnetic Applications*, vol. 16, no. 4, pp. 1012–1018, Dec. 2017.
[19] J. C. Narayana Swamy and D. Seshachalam, “Multiband microstrip patch antenna for IoT applications”, *International Journal of Innovative Technology and Exploring Engineering (IJITEE)*, vol. 9, no. 1, pp. 2455–2457, November 2019.
[20] S. K. Sharma and G. Goswami, “Frequency reconfigurable multiband antenna for IoT applications in WLAN, Wi-Max, and C-Band”, *Progress in Electromagnetics Research C*, vol. 102, pp. 149–162, 2020.
[21] M. Borhani, P. Rezaei and A. Valizade, “Design of a reconfigurable miniaturized microstrip antenna for switchable multiband systems”, *IEEE Antennas and Wireless Propagation Letters*, vol. 15, p. 882, 2016.
[22] M. Mudda, “Target tracking system: PAN TILT”, *International Journal of Engineering and Advanced Technology (IJEAT)*, vol. 8, no. 5, June 2019.
[23] M. Mudda “Brain tumor classification using enhanced statistical texture features”, *IETE Journal of Research*, https://doi.org/10.1080/03772063.2020.177550, PP-1-12, ©2020 IETE.
[24] S. Mudda, “Frequency reconfigurable ultra-wide band MIMO antenna for 4G/5G portable devices applications: Review”, *International Journal on Emerging Technologies*, vol. 11, no. 3, pp. 486–490, 2020.

17 Design of a Microstrip Patch Antenna Based on Fractal Geometry for IoT Applications

N. Koteswaramma and P. A. Harsha Vardhini

17.1 INTRODUCTION

The Internet of Things (IoT) interlinks various digital, mechanical, and computing systems with unique identifiers without any interaction between human and human–computer where enormous amounts of data are transferred over the network. In the modern paradigm of wireless communication, thousands of smart wireless devices communicate with each other uninterruptedly and IoT plays an important role in these communication networks. The design of efficient hardware plays a major role in wireless applications in addition to sophisticated communication protocols. The antenna is an essential hardware module at the front end of every wireless communication module. The antennas that are designed to suite IoT applications systems must meet the specific requirements. A miniaturized antenna with energy efficient design having the ability to function in a multiplexing environment is preferred among various designs. In the proposed design, the antenna is simulated using CST Microwave Studio and design is verified in MIMO.

DOI: 10.1201/9781003093558-23

17.1.1 Importance of Antenna Design for IoT-Based Wireless Applications

The goals of effective antenna design, testing, and inclusion should be clear. A well-designed and tested/verified antenna leads to the success of the system by:

- Providing gain which increases the radiated signal while consuming no additional battery power
- Ensuring relatively consistent signal propagation so that communication is not lost due to movement
- Maximizing the signal level while minimizing the reception of noise through optimum location or optimum level positioning (particularly to minimize noise reception from other EMF devices)

As IoT modules continue to occupy most of the wireless communication applications, it a major requisite for adaption of scaled antenna design for communication modules.

17.1.2 Different Types of Antennas

In radio frameworks, extensive categories of antennas are applied with unique houses for particular packages. It may be organized in a one-of-a-kind way. The rundown gatherings of antennas underneath primary running standards, following the way antennas are ordered in the literature mentioned by researchers. The monopole, dipole, huge loop, and array antenna types typically function as resonant antennas; waves of voltage and current bounce back and forth between the ends creating standing waves alongside the elements.

17.1.3 Isotropic Antenna

The isotropic antenna is a theoretical antenna that transmits identical electricity of the signal in each path, regularly contrasted with a glowing light. It is a numerical version that is applied as the base of correlation for computing the directionality or gain of genuine receiving antennas. No real radio antenna can have an isotropic radiation pattern, but the isotropic radiation design serves as a reference for contrasting the radiation force of different receiving antennas, regardless of the type. It should not be mistaken for an "omnidirectional radio antenna"; the isotropic antenna transmits equal power in all three measurements, while the last emanates parallel power in every single even heading; however, the power transmitted differs with the rising edge, diminishing to zero on the receiving antenna's pivot point. Isotropic radio antennas can be built utilizing numerous small components and are used as reference reception apparatuses for testing different receiving antennas and field quality estimations, and for reinforcement receiving antennas on satellites which work without the satellite being focussed towards a communication station.

17.1.4 Dipole Antenna

The dipole antenna is one of the most common antennas. They are easy to make, most simple, and widely used in telecommunications. They are made to work either in horizontal or in vertical form. Radio frequency voltages are directly applied to the centre of this antenna, between the two conductors. They are especially used in rabbit-ear television antenna and have driven elements in some other antennas. The receiving twin emanates maximum ally heading opposite reception equipment's pivot point, giving it a chunk order gain of 2.15 dB. Although half of wave dipoles are applied as omnidirectional receiving antennas, they are a constructing square of several unique grade-by-grade muddled directional reception apparatuses.

17.1.5 Monopole Antenna

This incorporates a solitary conduit, for example, a metal pole, typically installed over the floor or a counterfeit directing floor. One side of the feed line from the receiver is connected with the conveyor. These radio waves of beginning at the earliest degree may likely originate the reception of image equipment beneath, therefore the receiving cord of monopole has a radiation layout indistinguishable great 50% to instance comparative receiving wire of dipole. Because the majority radiation of dipole is proportional gathered at 1/2-space area and reception apparatus may have two times the growth of a comparative dipole, it is no longer considered a misfortune inside the floor plane.

17.1.6 Array Antenna

This comprises numerous antennas working as a solitary receiving wire. Broadside clusters comprise numerous indistinguishable driven components, normally dipoles, encouraged in stage, emanating a bar opposite to the radio wire plane. Parasitic exhibits comprise different reception apparatuses, generally dipoles, with one driven component and the rest parasitic components, which transmit a bar along the line of the receiving wires.

17.2 MICROSTRIP PATCH ANTENNAS FOR IOT APPLICATIONS

The antenna that can work in multiband frequencies has been guided by the various frequency bands used in mobile and IoT as presented in Table 17.1. Though, there is a fact that the size reduction levels remain unsatisfactory to the expectation, a

TABLE 17.1
Operational frequencies of various IoT technologies

Application	IoT technology	Frequency
General (smart home, smart building)	Zigbee	915 MHz, 2.4 GHz
	Z-wave	2.4 GHz
	Bluetooth	2.4 GHz
	Wi-Fi	2.4 GHz, 3.6 GHz, 4.9 GHz, 5 GHz, and 5.9 GHz
	GPS	1575.42 MHz, 1227.6 MHz, 1176.45 MHz
IIoT	Wireless HART	2.4 GHz
	Isa 100.11a	2.4 GHz

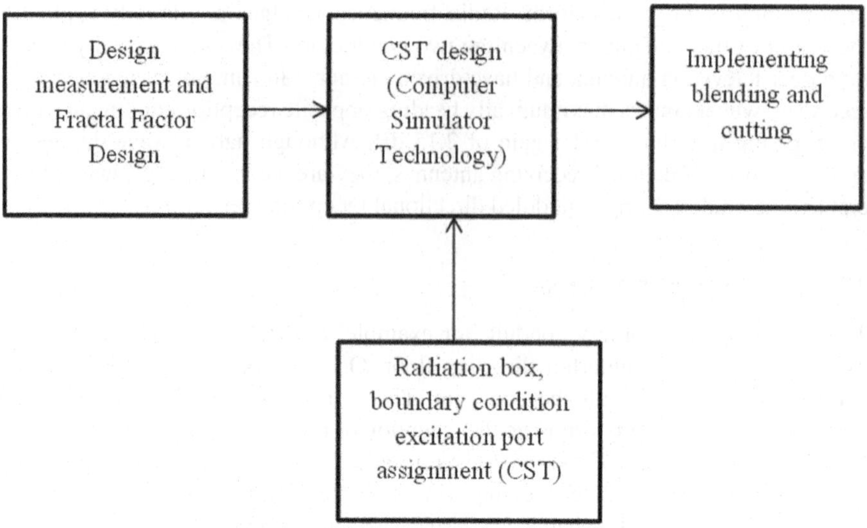

FIGURE 17.1 Proposed antenna design.

microstrip patch antenna is one of the antennas which can overcome this kind of demand due to its inherent capabilities such as low cost, less weight, low profile, and multiband support. Since space is very limited in recent IoT devices, a compact antenna is needed. Thus, the type of microstrip patch antenna design as depicted in Figure 17.1 is almost suitable for the purpose.

Many attempts to design small and compact antennas in the past have been endeavoured through slots, shorting, and folded geometries, but it was not until the introduction of fractals in antenna engineering that this could be done in a most efficient and sophisticated manner. With fractal antenna physical geometry, it facilitates miniaturization. Fractal geometry involves self-resonant radiators and electrical miniaturization. Minkowski fractal geometry, being the size of the conventional loop antenna, leads to compactness and miniaturization. With the same bandwidth and using tight packing of Minkowski Island loop antenna elements in the array, reduced mutual coupling was achieved.

17.2.1 Antenna Design

Numerous efforts have been made in the intervening decades to develop lightweight and portable antennas using slots, shorting, and folded configurations; it was not until the incorporation of fractals in antenna engineering that this could be achieved in the most productive and sophisticated way.

The ground plane is initially 100×100 mm^2 in size, from which the measurement of the circle opening is determined using lambda/4 where the receiving wire is formed. Two copper plates are taken here for two data sources with the aim of using MIMO design to separately accomplish the first focus and second loop as illustrated in Figure 17.2.

1 First Indentation Analysis

Feed is given to the copper plate prior to the fractal test. The fixed radio wire of dimension 25 × 25 mm^2 is taken for this process. Each side is then separated to equal proportion of 1/3, which gives each side three 8.33. Therefore, the space factor is 0.9 for the first loop and indentation width = 0.9 = 7.4977.

2 Second Indentation Analysis

In this method, the fixed reception apparatus with a dimension of 25 × 25 mm^2 is taken. Then each side is divided into a proportion of 1/3, giving each side three 8.33. The space factor for the first period is 0.9 and then each of the square-carved sides undergoes another focus in the stage again using the space factor of 0.5. The second space factor estimation is therefore given by indentation width = 0.5 × 2.499 = 1.2495. In addition, the focus phase continues to use this fractal plan.

In order to operate at several frequency bands, the existing structure should be reconfigured by modifying each strip as well as its ground plane. Slot antenna comprises a metal surface more often than not a level plate with a gap or space cut out.

Design parameters and relating values are specified in Table 17.2.

Step 1: For the design of microstrip antenna, the choice of frequency is the first step. In the proposed design, six frequency bands are chosen in the bandwidth from 1 GHz to 6 GHz.

Step 2: The next step in the design process is the measurement of the notch band.

Step 3: After the notch-band design, simulation is performed in an EM simulator.

Step 4: In the next step, iterations are performed to achieve the optimum design frequency and dimensions of the patch.

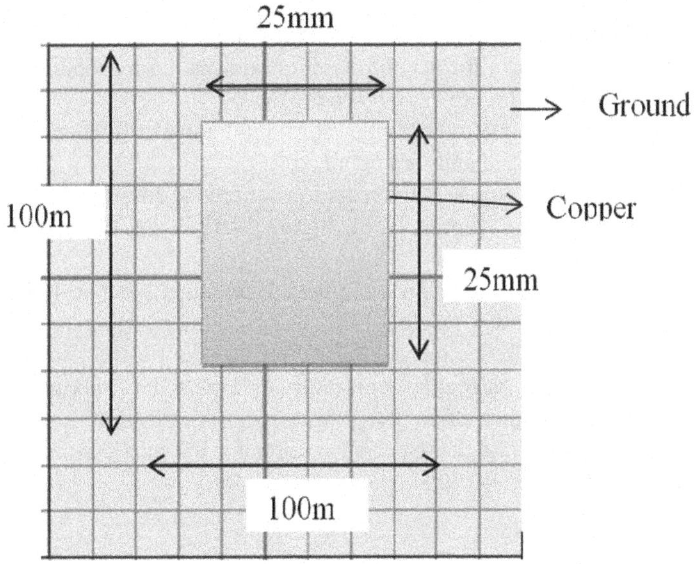

FIGURE 17.2 Initial copper plate.

TABLE 17.2
Design parameters

Parameter	Values
Ground plane size (L × W)	100 mm × 100 mm
Loop slot dimension (L0 × W0)	25 mm × 25 mm
Stub length (Ls)	3 mm
Slot width (Sw)	0.4 mm
Second space factor (i2)	0.5 mm
Dielectric slab dimensions (A × B × H)	33 mm × 30 mm × 5 mm

Step 5: The EM simulator chosen is CST Microwave Studio in which the tetrahedral patch is designed by choosing the optimum patch dimensions.
Step 6: Radiation patterns are drawn and the return loss graph is plotted in CST microwave studio.

Minkowski fractal configuration-based receiving wire and MIMO design for subjective radio applications are presented in Figures 17.3 and 17.4.

17.3 SIMULATION RESULTS OF MICROSTRIP ANTENNAS FOR IOT APPLICATIONS

In this research work, utilizing Minkowski fractal configuration-based receiving wire and MIMO design for subjective radio applications presents an increase of 3.5 dBi. Simulation is performed using CST Microwave Studio and radiation patterns are indicated in the below figures. The design of antenna in the first iteration and the corresponding output is depicted in Figure 17.5 and Figure 17.6, respectively.

The design of antenna in the second iteration and the corresponding output are shown in Figure 17.7 and Figure 17.8, respectively.

Figure 17.9 represents the final Minkowski fractal antenna frequency plot and it illustrates that six frequency bands are obtained.

Figures 17.10 to 17.13 represent the radiation patterns at different frequencies i.e., Figure 17.10, Figure 17.11, Figure 17.12, Figure 17.13 at 1 GHz, 2.61 GHz, 4.93 GHz, and 6 GHz, respectively.

Figure 17.14 and Figure 17.15 illustrate fractal LTE Tablet 2 and far-field of second table, respectively. Simulation results of the proposed microstrip patch antenna are tabulated in Table 17.3. Reflection coefficient/return loss (i.e., S_{11}) of −25 dBi at 2.4 GHz is achieved (i.e., very minimum return loss) and thus maximum power is transmitted exhibiting higher efficiency. With minimum reflection, high impedance matching is achieved that makes the design suitable for IoT applications.

17.4 CONCLUSION

Anticipation is undertaken in fluctuating multiband recurrence of the Minkowski shape expected reception system that offers an example of 11 classes of MIMO

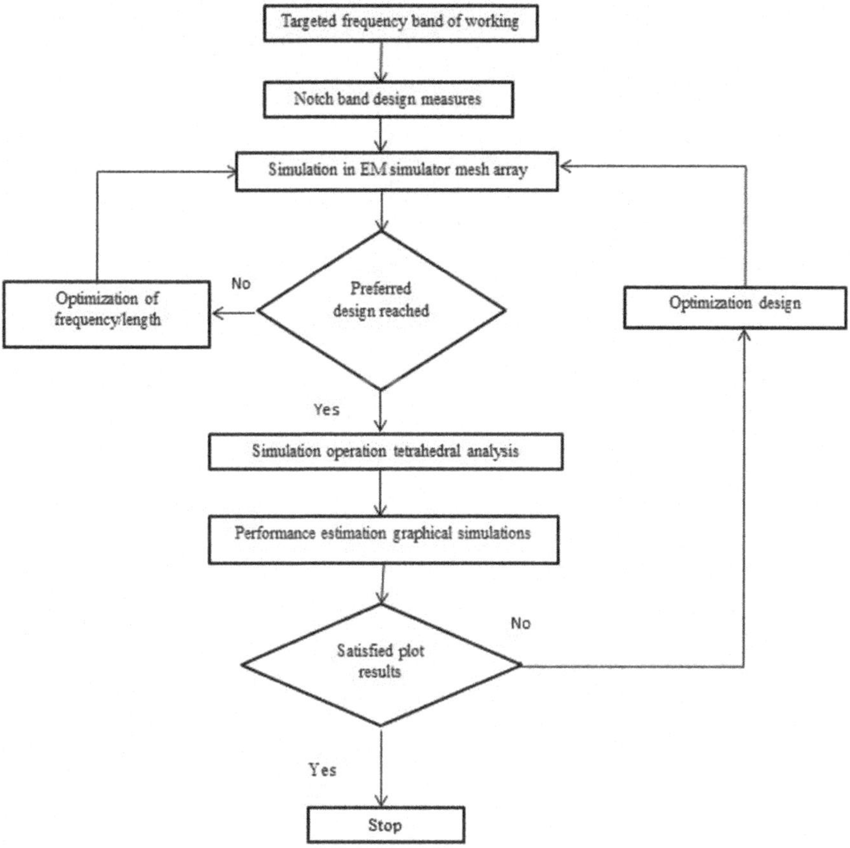

FIGURE 17.3 Design process.

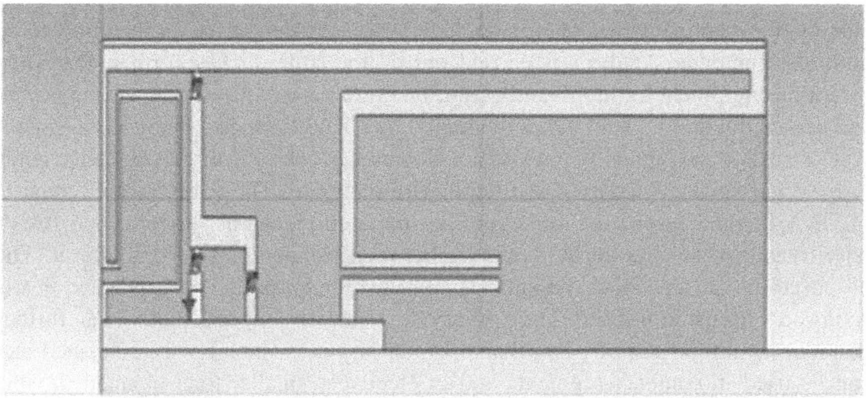

FIGURE 17.4 Lower band frequency antenna.

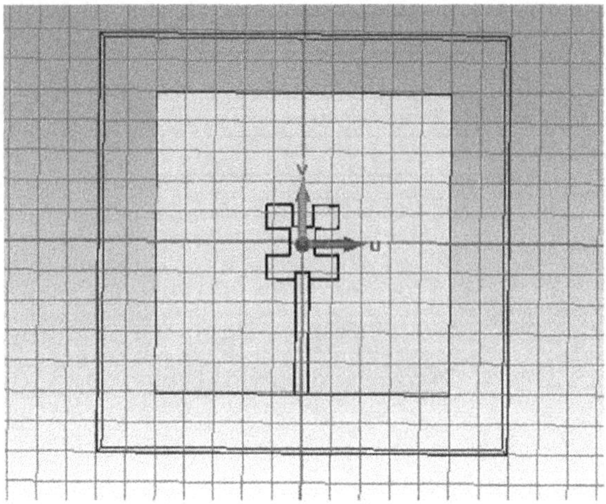

FIGURE 17.5 First iteration.

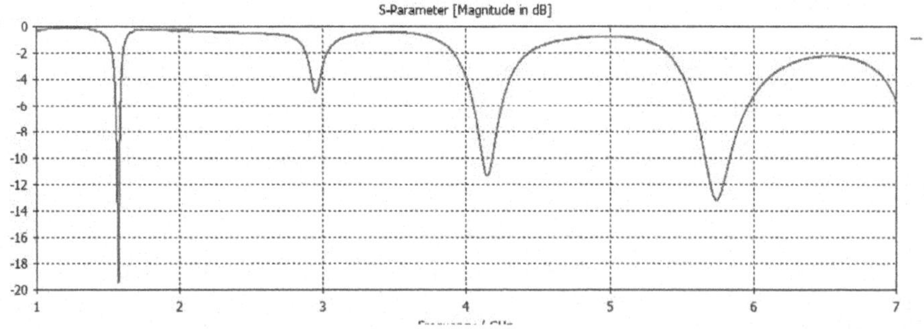

FIGURE 17.6 Frequency plot of first iteration.

manipulation and stuff resonator stacking. This work demonstrates the anticipated multiband antenna. Analysis is carried out on the stuff stacked form space circle reception equipment for multiband execution. Minkowski limit space cut is expected and depicted with FP4 substrate encouraged by CPW. To update the receiving wire-type criteria, steady quantity reviews are allocated. Consistent circuit evidence tends to have partner knowledge of the radio wire's service. The entire circuit consists of DRA lumped resonators, opening line-scattered resonator sections, and resistivity transformers to point the coupling between isolated parts of the circuit. The proposed imaginary model presents 10 dB reflection co-proficient bands and this is achieved with the multiband. The primary drawback of this undertaking can further be reduced in the future by less than 100 × 100 mm^2, which can be relocated and can be used in compact segments and in electively smaller than planned flexible applications.

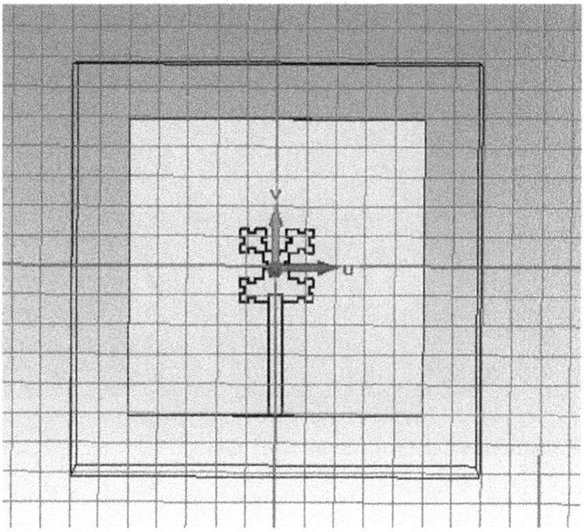

FIGURE 17.7 Second iteration.

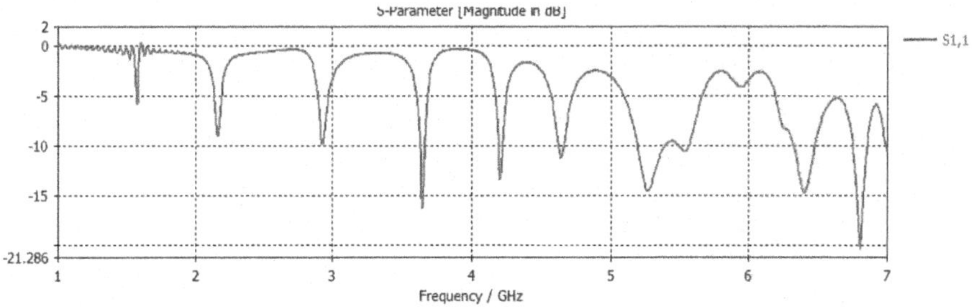

FIGURE 17.8 Frequency plot of second iteration.

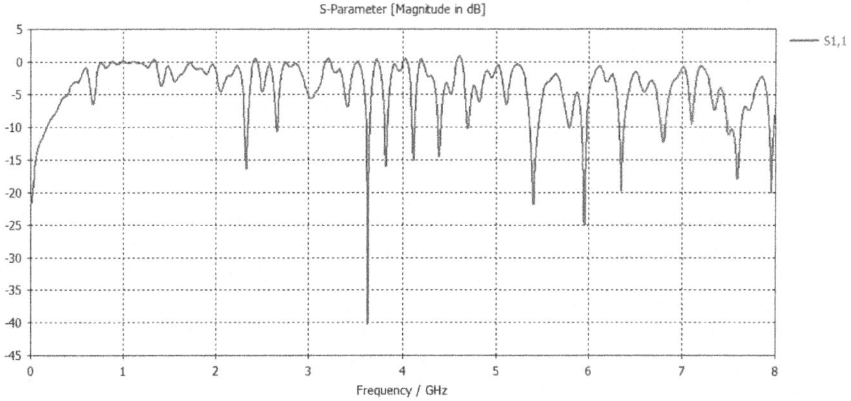

FIGURE 17.9 Final Minkowski fractal antenna MIMO frequency plot output S11.

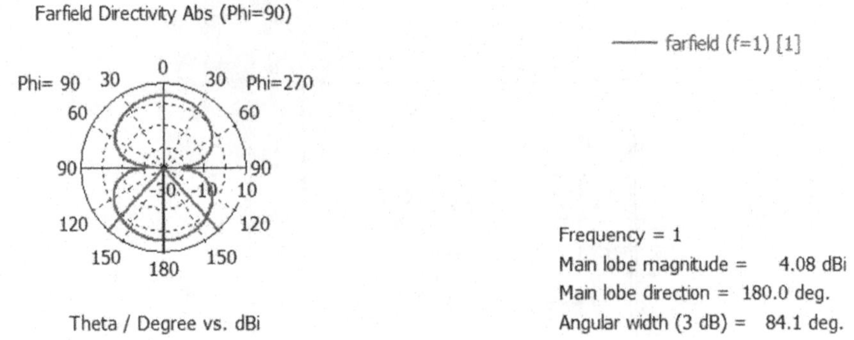

FIGURE 17.10 Radiation pattern for frequency 1 GHz.

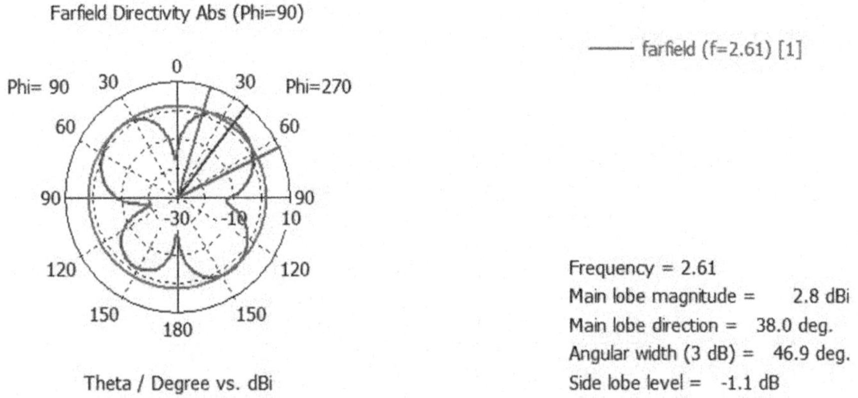

FIGURE 17.11 Radiation pattern for frequency 2.61 GHz.

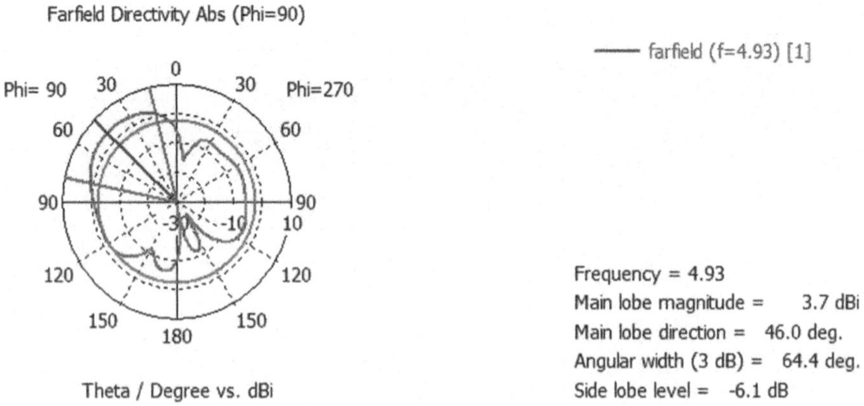

FIGURE 17.12 Radiation pattern for frequency 4.93 GHz.

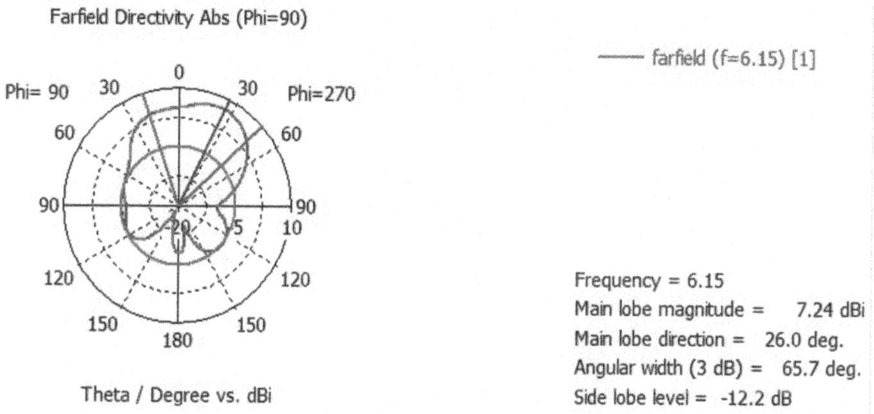

FIGURE 17.13 Radiation pattern for frequency 6 GHz.

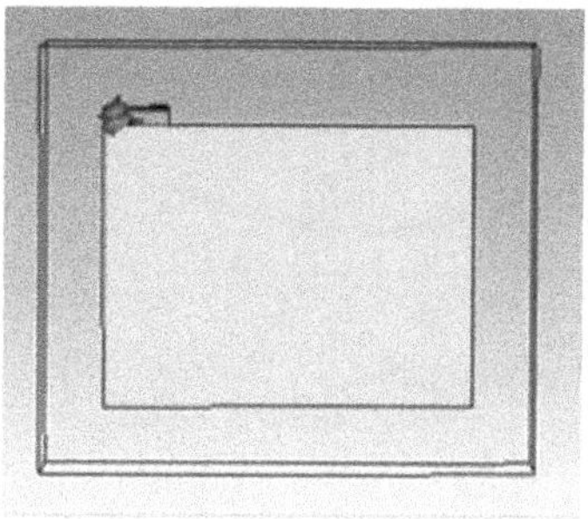

FIGURE 17.14 Fractal LTE Tablet 2.

TABLE 17.3
Simulation results

Frequency (GHz)	Main lobe magnitude (dBi)	Main lobe direction (deg)	Angular width (3 dB)	Side lobe level (dB)
1	4.8	180	46.9	−1.1
1.4	5.71	147	56.9	−1.2
2.61	2.8	38	46.9	−1.1

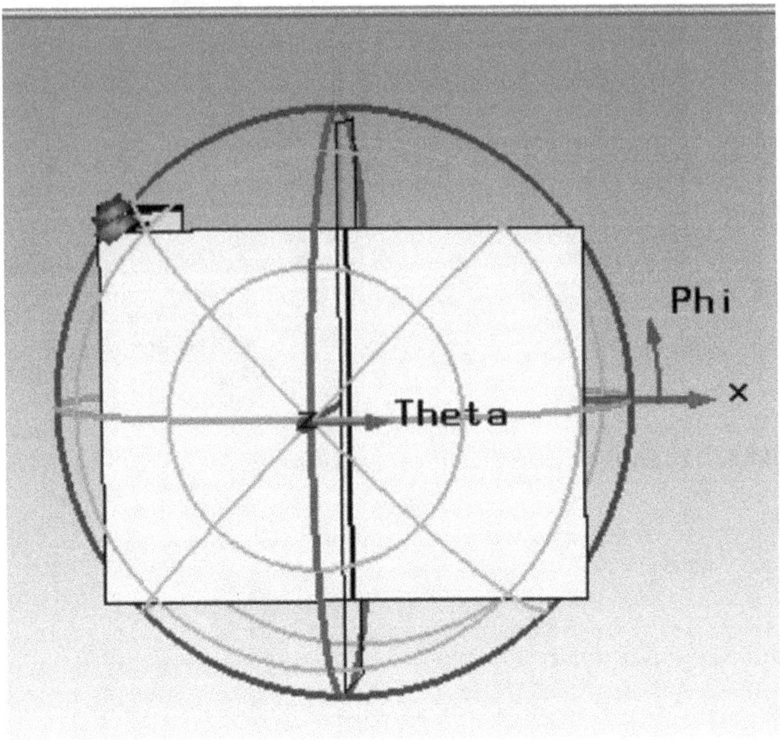

FIGURE 17.15 Far-field of second tablet.

Part VII

Ultra-Wideband Antenna Design for Wearable Applications

18 UWB and NB Performance Integrated Antennas for Cognitive Radio Applications

A Survey

G. Shine Let, G. Josemin Bala and C. Benin Pratap

18.1 INTRODUCTION

Cognitive radio technology utilizes the unused spectrum in the licensed band without interfering with users in the licensed spectrum band. Primarily, cognitive radio research is initiated to utilize the unused portion of the television band in the ultra-high frequency range of 470 MHz to 890 MHz. Since most of the communication systems are operating in the ultra-wideband frequency range, the research started to

DOI: 10.1201/9781003093558-25

focus on utilizing the ultra-wideband of 3 GHz to 10 GHz. The key problem in the design of cognitive radio antennas is that the same antenna has to detect the vacant spectrum and communicate in the available vacant spectrum. The designed cognitive radio antenna should have ultra-wideband spectrum sensing and narrowband communication competence. Various literature has suggested diverse antenna structures with different materials possessing one or many of the following design considerations: occupying less space, high spectrum efficiency, high radiation efficiency and high gain. This chapter provides a comprehensive survey of different reconfigurable and non-reconfigurable antennas designed for cognitive radio applications. It also acts as an eye-opener for the new researchers to have a quick start in this research area for a novel design. Nowadays all the communication networks operate based on a fixed spectrum access (FSA) policy. The Federal Communications Commission (FCC) is an independent regulatory agency that regulates the use of 9 KHz to 275 GHz under specific conditions to terrestrial communication, space radio communication and radio astronomy. As per the 2002 report by the FCC Spectrum Policy Task Force, most of the fixed allocated spectrum for specific usages are underutilized [1]. The maximum amount of usable data/information transmitted within a given spectrum (i.e. spectrum efficiency) is found to be less. In 1999, Joseph Mitola III introduced the concept of cognitive radio (CR) for mobile communication as provided in Ref. [2]. Using dynamic spectrum access mechanism, cognitive radio network can dynamically access the available spectrum band for communication without affecting the already existing users in that particular band. Therefore, by adapting the cognitive radio concept, the spectrum efficiency can be greatly increased. This requirement can be achieved by sensing the wide frequency band and transmitting the data in the available vacant narrow spectrum band. Thus, the design of antenna plays a major role in any communication system to transmit and receive signals. The transceiver with cognitive radio capability has to be supported with 'n' different bands for spectrum sensing and to select the unused band for communication [3]. Thus, the antenna designed for cognitive radio applications should have an ultra-wideband (UWB) antenna for wide-range spectrum sensing and a narrowband (NB) antenna for communication in the desired vacant band. In this regard for CR applications, many authors have proposed an integrated antenna with wideband and narrowband functionalities.

The basic functionalities of antenna used in interweave and underlay cognitive radio RF system (radio frequency system) is given in Ref. [4]. The cognitive radio engine has to self-configure the hardware based on the closed-loop cognitive cycle: observe, orient, decide, learn, and act. To have better spectrum efficiency, cognitive radio device operates in overlay, underlay, or interweave modes. Based on the cognitive radio operation, the device has to find the vacant spectrum (also called white space) in the licensed band and dynamically utilize that band without disturbing the already existing users in that band. In underlay mode, power constraint transmission can be done by unlicensed users.

A wide variety of UWB antennas and NB antennas have been designed and investigated in the literature by providing microstrip feeds or coplanar waveguide feeds to antennas. Most of the designed UWB antenna can sense the frequency range from 3.1 GHz to 10.6 GHz. This range overlaps with the existing communication systems such as Wireless Local Area Network (WLAN) from 5.15 GHz to 5.725 GHz, military X-band satellite communication downlink from 7.25 GHz to 7.745 GHz and uplink from 7.9 GHz to 8.4 GHz, worldwide inter-operability for microwave access (Wi-Max) from 3.3 GHz to 3.7 GHz, HiperLAN2, C-band, radio frequency identification (RFID) band and so on. To avoid interference with the existing communication systems, notch functionality is integrated with microstrip UWB antenna using slot structures, stub or resonators. The technique such as slot in the patch called a defected microstrip structure (DMS), slot in ground plane referred to as defected ground structure (DGS), metamaterial designs, addition of various shape radiating strips or incorporation of dielectric resonators, to render microstrip fed microstrip patch antenna to function in various wireless frequency bands. Based on the dimension and location of slots/stubs in the ground plane or radiating surface of the antenna, there is a change in the analyzing parameters such as operational bandwidth, return loss, size, etc. [5]. To make the antenna operate in different vacant spectrum bands or to switch from one frequency band to another, reconfigurability function is added to the antenna design. Reconfigurability function in an antenna can also vary the radiation field, current distribution and polarization behavior [6–8]. Using switches or non-switches reconfigurability function is incorporated into the antenna [9]. The three commonly used softwares for antenna design are Ansys-High Frequency Structure Simulator (HFSS), Computer Simulation Technology (CST) and Keysight Technologies-Advanced Design System (ADS). Figure 18.1 shows the design flow of cognitive radio application antenna.

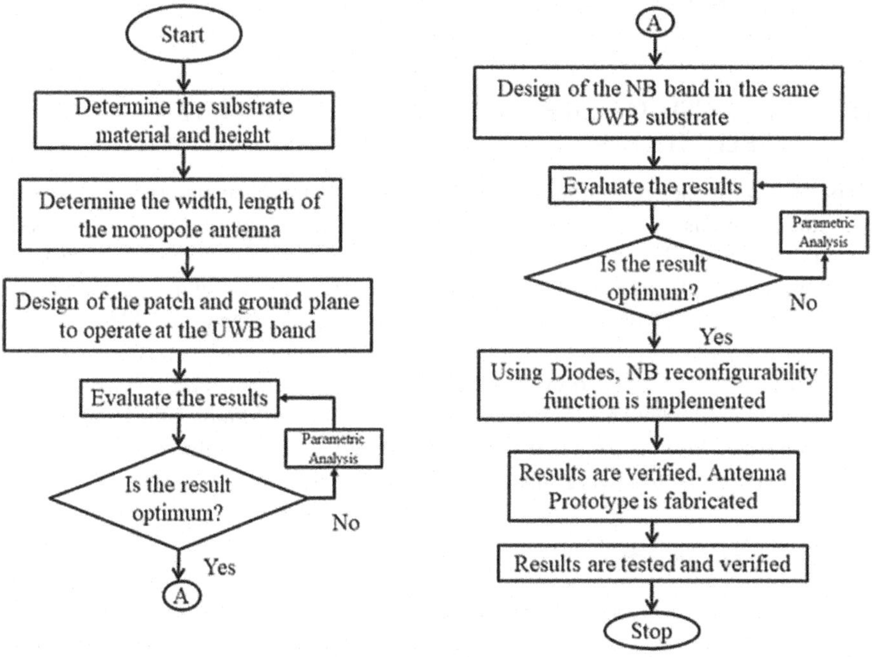

FIGURE 18.1 Cognitive radio application antenna design flow.

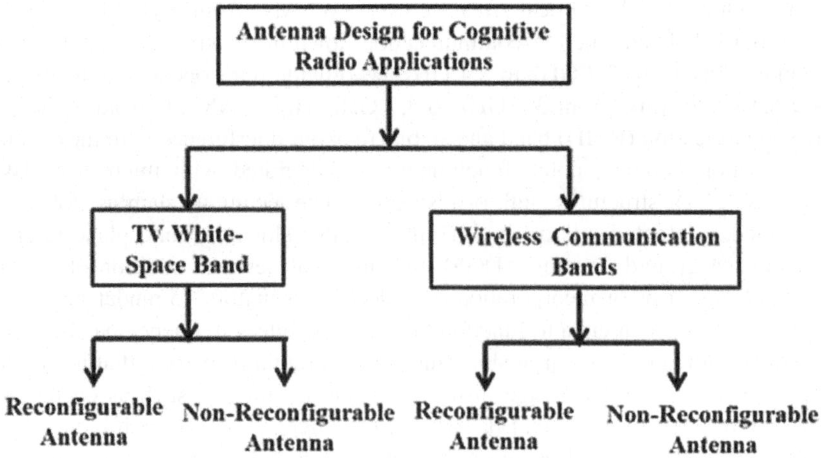

FIGURE 18.2 Taxonomy of CR application antenna design.

Cognitive radio research is initially initiated to utilize the unused portion of the television band in the ultra-high frequency range from 470 MHz to 890 MHz. Since most of the communication systems are operating in the ultra-wideband frequency range, the research started to focus on utilizing the ultra-wideband of 3.1 GHz to 10.6 GHz. Figure 18.2 shows the taxonomy of CR application antenna design. The following sections discuss about the various reconfigurable and non-reconfigurable antennas used for cognitive radio applications.

18.2 RECONFIGURABLE ANTENNA FOR COGNITIVE RADIO APPLICATIONS

To provide frequency reconfigurability, the antenna physical and electrical dimensions are altered by using switches. The switches act as a circuit breaker or circuit maker. The ON/OFF functioning of the switch changes the antenna characteristics/parameters and interconnect different antenna elements. Ideal switches, pin diodes, varactor diodes, FET devices, micro-electro mechanical system (MEMS) switches and optical switches are several types of switches used for antenna reconfigurability. The current flow is regulated in the desired direction using switches. The placement of switch, isolation between switches and insertion loss determines the performance of the antenna. Various switches considered for the reconfigurable antenna designs are discussed in the following subsections.

18.2.1 Ideal Switches

This does not dissipate or consume any input power. The ON state of an ideal switch resembles integrating a metal copper bridge. In OFF state, the integration has to be removed. The tuning of ideal switch is practically difficult [10].

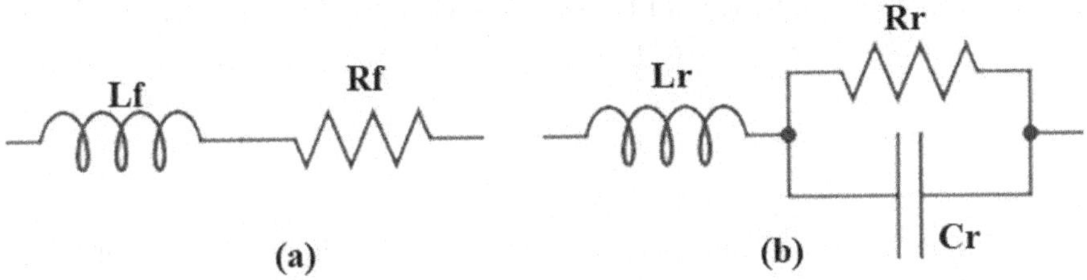

FIGURE 18.3 Pin diode equivalent circuit (a) forward bias (b) reverse bias.

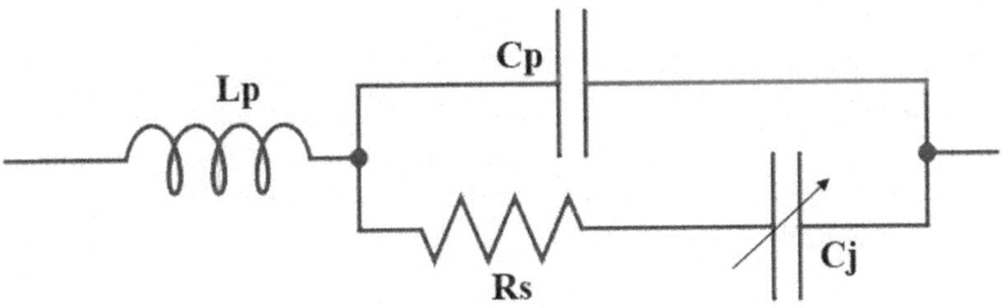

FIGURE 18.4 Equivalent circuit of varactor diode.

18.2.2 Pin Diodes

Most applications use pin diodes to provide reconfigurability function because of its following characteristics. This device is more reliable, compact in size, switching ON and OFF offers only small variation of resistance (1 Ω to 10 kΩ) and capacitance and switching speed is less than 100 ns. For a pin diode ON state, 10 mA bias current is required. The equivalent circuit of pin diode in forward bias and reverse bias is shown in Figure 18.3.

18.2.3 Varactor Diodes

It is also called a variable capacitance diode. The antenna can be tuned to various frequencies by varying the capacitance of the diode. This provides continuous change in the resonator length. Figure 18.4 shows the equivalent circuit of a varactor diode. The capacitance of the varactor diode varies for a definite range as per the applied bias [11]. Larger capacitance (equivalent to short circuit) makes the antenna to operate in the lowest operating frequency. Low capacitance in varactor diode (open circuit) tunes the antenna to operate in highest operating frequency. Broad frequency range operation is offered with low insertion loss.

18.2.4 FET Switches

FET device switching is controlled by the voltage signal and hence the device has zero DC power consumption. At high frequencies, due to drain-source capacitance effect, isolation gets degraded in FET switches. GaAs FET switches have very high switching speed, i.e. less than 10 ns [10].

18.2.5 MEMS Switches

This is a mechanical switch. Compared to other switches, insertion loss is very less (<0.2 dB) and power consumption is in microwatts. The main disadvantage of MEMS switch is that it has a slower switching speed. However, losses will be high at microwave and millimeter wave frequencies.

18.2.6 Optical Switches

Based on the wavelength of the light, optical switches become ON and OFF. An optical switch does not require any bias lines, so there is no interference in the radiation of the antenna [12, 13]. Laser diode activates the optical switch through optical fiber [13]. Good electrical and thermal isolation is provided between antenna and optical switch control circuitry.

18.2.7 Comparison of Different Switches

Pin diodes are commonly used for various switching purposes because of its high speed switching time and low power consumption [14]. Table 18.1 provides the comparison between various switches used in the antenna design.

Most of the constructed cognitive radio antennas take single-port excitation into account. In recent days, based on the demand of high data rate and to use multiple-input multiple-output (MIMO) technology, multi-port excitation is considered. Sections 18.3 and 18.4 provide different antenna designs for single-port and multi-port excitation, respectively.

18.3 SINGLE-PORT RECONFIGURABLE ANTENNA

Figure 18.5 shows the simple structure of a single-port reconfigurable antenna having a stub or slot in the design. The radiating patch structure or a ground plane structure is changed based on the frequency of operation and the author's interest. Table 18.2 shows the comparison of various single-port reconfigurable antenna used for CR applications.

Reconfigurability is achieved by integrating four ideal switches and the antenna can operate in eight different modes. These triple bands are notched 4.2 GHz to 6.2 GHz, 6.6 GHz to 7.0 GHz and 12.2 GHz to 14 GHz from the designed UWB frequency band 3.1 GHz to 14 GHz. In Ref. [14], the authors have proposed reconfigurability among the notch bands 3.2 GHz to 3.65 GHz, 4.8 GHz to 6 GHz and 8 GHz to 8.3 GHz using three pin diodes. The notch band functionality is obtained by inserting two slots and an L-shaped stub on the ground plane. The tuning range of the designed UWB dielectric resonator antenna is 82.75%. Usually, tunability of the antenna can be measured using various metrics such as tuning range, impedance bandwidth,

TABLE 18.1
Comparison of switches used in antenna design

Attributes switch	Power handling	Switching time/speed	Control current	Power consumption (mW)	Frequency range	Isolation	Sensitive to
Pin diode	>1 kW	<100 ns/Fast	Up to 100 mA	5–100	From MHz to GHz (Not DC)	Good at high frequencies	Temperature
FET switch	<10 W	<10 ns/ Average	<100 μA	~5	From DC	Good at low frequencies	Temperature
MEMS switch	–	<200 μs/ Slow	–	0.05–0.1	From DC	Good across broad frequencies	Vibrations
Optical switch	–	<9 μs/ Average	Up to 87 mA	0–50	From DC		Light intensity

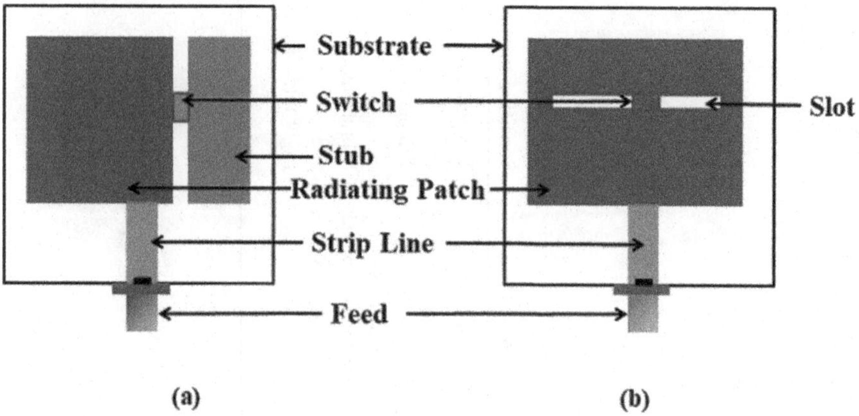

FIGURE 18.5 Single-port reconfigurable antenna (a) stub structure (b) slot structure.

bandwidth ratio, etc. [15]. Tuning range (*TR*) is one of the commonly used metric and is defined by Equation (18.1) based on the antenna operation at S_{11} minima.

$$TR(\%) = \frac{2\times(f_h - f_l)}{(f_h + f_l)} \times 100 \tag{18.1}$$

where f_h and f_l are the highest and lowest resonant frequencies, respectively. Impedance bandwidth (*BW*) of the designed antenna is given by Equation (18.2) [5].

$$BW = \frac{f_h - f_l}{f_c} \times 100 \tag{18.2}$$

where f_c is the center frequency in which antenna is tuned. Impedance bandwidth is also referred as fractional bandwidth. For any designed UWB antenna, the bandwidth ratio should be high and is given by Equation (18.3).

$$\text{Bandwidth ratio} = \frac{f_h}{f_l} : 1 \tag{18.3}$$

The monopole antenna is integrated with two double split ring resonators to provide dual band notches in the frequency bands 3.3 GHz to 3.7 GHz and 5.4 GHz to 5.7 GHz. Using two pin diodes, the reconfigurability is achieved between two notch bands [16]. The reconfigurability between notch bands 5.725 GHz to 5.825 GHz and 8.05 GHz to 8.4 GHz is achieved by changing the length of the resonators in the designed UWB antenna in the frequency range of 3.05 GHz to 12.15 GHz. The lengths of the resonators are controlled by using two ideal switches [17]. The authors have designed this antenna to work effectively for underlay cognitive radio

TABLE 18.2
Single-port reconfigurable antenna for CR applications

References	Substrate	Antenna dimension (mm^3)	UWB band of operation (GHz)	Tuning element	Reconfigurable frequency band (GHz)	Simulation software used
[3]	FR4 substrate $\varepsilon_r = 4.3$, $\gamma = 0.0018$	68 × 51 × 1.6	2.63 to 3.7	Pin diodes	Four sub-bands between 2.63 and 3.7	CST
[5]	FR4 substrate $\varepsilon_r = 4.4$, $\gamma = 0.02$	50 × 50 × 1.6	–	Pin diodes	2.46–3.90	HFSS
[10]	FR4 substrate $\varepsilon_r = 4.3$	50 × 50 × 1.52	2 to 10	GaAs FET	2.1–2.6, 3.6–4.6, 2.8–3.4 and 4.9–5.8	–
[14]	Rogers RT6010LM $\varepsilon r = 10.2$, mounted on an FR4 substrate $\varepsilon r = 4.4$	40 × 40 × 7.3	3 to 11	Pin diodes	3.2–3.65, 4.8–6 and 8–8.3	HFSS
[16]	Rogers RT/ Duroid 5880 $\varepsilon_r = 2.2$, $\gamma = 0.0009$	–	2.58 to 15.5	Pin diodes	3.3–3.7 and 5.4–5.7	CST
[17]	RO3003 substrate $\varepsilon_r = 3$, $\gamma = 0.0013$	36 × 30 × 0.762	3.05 to 12.15	Ideal Switches	5.725–5.825 and 8.05–8.4	CST
[18]	FR4 substrate $\varepsilon_r = 4.4$, $\gamma = 0.018$	–	3 to 14.5	Pin diodes	5–6, 5.8–6.4 and 8–8.4	HFSS
[20]	Foam substrate $\varepsilon_r = 1$, Rogers Duroid 5880 substrate $\varepsilon_r = 2.2$	100 × 120 × 11.574	2 to 3.2	RF MEMS switch	2.0–2.6 and 2.6–3.2	HFSS
[21]	RO3003 substrate, $\varepsilon_r = 3$, $\gamma = 0.0013$	40 × 40 × 0.762	–	Varactor diode	4.26–5.94, 3.85–5.58	CST Microwave Studio

(continued)

TABLE 18.2 (Continued)
Single-port reconfigurable antenna for CR applications

References	Substrate	Antenna dimension (mm^3)	UWB band of operation (GHz)	Tuning element	Reconfigurable frequency band (GHz)	Simulation software used
[22]	Rogers RT5880 $\varepsilon_r = 2.2$, $\gamma = 0.0004$	47 × 52 × 17.787	2.8 to 11	Varactor diodes	4.3–5.3	HFSS
[23]	FR4 substrate $\varepsilon_r = 4.4$	80 × 60 × 1.6	1.35 to 6.2	Pin diodes and Varactor diodes	2.55–3.2	–
[24]	FR4 substrate $\varepsilon_r = 4.4$, $\gamma = 0.02$	21 × 9 × 0.8	2.8 to 12	Pin diodes	3.2–4.5, 4.3–7.8, 7.4–11.2	HFSS
[25]	Rogers RT5880 $\varepsilon_r = 2.2$, $\gamma = 0.0004$	28 × 26 × 0.8	3.1 to 10.6	Pin diodes	5.15–5.725 and 7.25–7.745	CST
[26]	FR4 substrate $\varepsilon_r = 4.4$	26 × 23 × 1.6	3.14 to 12	Pin diodes	3.12–4.1, 3.6–6.9, 6.24–7.7, 8.15–10.3	CST
[28]	FR4 substrate $\varepsilon_r = 4.4$, $\gamma = 0.02$	40 × 40 × 1.6	3.13 to 12	MEMS	3.54–6.09, 4.81–6.46, 18.44–12, 6.35–11	HFSS

application [17]. For operating in underlay and overlay cognitive radio approach, a monopole antenna with C-shaped and L-shaped parasitic structures in the ground plane is suggested in Ref. [18]. To provide reconfigurability between three notch bands 5 GHz to 6 GHz (WLAN), 5.8 GHz to 6.4 GHz (ITS) and 8 GHz to 8.4 GHz (ITU) three pin diodes are used.

The authors have proposed a coaxial feed reconfigurable microstrip patch antenna for the frequency band of 0.8 GHz to 3.0 GHz. The reconfigurability is achieved by using seven pin diodes at different center frequencies 900 MHz, 1.57 GHz, 1.84 GHz and 2.45 GHz [19]. To provide reconfigurability, stub structures are used in the UWB monopole antenna [10]. The connection or disconnection of four stubs with the microstrip feed line is controlled by two GaAs FET switches. This antenna can provide reconfigurability in the following bands 2.1 GHz to 2.6 GHz, 3.6 GHz to 4.6 GHz, 2.8 GHz to 3.4 GHz and 4.9 GHz to 5.8 GHz. About 12 pin diodes are used to switch between four sub-bands in the frequency range 2.63 GHz to 3.7 GHz [3].

Using two RF MEMs switches in the E-shaped patch, frequency reconfigurability is achieved between two bands 2.0 GHz to 2.6 GHz and 2.6 GHz to 3.2 GHz [20]. Based on the particle swarm optimization algorithm, seven design variables, representing length and width of the E-shape patch, are optimized and used in the antenna design. Using six pin diodes at the perimeter of the triangular slot, the antenna is used to tune 12 different frequencies, ranging from 2.46 GHz to 3.90 GHz [5]. The length of the triangular slot is controlled by ON/OFF of the pin diodes. The radiation efficiency of the antenna at different frequency of operation varies from 76.27% to 88.58%. The authors in Ref. [21] have used T-shaped and H-shaped bandpass filters (resonators) connected on the stripline made the UWB antenna to operate in the 4.26 GHz to 5.94 GHz and 3.85 GHz to 5.58 GHz, respectively. The other end of T-shaped and H-shaped bandpass filters is connected to varactor diode. The operating frequency of the narrowband can be varied by varying the capacitance of the varactor diode [21, 22]. Nachouane, Najid, Tribak and Riouch proposed a tunable and reconfigurable quad band antenna using two varactor diodes and two pin diodes as switches [23]. In Ref. [24], the length and width of the rectangular slot is changed based on the state of five pin diodes attached to the antenna. This antenna can operate in three communication bands as the diode switching state changes.

The authors in Ref. [25] designed a compact antenna with dimension $28 \times 26 \times 0.8$ mm^3 for underlay cognitive radio applications. The antenna was able to operate in the UWB band and by using two pairs of pin diodes reconfigurability can be achieved between two notch bands 5.15 GHz to 5.725 GHz (WLAN) and 7.25 GHz to 7.745 GHz (military X-band satellite communication). The authors in Ref. [26] also designed a compact antenna with half-elliptical stub for tuning the UWB band. Reconfigurability between four narrowbands is achieved by using eight pin diodes. Using MEMS switch, the reconfigurability is done between two frequency bands 3.5 GHz and 5.5 GHz, which can be used for cognitive radio applications [27]. In Ref. [28], the reconfigurability is achieved between five NB communication using ON/OFF state of six pin diodes.

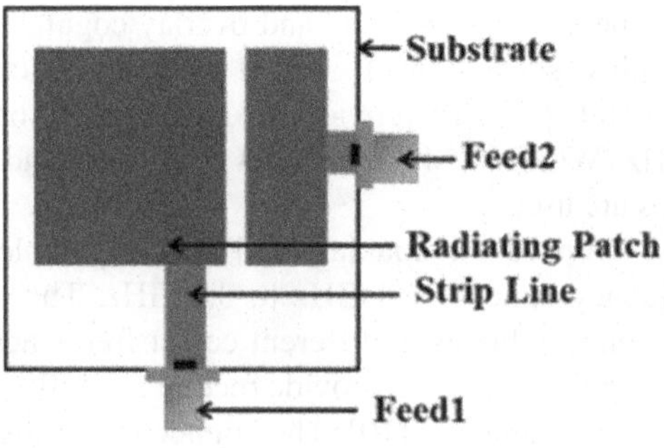

FIGURE 18.6 Dual-port antenna structure.

18.4 MULTI-PORT RECONFIGURABLE ANTENNA

Using single substrate and switchable dual ports, UWB antenna for spectrum sensing and reconfigurability in NB for communication is suggested in much of the literature [29, 30]. Figure 18.6 shows a simple dual-port antenna structure.

Table 18.3 shows the various multi-port reconfigurable antenna for CR applications. For CR applications, a dual-port UWB antenna was designed in Ref. [29]. One port was used for UWB spectrum sensing and the other port for NB communication. The reconfigurability was achieved by integrating two pin diodes within the antenna. Based on the different ON/OFF configuration of diodes, the antenna could operate in three different frequency ranges with a tuning range (TR) of 80.34%. The authors in Ref. [30] have designed a U-shape monopole antenna in one side of the substrate to operate in the UWB frequency range 2 GHz to 12 GHz. In the other side of the substrate, slot and stub structures were provided meticulously to make the antenna operate in six narrowbands. The reconfigurability of the NB was achieved by using five pin diodes. A half-elliptical patch designed in one side of the substrate operated as a UWB sensing antenna in the frequency range of 2.8 GHz to 11.6 GHz [31]. To provide reconfigurability in the NB communication, a varactor diode was used. As the capacitance of the varactor diode was varied, NB communication can be made at different bands between 5 GHz and 6 GHz. Combination of two pin diodes and two varactor diodes are used to provide reconfigurability among different bands in the frequency range of 0.573 GHz to 2.55 GHz [32]. To provide fine tuning in the lower frequency bands, varactor diodes are used and pin diodes are used for reconfigurability. The same antenna provides sensing in the UWB band of 0.72 GHz to 3.4 GHz [32].

Using four pin diodes and two ports, switching between narrowband and wideband was achieved in Ref. [33]. In a single metal layer, open annular ring-based topology is considered for the antenna design. The reconfigurability in narrowband is achieved by illuminating two photoconductive silicon switches using laser diodes [12]. When both the switches are in OFF condition, the NB operating frequency range is 4.15 GHz to 5.1 GHz. When the two switches are simultaneously ON, the operating ranges are 3.2 GHz to 4.3 GHz and 4.8 GHz to 5.7 GHz. In Ref. [34], a slotted elliptical monopole antenna was designed for UWB sensing, and reconfigurability in NB communication

TABLE 18.3
Multi-port reconfigurable antenna for CR applications

References	Substrate	Antenna dimension (mm^3)	UWB band of operation (GHz)	Tuning element	Reconfigurable frequency band (GHz)	Simulation software used
[12]	Taconic TLY substrate, $\varepsilon_r = 2.2$	50 × 45.5 × 1.6	3 to 11	Laser diodes integrated with photoconductive silicon switches	3.2–4.3, 4.8–5.7	HFSS
[29]	FR4 substrate, $\varepsilon_r = 4.4$, $\gamma = 0.02$	65 × 40 × 6.5	3 to 11	Pin diodes	3.25–3.7, 5.5–6.5 and 8–8.4	HFSS
[30]	-	60 × 60 × 1.6	2 to 12	Pin diodes	5.65, 3.6, 5, 2.94, 4.5 and 2.15	HFSS
[31]	FR4 substrate $\varepsilon_r = 4.3$, $\gamma = 0.025$	36 × 40 × 1.6	2.8 to 11.6	Varactor diode	Different bands between 5 and 6	CST
[32]	FR4 substrate $\varepsilon_r = 4.4$	65 × 120 × 1.56	0.72 to 3.4	Pin diode and varactor diode	Different bands between 0.573 and 2.55	HFSS
[33]	Rogers RT/Duroid 6035 HTC $\varepsilon_r = 3.6$, $\gamma = 0.0013$	80 × 80 × 1.52	3.4 to 8.0	Pin diodes	4.7–5.4	CST Microwave Studio
[34]	Rogers Duroid 5880 substrate $\varepsilon_r = 2.23$	50 × 30 × 0.8	3.1 to 10.6	Pin diodes	4.25–6, 6.6–7.35, 4.25–7.15, 5–6	CST
[35]	FR4 substrate $\varepsilon_r = 4.3$, $\gamma = 0.02$	80 × 65 × 1.6	2 to 5.5	Inductors	2.8	CST and HFSS
[36]	FR4 substrate $\varepsilon_r = 4.4$, $\gamma = 0.02$	75.6 × 66 × 1.6	1 to 12	Pin diodes	2.1, 2.96, 3.5, 5	HFSS
[37]	Rogers TMM6 laminate $\varepsilon_r = 6$, $\gamma = 0.0037$	70 × 70.5 × 0.762	2.6 to 11	Varactor diode	4.9–6.2	HFSS
[38]	Rogers Duroid 5880 substrate $\varepsilon_r = 2.2$	70 × 50 × 1.6	2 to 10	Rotational motion	3.4–5.56, 2.1–3, 3–3.4, 6.3–10, 5.4–6.2	HFSS
[39]	Taconic TLY and Rogers Duroid 5880 substrate $\varepsilon_r = 2.2$	50 × 4 5.5 × 1.6	3 to 11	n-type silicon switch	4.15–5.1, 4.8–5.7, 3.2–4.3	HFSS

was achieved using two pin diodes in triangular monopole antenna. In Ref. [35], to provide reconfigurability two inductors were incorporated in the antenna design. As the inductor value changed from 5.6 nH to 10 nH, the antenna could be tuned between 2.6 GHz and 2.8 GHz communication band. Sonia Sharma and Chandra presented a three pin diode configuration to provide reconfigurability between four NB communication bands [36–37].

18.4.1 Reconfigurability without Using Switches

To provide reconfigurability, various types of switches are used. Bias lines are required to activate/deactivate the switches. However, the use of a bias line affects the electromagnetic structure of the antenna and increases the complexity in antenna design [38]. Without using switches, the antenna radiation characteristics can be altered by rotating or moving the antenna parts. The rotational motion is provided by using actuator or stepper motor or FPGA or microcontroller [38, 39]. The actuator and other devices used for rotational motion consume more power than semiconductor switches used for reconfigurability. In Ref. [39], the reconfigurability was achieved by the rotation of antenna patch by 180°. The antenna was able to tune to 5.3 GHz to 18.15 GHz and 3.4 GHz to 4.85 GHz frequencies at two different positions. Between 3 GHz and 6 GHz, optical switches were used to provide reconfigurability. The authors in Ref. [38] designed an oval-shaped patch to operate in the UWB and five different-shaped patches to operate in five different narrowbands. The reconfigurability among NB was achieved by rotational motion. The five different patches were rotated by using a stepper motor. The rotational motion changed the feed given to different patches.

18.5 NON-RECONFIGURABLE ANTENNAS FOR COGNITIVE RADIO APPLICATIONS

This section compares the suggestions provided by multiple researchers on non-reconfigurable antennas as shown in Table 18.4. The antenna system built in a single substratum can function in various frequency bands, but in a non-reconfigurable antenna it is not feasible to switch the band.

TABLE 18.4
Non-reconfigurable antennas for CR applications

References	Substrate	Antenna dimensions (mm^3)	UWB band of operation (GHz)	Simulation software used
[41]	Roger substrate $\varepsilon_r = 2.5$, $\gamma = 0.0015$	35 × 52.2 × 1.57	1.7–10.6	HFSS
[42]	FR4 substrate $\varepsilon_r = 4.4$, $\gamma = 0.02$	40 × 40 × 1.6	3.0–12	HFSS

18.5.1 Single-Port Non-Reconfigurable Antennas

A very high frequency octagonal loop UWB antenna with a rectangular patch was designed for spectrum sensing at the frequency band 25 GHz to 45 GHz, 50 GHz to 60 GHz and 60 GHz to 80 GHz [40]. The antenna performance analysis was carried out with Advanced Design System (ADS) software.

18.5.2 Multi-Port Non-Reconfigurable Antennas

To increase the system capacity, the multiple-input multiple-output (MIMO) concept is introduced. Hence, MIMO antenna design has become the research frontier. The authors in Ref. [41] have designed a four port MIMO integrated antenna for CR applications. The UWB and NB functionalities can be performed by the same antenna with total substrate area as $0.18 * \lambda_0^2$, where λ_0 is the highest operating wavelength. The antenna can operate at different NB frequencies such as 5.1 GHz to 5.5 GHz, 6.6 GHz to 7.2 GHz and 18.7 GHz to 10.2 GHz. The triband operation is obtained by exciting cylindrical dielectric resonator antenna with orthogonal T-feed microstrip line. Envelope correlation coefficient, diversity gain and mean effective gain parameters are evaluated to analyze the performance of MIMO antenna. In Ref. [42], the authors have also proposed a dual-port UWB and NB functionalities antenna in the same substrate. The designed antenna can operate in the UWB range 3.0 GHz to 12 GHz and the NB frequency ranges are 4.93 GHz to 5.4 GHz, 5.9 GHz to 6.7 GHz and 18.28 GHz to 10.2 GHz. The UWB antenna is fed through coplanar waveguide and microstrip feed line is given to NB antenna. Dielectric resonator antenna (DRA) made of eccostock dielectric material (HIK) with $\varepsilon_r = 10$, $\gamma = 0.003$ is used for NB communication. This integrated antenna ensures good isolation below −10 dB throughout the operating band.

18.6 TV WHITE SPACE FOR COGNITIVE RADIO APPLICATIONS

In 2008, FCC issued an order to utilize the unutilized television (TV) band below 900 MHz. Ultra-high frequency (UHF) range of 470 MHz to 890 MHz has been allocated for TV broadcast in many countries. As per FCC, the underutilized UHF band can be utilized for wireless communication. Many authors have designed antennas to utilize the vacant spectrum in the UHF band. Table 18.5 shows the comparative study of various antenna designs used to utilize TV white space (TVWS). To utilize TVWS, a strip profile based on an omnidirectional antenna is designed in UHF range 460 MHz to 870 MHz with the overall antenna dimensions of $261 \times 30 \times 0.8$ mm^3 [43].

To have wide bandwidth operation, a defected ground structure (DGS) is used in Ref. [44]. The size of the antenna is comparatively well reduced and simple rectangular patches are used as the radiating element. The two U-shaped arms, which behave as capacitance, make the antenna operate in a wide bandwidth of 414 MHz [45]. Effective power radiation for the entire bandwidth is ensured using characteristics mode analysis. By considering the dimensions of the smartphone, a compact antenna of dimension $59 \times 115 \times 1$ mm^3 is designed to utilize the TV UHF in Ref. [46]. The two trapezoidal sections joined with a rectangular section result

TABLE 18.5
Antenna designs used to utilize TVWS

References	Substrate	Antenna dimension (mm³)	UWB band of operation (MHz)	Simulation software used
[43]	FR4 substrate $\varepsilon_r = 4.4$, $\gamma = 0.02$	261 × 30 × 0.8	460–870	–
[44]	FR4 substrate $\varepsilon_r = 4.4$, $\gamma = 0.02$	128 × 64 × 1.6	623–860	HFSS
[45]	FR4 substrate $\varepsilon_r = 4.3$	231 × 35 × 0.8	470–884	CST
[46]	FR4 substrate	59 × 115 ×x 1	490–740	–
[47]	FR4 substrate $\varepsilon_r = 4.4$, $\gamma = 0.02$	198.32 × 111.11 ×x 1.6	420–1040	CST
[48]	FR4 substrate $\varepsilon_r = 4.4$, $\gamma = 0.02$	198.9 × 150 × 1.6	470–700	CST
[49]	FR4 substrate $\varepsilon_r = 4.3$, $\gamma = 0.019$	280 × 305 × 1.52	470–790	–
[50]	FR4 substrate $\varepsilon_r = 4.4$	105 × 40 × 0.8	600–800	HFSS

in an M-shaped monopole antenna capable of operating in wide frequency range of 420 MHz to 1040 MHz [47]. The same authors in Ref. [47] designed a planar trapezoidal antenna to utilize the TVWS for customer premise equipment [48]. The antenna with high gain is designed using rectangular slots in the log-periodic fractal Koch antenna [49].

18.6.1 Reconfigurable Antennas in TVWS

The authors have designed a 1 mm thin spiral monopole antenna [50-51]. A tunable inductor is inserted in the spiral monopole. As the inductor value changes from 1 nH, 5 nH, 10.5 nH and 27 nH, the antenna is tuned to 750 MHz, 670 MHz, 682 MHz and 720 MHz, respectively with bandwidth between 15 MHz and 20 MHz.

18.7 CONCLUSION

The spectrum efficiency can be improved and the vacant spectrum can be effectively utilized without any interference using cognitive radio technology. This chapter provides a comprehensive survey of antenna design used for cognitive radio applications. Different single-port and multi-port antennas are designed to have cognitive radio functionality. Switches are incorporated in the antenna design to provide reconfigurability between different narrowbands. Furthermore, this chapter provides great insights to budding researchers in the field of CR antenna design.

REFERENCES

[1] "Federal Communications Commission Spectrum Policy Task Force," 2002.

[2] J. Mitola III, "Cognitive radio for flexible mobile multimedia communications," in *1999 IEEE International Workshop on Mobile Multimedia Communications (MoMuC'99)*, 1999, pp. 3–10.

[3] A. Mansoul, F. Ghanem, M. R. Hamid, and M. Trabelsi, "A selective frequency-reconfigurable antenna for cognitive radio applications," *IEEE Antennas Wirel. Propag. Lett.*, vol. 13, pp. 515–518, 2014.

[4] Y. Tawk, J. Costantine, and C. G. Christodoulou, "Cognitive-radio and antenna functionalities: A tutorial," *IEEE Antennas Propag. Mag.*, vol. 56, no. 1, pp. 231–243, 2014.

[5] R. Kumar and R. Vijay, "Frequency agile quadrilateral patch and slot based optimal antenna design for cognitive radio system," *Int. J. RF Microw. Comput. Eng.*, vol. 28, no. 2, pp. 352–356, 2017.

[6] J. Costantine, Y. Tawk, S. E. Barbin, and C. G. Christodoulou, "Reconfigurable antennas: Design and applications," *Proc. IEEE*, vol. 103, no. 3, pp. 424–437, 2015.

[7] M. Kamran Shereen, M. I. Khattak, and G. Witjaksono, "A brief review of frequency, radiation pattern, polarization, and compound reconfigurable antennas for 5G applications," *J. Comput. Electron.*, vol. 18, no. 3, pp. 1065–1102, 2019.

[8] M. S. Shakhirul, M. Jusoh, Y. S. Lee, and C. R. Nurol Husna, "A review of reconfigurable frequency switching technique on microstrip antenna," *J. Phys. Conf. Ser.*, vol. 1019, no. 1, pp. 1–7, 2018.

[9] N. O. Parchin, H. J. Basherlou, Y. I. A. Al-Yasir, A. M. Abdulkhaleq, and R. A. Abd-Alhameed, "Reconfigurable antennas: Switching techniques – a survey," *Electron.*, vol. 9, no. 2, p. 336, 2020.

[10] T. Aboufoul, A. Alomainy, and C. Parini, "Reconfiguring UWB monopole antenna for cognitive radio applications using GaAs FET switches," *IEEE Antennas Wirel. Propag. Lett.*, vol. 11, pp. 392–394, 2012.

[11] S. Ramya and M. Maheswari, "Design of Frequency Reconfigurable Antenna using Varactor Diode," in *International Conference on Engineering Innovations and Solutions Design*, 2016, pp. 136–139.

[12] Y. Tawk, J. Costantine, S. Hemmady, G. Balakrishnan, K. Avery, and C. G. Christodoulou, "Demonstration of a cognitive radio front end using an optically pumped reconfigurable antenna system (OPRAS)," *IEEE Trans. Antennas Propag.*, vol. 60, no. 2, pp. 1075–1083, 2012.

[13] Y. Tawk, A. R. Albrecht, S. Hemmady, G. Balakrishnan, and C. G. Christodoulou, "Optically pumped frequency reconfigurable antenna design," *IEEE Antennas Wirel. Propag. Lett.*, vol. 9, pp. 280–283, 2010.

[14] M. Abioghli, A. Keshtkar, M. Naser-Moghadasi, and B. Ghalamkari, "A frequency reconfigurable band-notched UWB dielectric resonator antenna with a wide-tuning range for cognitive radio systems," *IETE J. Res.*, pp. 1–10, 2018.

[15] S. Rao and N. Llombart, "An overview of tuning techniques for frequency-agile antennas," *IEEE Antennas Propag. Mag.*, vol. 54, no. 5, pp. 271–296, 2012.

[16] A. Alhegazi, Z. Zakaria, N. A. Shairi, I. M. Ibrahim, and S. Ahmed, "A novel reconfigurable UWB filtering-antenna with dual sharp band notches using double split ring resonators," *Prog. Electromagn. Res. C*, vol. 79, pp. 185–198, 2017.

[17] H. A. Atallah, A. B. Abdel-Rahman, K. Yoshitomi, and R. K. Pokharel, "Reconfigurable band-notched slot antenna using short circuited quarter wavelength microstrip resonators," *Prog. Electromagn. Res.*, vol. 68, no. October, pp. 119–127, 2016.

[18] N. Ojaroudi, N. Ghadimi, Y. Ojaroudi, and S. Ojaroudi, "A novel design of microstrip antenna with reconfigurable band rejection for cognitive radio applications," *Microw. Opt. Technol. Lett.*, vol. 56, no. 12, pp. 2998–3003, 2014.

[19] S. Genovesi, A. Monorchio, and G. Manara, "Microstrip reconfigurable antenna for cognitive radio systems," in *International Symposium on Electromagnetic Theory*, 2013, pp. 429–431.

[20] H. Rajagopalan, J. M. Kovitz, and Y. Rahmat-Samii, "MEMS reconfigurable optimized E-shaped patch antenna design for cognitive radio," *IEEE Trans. Antennas Propag.*, vol. 62, no. 3, pp. 1056–1064, 2014.

[21] H. A. Atallah, A. B. Abdel-Rahman, K. Yoshitomi, and R. K. Pokharel, "Compact frequency reconfigurable filtennas using varactor loaded T-shaped and H-shaped resonators for cognitive radio applications," *IET Microwaves, Antennas Propag.*, vol. 10, no. 9, pp. 991–1001, 2016.

[22] S. Keerthipriya and C. Saha, "Tunable multifunctional reconfigurable step profiled dielectric resonator antenna for cognitive radio applications," in *Proceedings of the 2019 TEQIP – III Sponsored International Conference on Microwave Integrated Circuits, Photonics and Wireless Networks, IMICPW 2019*, 2019, pp. 440–442.

[23] H. Nachouane, A. Najid, A. Tribak, and F. Riouch, "Reconfigurable and tunable filtenna for cognitive LTE femtocell base stations," *Int. J. Microw. Sci. Technol.*, vol. 2016, 2016.

[24] G. Srivastava, A. Mohan, and A. Chakrabarty, "Compact reconfigurable UWB slot antenna for cognitive radio applications," *IEEE Antennas Wirel. Propag. Lett.*, vol. 16, no. c, pp. 1139–1142, 2017, doi: 10.1109/LAWP.2016.2624736.

[25] A. S. A. Falih Mahdi Alnahwi, A. Abdalhameed, and H. L. Swadi, "A compact wide-slot UWB antenna with reconfigurable and sharp dual band notches for underlay cognitive radio applications," *Turkish J. Electr. Eng. Comput. Sci.*, vol. 27, pp. 11–13, 2018.

[26] S. Koley, D. Bepari, and L. Murmu, "Compact coplanar waveguide-fed ultra wide band and band switchable slot antenna for cognitive radio front-end system," *J. Commun. Technol. Electron.*, vol. 63, no. 12, pp. 1373–1378, 2018.

[27] T. Singh, K. A. Ali, H. Chaudhary, D. R. Phalswal, V. Gahlaut, and P. K. Singh, "Design and analysis of reconfigurable microstrip antenna for cognitive radio applications," *Wirel. Pers. Commun.*, vol. 98, no. 2, pp. 2163–2185, 2018.

[28] R. M. C. Cleetus and G. J. Bala, "Wide-narrow switchable bands microstrip antenna for cognitive radios," *Prog. Electromagn. Res. C*, vol. 98, no. January, pp. 225–238, 2020.

[29] M. Abioghli, A. Keshtkar, M. Naser-Moghadasi, and B. Ghalamkari, "UWB rectangular DRA integrated with reconfigurable narrowband antenna for cognitive radio applications," *IETE J. Res.*, vol. 67, no. 1, pp. 139–147, 2018, doi: 10.1080/03772063.2018.1530075.

[30] S. Sharma and C. C. Tripathi, "An integrated frequency reconfigurable antenna for cognitive radio application," *Radioengineering*, vol. 26, no. 3, pp. 746–754, 2017.

[31] F. M. Alnahwi, A. A. Abdulhameed, A. S. Abdullah, and S. Member, "A compact integrated UWB / reconfigurable microstrip antenna for interweave cognitive radio applications," *Int. J. Commun. Antenna Propag.*, vol. 8, pp. 81–86, 2018.

[32] R. Hussain and M. S. Sharawi, "A cognitive radio reconfigurable MIMO and sensing antenna system," *IEEE Antennas Wirel. Propag. Lett.*, vol. 14, pp. 257–260, 2015.
[33] B. P. Chacko, G. Augustin, and T. A. Denidni, "Electronically reconfigurable uniplanar antenna with polarization diversity for cognitive radio applications," *IEEE Antennas Wirel. Propag. Lett.*, vol. 14, pp. 213–216, 2015.
[34] M. Y. Li et al., "Eight-port orthogonally dual-polarized antenna array for 5G smartphone applications," *IEEE Trans. Antennas Propag.*, vol. 64, no. 9, pp. 3820–3830, 2016.
[35] H. Nachouane, A. Najid, A. Tribak, and F. Riouch, "Dual port antenna combining sensing and communication tasks for cognitive radio," *Int. J. Electron. Telecommun.*, vol. 62, no. 2, pp. 121–127, 2016.
[36] S. Sharma and C. C. Tripathi, "A wide spectrum sensing and frequency reconfigurable antenna for cognitive radio," *Prog. Electromagn. Res. C*, vol. 67, no. July, pp. 11–20, 2016.
[37] G. Augustin and T. A. Denidni, "An integrated ultra wideband / narrow band antenna in uniplanar configuration for cognitive radio systems," *IEEE Trans. Antennas Propag.*, vol. 60, no. 11, pp. 5479–5484, 2012.
[38] Y. Tawk, J. Costantine, K. Avery, and C. G. Christodoulou, "Implementation of a cognitive radio front-end using rotatable controlled reconfigurable antennas," *IEEE Trans. Antennas Propag.*, vol. 59, no. 5, pp. 1773–1778, 2011.
[39] Y. Tawk, M. Bkassiny, G. El-Howayek, S. K. Jayaweera, K. Avery, and C. G. Christodoulou, "Reconfigurable front-end antennas for cognitive radio applications," *IET Microwaves, Antennas Propag.*, vol. 5, no. 8, pp. 985–992, 2011.
[40] S. S. S. Pandian and C. D. Suriyakala, "A new UWB tri-band antenna for cognitive radio," *Procedia Technol.*, vol. 6, pp. 747–753, 2012.
[41] S. Pahadsingh and S. Sahu, "Four port MIMO integrated antenna system with DRA for cognitive radio platforms," *AEU – Int. J. Electron. Commun.*, vol. 92, pp. 98–110, 2018.
[42] S. Pahadsingh and S. Sahu, "Planar UWB integrated with multi narrowband cylindrical dielectric resonator antenna for cognitive radio application," *AEU – Int. J. Electron. Commun.*, vol. 74, pp. 150–157, 2017.
[43] N. Wang, Y. Gao, and Q. Zeng, "Compact wideband omnidirectional UHF antenna for TV white space cognitive radio application," *AEU – Int. J. Electron. Commun.*, vol. 74, pp. 158–162, 2017.
[44] S. Gite and D. Niture, "A compact printed monopole antenna for TV white space applications using defected ground structure," in *IEEE International Conference on Information Processing, ICIP 2015*, 2015, pp. 198–200.
[45] Q. Zhang, Y. Gao, and C. G. Parini, "Compact U-shape ultra-wideband antenna with characteristic mode analysis for TV white space communications," in *2016 IEEE Antennas and Propagation Society International Symposium, Proceedings*, 2016, pp. 17–18.
[46] J. Bauer and M. Schuhler, "Compact wideband antenna for TV white spaces," in *8th European Conference on Antennas and Propagation*, 2014, vol. 1, pp. 2894–2896.
[47] A. K. Singh and M. Ranjan Tripathy, "M shaped ultra wide band monopole antenna for TVWS CPE," in *2018 5th International Conference on Signal Processing and Integrated Networks (SPIN)*, 2018, pp. 240–243.
[48] A. K. Singh and M. R. Tripathy, "Planar trapezoidal monopole antenna for TVWS CPE," in *2018 3rd International Conference for Convergence in Technology, I2CT 2018*, 2018, pp. 6–8.

[49] N. A. A. Rahman, M. F. Jamlos, H. Lago, M. A. Jamlos, P. J. Soh, and A. A. Al-Hadi, "Reduced size of slotted-fractal Koch log-periodic antenna for 802.11af TVWS application," *Microw. Opt. Technol. Lett.*, vol. 57, no. 12, pp. 2732–2736, 2015.

[50] M. Y. Abou Shahine, M. Al-Husseini, Y. Nasser, K. Y. Kabalan, and A. El-Hajj, "A reconfigurable miniaturized spiral monopole antenna for TV white spaces," in *PIERS Proceedings*, 2013, pp. 1026–1029.

[51] Y. L. Ban, C. Li, C. Y. D. Sim, G. Wu, and K. L. Wong, "4G/5G multiple antennas for future multi-mode smartphone applications," *IEEE Access*, vol. 4, pp. 2981–2988, 2016.

19 Ultra-Wideband Wearable Vivaldi Antennas for Biomedical Applications

K. Lalitha

19.1 INTRODUCTION

This chapter presents a Vivaldi antenna for wearable applications in the field of biomedical imaging. The structure of this proposed antenna is designed on the FR4 substrate (dielectric constant = 4.3) having a thickness of 0.018 mm with a Vivaldi structure for high gain and high bandwidth applications. The antenna is designed and optimized using computer simulation software. The simulated results give an S11 value of less than −10 dB over the ultra-wideband region. The structure of this proposed antenna structure can also be employed for imaging hidden or shielded objects. This Vivaldi ultra-wideband antenna produces non-ionized radiation that makes them suitable for wearable healthcare applications. The telemedicine industry and developments in radio wave communications have been working together to provide better medical facilities to people in the remote areas at an affordable cost. Wearable antennas with smaller sizes are generally employed for the transmission of patient's data to a medical expert through the IoT device for early diagnosis and continuous monitoring. The antenna requirement for wearable technology is quite different. In general, the wearable antenna interacts with human body to provide remote monitoring. The UWB frequency spectrum lies in the frequency range of 3.1 to 10.6 GHz, as described by the Federal Communications Commission (FCC). The UWB antennas for wearable applications should be simple and have compact design, low profile, and high-speed data transfer rate for short-range communication with low power consumption. Vivaldi antenna is the best solution for satisfying the aforementioned needs of the UWB antenna. This is a linearly polarized antenna and capable

DOI: 10.1201/9781003093558-26

of providing wide bandwidth and high gain [1]. There are three types of tapered slot antennas: (i) Linearly Tapered Slot Antenna (LTSA), (ii) Vivaldi or Exponentially Tapered Slot Antenna (ETSA), and (iii) Constant Width Slot Antenna (CWSA) [2]. Amplifiers may be included to improve the dynamic range of the system. The microwave imaging is an imaging technique which has been evolved to evaluate embedded objects using an antenna or array of antennas that radiate electromagnetic waves close to the target. The reflections or backscattered arrays are analyzed to detect the presence of tumors, blood clots, stroke, and so on in various parts of the human body. The microwave spectrum band of 1–6 GHz is suggested for the human head since it offers enough penetration into the tissue and provides better resolution. A wearable microwave head imaging system is proposed with a hat-like structure using an ultra-wideband antenna array. This antenna array is designed with FR4 substrate in the frequency range of 1.5 GHz to 4 GHz [3]. Wearable antennas are designed with textile material to monitor vital signs of the patient [4].

19.2 ANTENNA DESIGN AND CONFIGURATION

The proposed antenna is designed for microwave imaging applications such as stroke detection, tumor detection, and so on. [5]. An antenna is used for sensing changes in the dielectric permittivity of biological tissues. Transmitting antenna is used to send short UWB pulse, and other antennas are used for receiving the reflected signals [6]. The matching medium is to be chosen for better impedance matching [7]. On-body wearable antennas are designed with unidirectional radiation characteristics and high directivity. The curve design equation for the Vivaldi antenna is given by the following equations:

$$f = se^{at} \tag{19.1}$$

$$g = -se^{at} \tag{19.2}$$

The various parameters with expressions related to the design of an antenna are indicated in Table 19.1.

In this chapter, the Vivaldi antenna has been designed for the UWB frequency range. The antenna is designed and simulated using CST microwave studio software (2020 version). The frequency range of this antenna is from 3.1 GHz to 10 GHz having a notch at 3.5 GHz and 2.2 GHz. The geometric structure of the antenna is shown in Figure 19.1.

This antenna is designed on an FR4 substrate material with 0.018 mm thickness and a dielectric constant of 4.3. There are several choices of substrate material available for designing an antenna. A substrate with low relative permittivity is suitable for improving the characteristics [8].

19.2.1 Simulation Results of Vivaldi Antennas

The antenna designed is simulated in CST microwave studio software and various results are analyzed. Figure 19.2 shows the S-parameter curve of the designed antenna with reflection coefficient s11 from 2 GHz to 10 GHz.

TABLE 19.1
Various design parameters of the designed antenna

Parameters	Expression	Dimension (mm)
ML1	22	22
ML2	14	14
MW	0.54688	0.54688
R1	7	7
R2	8	8
subH	0.5	0.5
subL	TL + sL + ext	60
subW	42	42
Thk	0.018	0.018
Ext	3	3
QWM	4.3	4.3
QMW	14	14
R	log(0.8 × subW/(s × 2))/TL	0.053
S	0.6	0.6
sL	13	23
TL	54	54

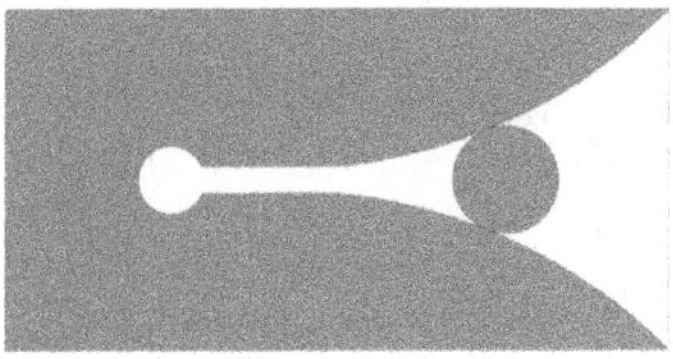

FIGURE 19.1 Vivaldi antenna (designed).

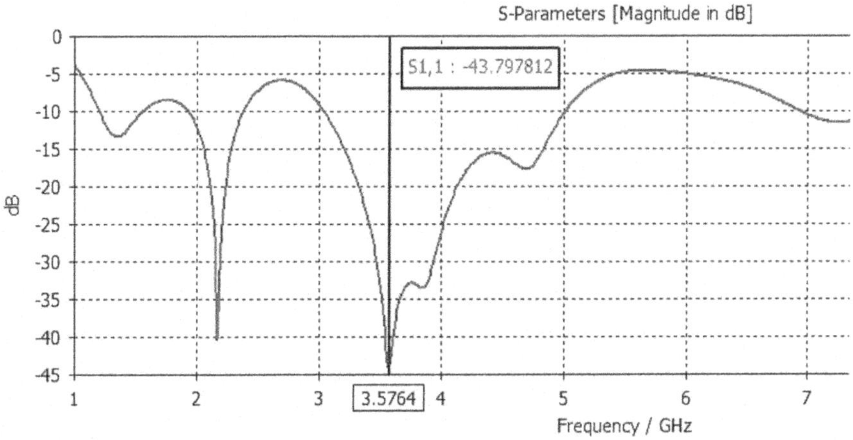

FIGURE 19.2 The S-parameter curve of the designed antenna.

It provides s11 value below 10 dB over a specified frequency range. The far-field pattern at 4 GHz, 5.5 GHz, and 9 GHz are represented in Figure 19.3, Figure 19.4, and Figure 19.5, respectively. The far-field pattern indicates high gain and directivity of nearly 10 dBi.

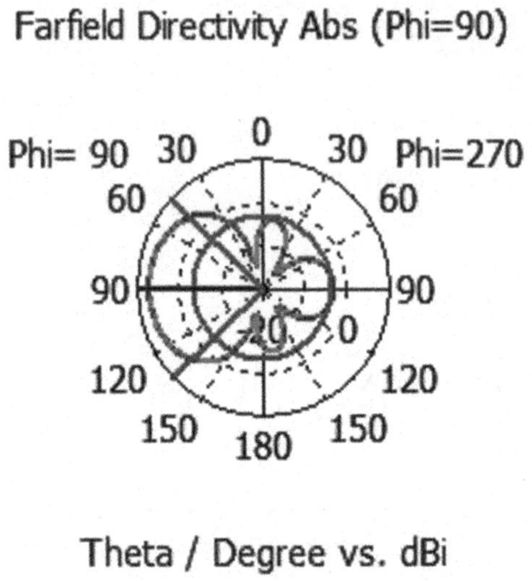

FIGURE 19.3 The radiation pattern and directivity at 4 GHz.

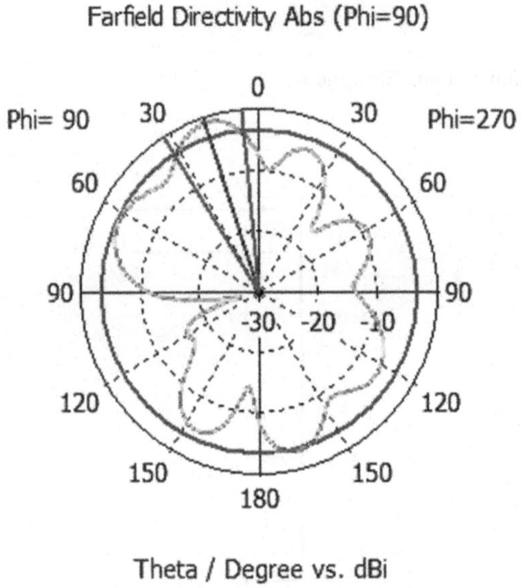

FIGURE 19.4 The radiation pattern and directivity at 5.5 GHz.

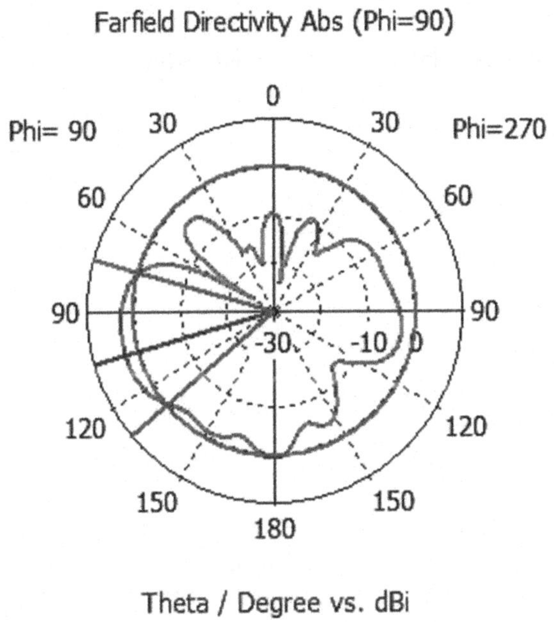

FIGURE 19.5 The radiation pattern and directivity at 9 GHz.

19.3 ANALYSIS AND COMPARISON OF VARIOUS ANTENNA PARAMETERS (CALCULATED AND SIMULATED)

The formula for received power is written using the Friis equation as shown in Equation (19.3). It is assumed that the gain of transmitter and receiver is the same. The s_{21} formula is represented by Equations (19.4) and (19.5). The value of R is 0.503 mm which is the distance between transmitting and receiving antenna. Equations (19.7) and (19.8) represent the gain formula logarithmic scale. Table 19.2 provides comparison of calculated and simulated results.

$$\mathrm{Pr} = \frac{\lambda^2 G_r G_t}{(4\Pi R)^2} P_t \tag{19.3}$$

$$s_{21} = \frac{\mathrm{Pr}}{P_t} = \frac{\lambda^2 G^2}{(4\Pi R)^2} \tag{19.4}$$

$$S_{21}(\mathrm{d}B) = 10\log \mathrm{Pr} - 10\log P_t = 10\log\left((\lambda G)^2\right) - 10\log((4\Pi R)^2) \tag{19.5}$$

$$S_{21}(\mathrm{d}B) = 20\log(\lambda G) - 20\log(4\Pi R) = 20\log(\lambda) + 20\log(G) - 20\log(4\Pi R). \tag{19.6}$$

TABLE 19.2
Comparison table (calculated and simulated results)

Frequency (GHz)	Calculated gain	Simulated gain
5	9.08	8.58
6	10.18	9.86
9	10.78	10

TABLE 19.3
Conclusion

Parameters	Result values
Return Loss (dB)	Less than −10 dB
VSWR	1.2
Resonant Frequency (GHz)	3.5 GHz
Bandwidth (%)	69%
Maximum Gain (dBi)	10 dBi

$$20\log(G) = S_{21}(dB) + 20\log(4\Pi R) - 20\log(\lambda) \tag{19.7}$$

$$G(dB) = \log(G) = \frac{S_{21}(dB) + 20\log(4\Pi R) - 20\log(\lambda)}{20} \tag{19.8}$$

19.4 CONCLUSION

The time-domain response is analyzed at various points using CST microwave simulation software. It can be observed from Table 19.3 that the antenna provides enhanced gain, wide bandwidth, and good return loss. In future, Vivaldi antenna array can be designed with the help of corporate feed power divider to increase the gain.

REFERENCES

[1] G. K. Pandey, H. S. Singh, and M. K. Meshram, "Meander-line-based inhomogeneous anisotropic artificial material for gain enhancement of UWB Vivaldi antenna," *Applied Physics A*, vol. 122, p. 134, 2016. https://doi.org/10.1007/s00339-015-9569-2.

[2] P. J. Gibson, "The Vivaldi aerial," in *Proceedings of the 9th European Microwave Conference*, Brighton, UK, 1979, pp. 101–105.

[3] S. R. Bashri, T. Arslan, W. Zhou, and N. Haridas, "Wearable device for microwave head imaging," *2016 46th European Microwave Conference (EuMC)*, London, 2016, pp. 671–674.

[4] X. Lin, Y. Chen, Z. Gong, B. Seet, L. Huang, and Y. Lu, "Ultra wideband textile antenna for wearable microwave medical imaging applications," *IEEE Transactions on Antennas and Propagation*, vol. 68, no. 6, pp. 4238–4249, June 2020.

[5] J. Bai, S. Shi and D. W. Prather, "Modified compact antipodal Vivaldi antenna for 4–50-GHz UWB application," *IEEE Transactions on Microwave Theory and Techniques*, vol. 59, no. 4, pp. 1051–1057, April 2011.

[6] A. M. Abbosh, "Compact antenna for microwave imaging systems," *2008 Cairo International Biomedical Engineering Conference*, Cairo, 2008, pp. 1–4.

[7] B. J. Mohammed, A. M. Abbosh, D. Ireland, and M. E. Bialkowski, "Compact wideband antenna immersed in optimum coupling liquid for microwave imaging of brain stroke," *Progress In Electromagnetics Research C*, vol. 27, pp. 27–39, 2012.

[8] D. Yang, S. Liu, and D. Geng, "A miniaturized ultra-wideband Vivaldi antenna with low cross polarization," *IEEE Access*, vol. 5, pp. 23352–23357, 2017.

20 Ultra-Wideband Antenna Design for Wearable Applications

Shailesh and Garima Srivastava

20.1 INTRODUCTION

Ultra-wideband (UWB) innovation can give short distance, high transfer speed communication systems with less energy utilization, which is very alluring for Wireless Body Area Networks (WBAN). The nearness of the human body to the antenna produces tremendous difficulties in designing wearable antennas in these systems. In addition, at the point when wearable (textile) antennas are utilized for on-body applications, they can deteriorate the performance of the antenna, and they can also be harmful to the human body. The Specific Absorption Rate (SAR) value is calculated to avoid the risk for the human body. Researchers are interested in this research area and some major developments have been attained lately. This chapter presents a description, results, substrate, and conductive layers of the latest UWB wearable antennas with a discussion of different analyses based on bending,

DOI: 10.1201/9781003093558-27

crumpling, antenna misalignment, SAR, and change in substrate parameters, which affects the antenna performance.

Currently, antennas are an essential part of WBAN applications such as surveillance, emergency services, and health monitoring. These communication systems need compact wideband antennas with developing technologies. Furthermore, the antennas that have new, good design arrangements with decent radiation performance are desirable for wearable/textile antennas (Nadh, Madhav, Kumar, Rao, and Anilkumar, 2018). The UWB antennas can have applications like target positioning, radar imaging, remote sensing, target positioning, and so on. The UWB is also appropriate for essential modifications in the monitoring, GPS observing, and WLAN low power transmission. These applications require less transmission power; consequently, they also do not harm human body tissues and meddle with other clinical signs (Nadh, Madhav, Kumar, Rao, and Anilkumar, 2018).

When an antenna is printed on a thin and flexible substrate, it provides different pros like bendability, inexpensive, and lightweight. Their advantages offer new opportunities in entertainment, communication, aerospace, and health. Wearable antennas are used in bendable wearable/implantable medical sensors, WBAN devices, Wireless Wide Area Network (WWAN) terminals, and so on. The UWB technology has an unlicensed 3.1–10.6 GHz frequency band, which is permitted by the Federal Communications Commission (FCC). The different merits of it are wider frequency range, less power utilization, and huge throughput which finds application in BAN, wireless communication, indoor location, and so on (Zhang, Shandong, Yang, Qu, and Zong, 2020).

On-body wearable antennas placed on different body parts or integrated into cloths are extensively used in numerous regions. The microstrip antennas are the best option chosen for wearable antennas integrated into cloths. The designing of such antennas is a challenging task. These antennas are needed to confirm some important demands like cheap, compact, improved impedance, and far-field radiation properties, less energy absorption, and bendability when placed on the different parts of the human body. To design the microstrip antenna having a planar arrangement, flexible substrates, and conductors, Electrical Technical Textiles (ETTs) demand rise (Ali, Mansour, and Mohamed, 2016). In recent times, the application of UWB has been increasing in the WBAN area because of the worn communication systems (Ur-Rehman, Abbasi, Akram, and Parini, 2015). The characteristics of the wearable antenna, user comfort, and no one body detuning are important to enable trustworthy communication connections in such environmental sceneries (Samal, Soh, and Vandenbosch, 2014).

This chapter presents the different types of UWB wearable antennas. Section 20.2 discusses some of the important challenges while designing the antenna. In Section 20.3, various wearable UWB antennas are described. Different analyses based on bending, crumpling, antenna misalignment, specific absorption rate (SAR), and change in substrate parameters, which affects the antenna performance are discussed in the subsections. The results of UWB wearable antennas are presented in Section 20.4. In Section 20.5, substrates and conductive layer specifications are mentioned, which are used for designing the UWB wearable antenna.

20.2 CHALLENGES IN DESIGNING WEARABLE UWB ANTENNAS

In this section, some essential challenges that are needed to be considered while designing UWB antenna for wearable applications are discussed.

- Antenna substrates are too thick (Zhang, Shandong, Yang, Qu, and Zong, 2020).
- Antennas are not able to cover the full UWB band for both flat and bend cases (Zhang, Shandong, Yang, Qu, and Zong, 2020).
- The small bending radius should be analyzed (Zhang, Shandong, Yang, Qu, and Zong, 2020).
- The frequency performance of antennas with bending situation should be improved (Zhang, Shandong, Yang, Qu, and Zong, 2020).
- Unidirectional radiation patterns are extremely suited for wearable applications (Poffelie, Soh, Yan, and Vandenbosch, 2015).

20.3 ANTENNA DESIGN

In this section, various UWB wearable antennas such as the electromagnetic bandgap (EBG)–based UWB antenna, CPW fed tapered slot UWB antenna, full ground plane UWB antenna, UWB antenna with different heights of the ground plane of CPW feeding, modified circular UWB antenna with DGS, UWB antenna with Ashoka Chakra–shaped radiator, and circular UWB antenna are discussed.

20.3.1 EBG-Based UWB Antenna

The UWB antenna consists of a beveled y-shaped patch fed by a feed line (microstrip) and a finite ground surface as shown in Figure 20.1. An inverted bell slot is added in

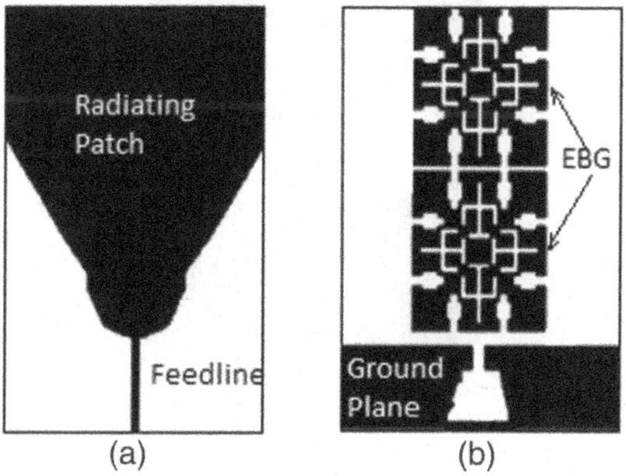

FIGURE 20.1 EBG-based UWB wearable antenna (a) top view, (b) bottom view (Sambandam, Kanagasabai, Ramadoss, Natarajan, Alsath, Shanmuganathan, Sindhadevi, and Palaniswamy, 2019).

the ground to improve the impedance matching. The lower part of the patch is altered like truncated edges to enhance bandwidth. By increasing the electrical length of the radiators, a low-frequency response is improved. Fork-slotted EBG design above ground is used for minimizing the frequency detuning, back radiation effects, and impedance mismatch because of the placement of the wearable antenna on the body parts (Sambandam, Kanagasabai, Ramadoss, Natarajan, Alsath, Shanmuganathan, Sindhadevi, and Palaniswamy, 2019).

Advantages

- Cover complete UWB frequency range
- Less chip area
- SAR within the limit
- Bend analysis does not affect the antenna performance

Limitations

- Complicated EBG structure
- EBG structure has some limitations such as narrow bandwidth, low gain, and excitation of surface waves

20.3.2 CPW Fed Tapered Slot UWB Antenna

The antenna contains a dual-tapered curved slot that acts as a ground surface to have a smooth impedance modification as presented in Figure 20.2. The impedance

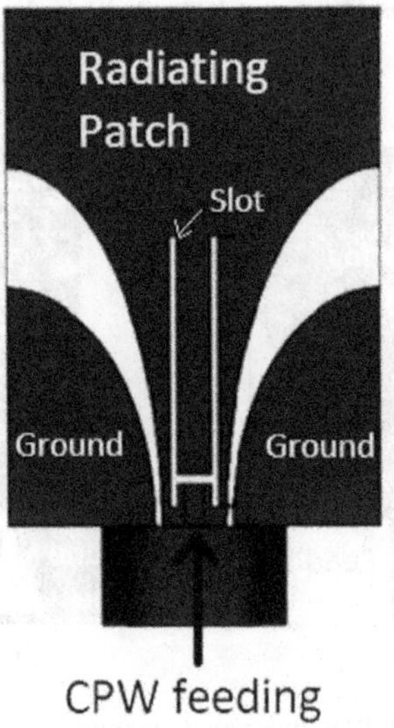

FIGURE 20.2 CPW-fed tapered slot UWB antenna (Abbasi, Ur-Rehman, Yang, Alomainy, Qaraqe, and Serpedin, 2013).

matching depends on the distance between the ground and the patch. Semi-major to the semi-minor axis ratio of both ellipses improves impedance matching. An H-slot in the patch, which is placed near the feed line, is used to reject the WLAN band (Abbasi, Ur-Rehman, Yang, Alomainy, Qaraqe, and Serpedin, 2013).

Advantages

- Simple structure
- Full UWB range is covered
- Notched band for WLAN band
- Small size
- Good bend and adverse condition analysis

Limitations

- The notched band becomes wider for in-body case
- Low gain

20.3.3 Full Ground Plane UWB Antenna

A full ground plane (FGP) UWB antenna is the newly introduced wearable antenna for WBAN applications such as monopole and Vivaldi antennas, which are fed by a coplanar waveguide (CPW) or microstrip line. The main problem with these antennas is that they produce radiating fields in both front and back directions. This shows that they are tightly coupled to the body (human). All wearable antennas with FGP can be operated within the complete UWB frequency band and have better insensitivity to the human body (Samal, Soh, and Vandenbosch, 2014). An arc-shaped UWB antenna and two arc-shaped radiating patches are positioned on the front side of the substrate as shown in Figure 20.3a. FGP is designed on the back plane of the substrate. Each radiating patch has two small transmission lines coupled to a coaxial connector on the substrate's back plane. Dual same T-shaped slots are added to further increase the impedance matching at low-frequency range (Simorangkir, Kiourti, and Esselle, 2018). In the UWB antenna with four radiators, probe feeds are used to feed three main radiating patches A, B, and C as presented in Figure 20.3b. Patch B and patch C has to feed with parallel topology and its output is fed in parallel with A. The additional radiator close to these main radiators is activated by capacitive coupling. The capacitive coupling among A, B, and C and nearer radiators can increase the bandwidth. The bends and steps are added in the radiators A, B, C, and additional radiating patch, that are used for tuning frequency response and matching for getting better performance in full UWB band (Samal, Soh, and Vandenbosch, 2014). Triple-layer UWB antenna design contains dual substrate layers and triple conductive layers as shown in Figure 20.3c. The top conductive layer has an octagonal shape radiator. This radiator is fed using a grounded CPW feeding. Steps are included in both ground surfaces, which enhance the bandwidth. The middle layer is a parasitic circular patch. This radiator patch is linked to the ground by a thin strip. On the bottom layer of the antenna, a full metallic plane (reflector) is located (Poffelie, Soh, Yan, and Vandenbosch, 2015).

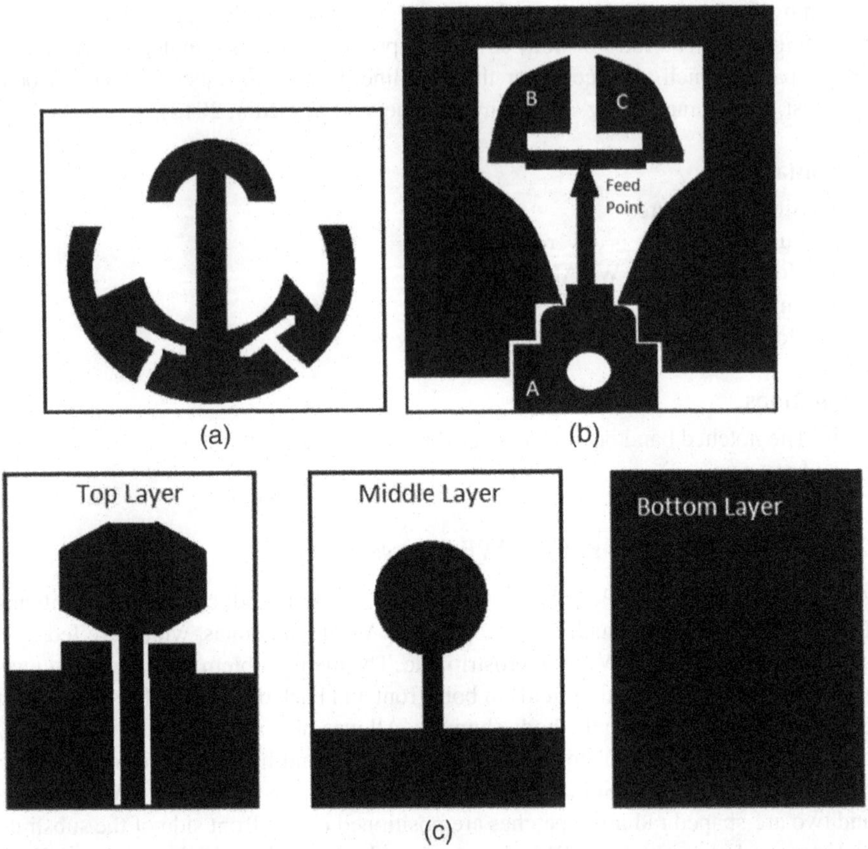

FIGURE 20.3 (a) Arc-shaped UWB antenna (Samal, Soh, and Vandenbosch, 2014), (b) UWB wearable antenna with four radiators (Simorangkir, Kiourti, and Esselle, 2018), (c) triple-layer UWB antenna (Poffelie, Soh, Yan, and Vandenbosch, 2015).

The advantages and disadvantages of full ground UWB antennas are presented in Table 20.1.

20.3.4 UWB Antenna with Different Heights of the Ground Surface of CPW Feeding

This antenna has CPW feeding with hybrid-shaped radiator patch (circle plus rectangle) as shown in Figure 20.4. The two ground planes have curved corners and different height for enhancing bandwidth. Here, CPW feeding is used because a thin microstrip feed line is not suitable for a wearable antenna and it is also difficult to weld an SMA connector on this type of feeding (Zhang, Shandong, Yang, Qu, and Zong, 2020).

TABLE 20.1
Pros and cons of full ground UWB antennas

Reference	Pros	Cons
Samal, Soh, and Vandenbosch, 2014	FGP can preserve the antenna performance by protecting it against connection to and power absorption by the human body	• Large chip size • Full UWB range is not covered • Complex structure
Simorangkir, Kiourti, and Esselle, 2018	FGP suppresses the loading of the antenna due to fundamental biological tissues and back radiation absorbed by the human body	• Low efficiency
Poffelie, Soh, Yan, and Vandenbosch, 2015	• Backward radiation reduction • High fidelity prerequisite for UWB pulse transmission/ reception	• Large antenna size • Slight variations in the measured fidelity factor

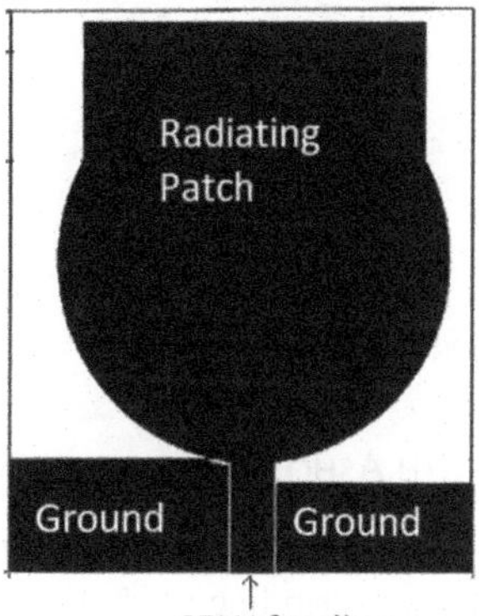

FIGURE 20.4 UWB antenna with different heights of the ground plane of CPW feeding (Zhang, Shandong, Yang, Qu, and Zong, 2020).

Advantages

- Simple structure
- Full UWB range is covered
- Small size
- Good bend and adverse condition analysis

Limitations

- No SAR analysis

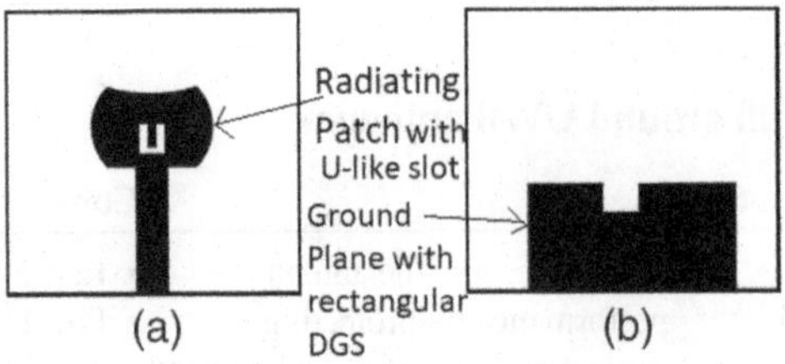

FIGURE 20.5 Modified circular UWB antenna with DGS (a) top view, (b) bottom view (Ali, Mansour, and Mohamed, 2016).

20.3.5 Modified Circular UWB Antenna with DGS

This antenna has a modified circle like a patch with a notch etched on it and an arc-like slot on the front plane. Also, a ground surface with rectangular-like defected ground structure (DGS) on the backside is presented in Figure 20.5. A U-like slot and ground surface are utilized for obtaining wide impedance bandwidth (Ali, Mansour, and Mohamed, 2016).

Advantages

- Simple structure

Limitations

- No SAR analysis
- No bending analysis
- Full UWB range is not covered

20.3.6 UWB Antenna with Ashoka Chakra–Shaped Radiator

This antenna has a circular ring like a radiating patch with 24 spokes, which look like Ashoka Chakra and fed by a microstrip line as shown in Figure 20.6. The triangular slotted ground with a rectangular slot disturbs the path of the current, which results in a change in inductance and capacitance. This change helps in the miniaturization of the antenna size (Nadh, Madhav, Kumar, Rao, and Anilkumar, 2018).

Advantages

- Simple structure

Limitations

- Efficiency is not mentioned

20.3.7 Circular UWB Antenna

This textile antenna contains a circular patch with a tapered feed line and ground surface with a small rectangular slot for improved impedance matching as presented

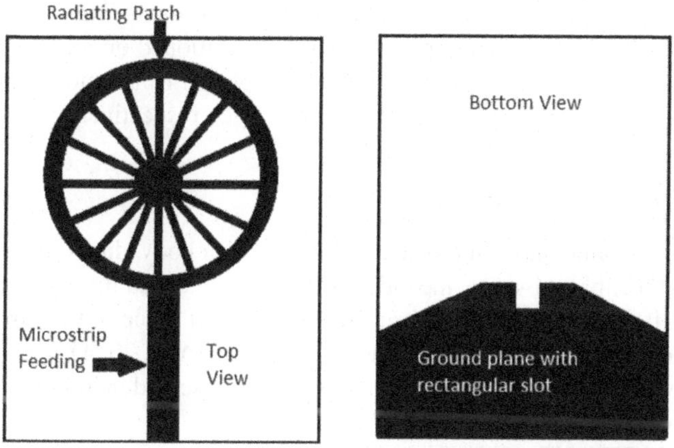

FIGURE 20.6 UWB antenna with Ashoka Chakra–shaped radiator (top and bottom view) (Nadh, Madhav, Kumar, Rao, and Anilkumar, 2018).

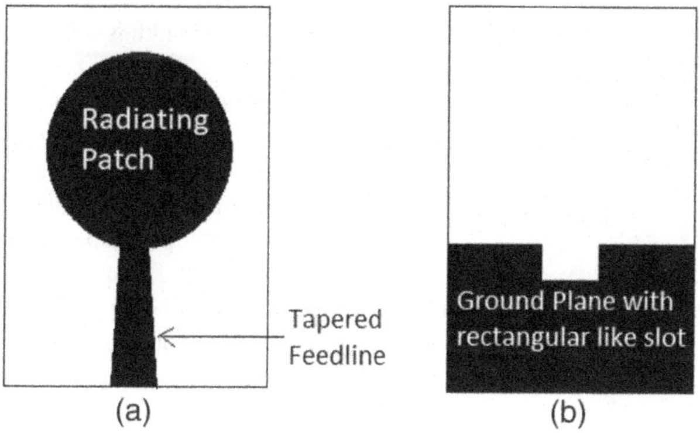

FIGURE 20.7 Circular UWB wearable antenna (a) top view, (b) bottom view (Sun, Cheung, and Yuk, 2014).

in Figure 20.7. The inductive and capacitive effects due to the radiating patch can change the radius of the circle (Sun, Cheung, and Yuk, 2014).

Advantages

- Simple structure

Limitations

- No SAR analysis

20.3.8 Bending and Crumpling Analysis

The wearable antenna is placed on the body (human) at different bending angles and radii, concerning the position where the antenna is placed. For instance, the bending angle on arm is greater than the bending angle of the torso. The wearable antenna can also be crumpled in any direction when it is positioned on a human body (Sun, Cheung, and Yuk, 2014). The bending and crumpling of the antenna affect its performance, which is displayed in Tables 20.2 and 20.3, respectively.

20.3.9 Antenna Misalignment

When the textile antenna is in contact with the human body, due to the movement of the body and flexible substrate, the radiator patch and the feed line on one plane of the substrate shift concerning a ground surface on the other plane of the substrate, which causes misalignment between them (Sun, Cheung, and Yuk, 2014). The impact of the misalignment of the antenna on its performance is presented in Table 20.4.

TABLE 20.2
Effects of bending

Reference	Bending (mm/deg)	Inferences
Zhang, Shandong, Yang, Qu, and Zong, 2020	10 and 20 mm	• UWB bandwidth is not affected by bending. • The bent antenna affects the radiation pattern. This effect develops stronger with a smaller bending radius. • Radiation patterns maintain omnidirectional shape due to the monopole patch in both flat and bend scenario.
Nadh, Madhav, Kumar, Rao, and Anilkumar, 2018	15, 30, 45, 60 deg	• In starting bending angles, the antenna produces a notched band. • With a further increment in the bending angle, the antenna works in the complete UWB range without a notched band.
Sun, Cheung, and Yuk, 2014	Along the y-axis (bending radius = 30, 40, and 50 mm)	• Higher resonance merges and lower resonance frequency shifts slightly. • Stable radiation patterns.
Sun, Cheung, and Yuk, 2014	Along the x-axis (bending radius = 30, 40, and 50 mm)	• Impedance bandwidth is slightly affected. • Quite stable radiation patterns.

TABLE 20.3
Effects of crumpling

Reference	Crumpling	Inferences
Sun, Cheung, and Yuk, 2014	Along the y-axis	• Impedance matching is slightly affected. • Omnidirectional radiation patterns are obtained.
Sun, Cheung, and Yuk, 2014	Along the x-axis	• Affects impedance matching more severely. • Stable radiation patterns.

TABLE 20.4
Antenna misalignment

Reference	Misalignment	Inferences
Sun, Cheung, and Yuk, 2014	Along the x-axis or in the horizontal direction	• Due to change in the location of the rectangular ground slot, matching at higher frequencies becomes worse. • At the low frequency range, the stable radiation pattern is obtained. At the higher end of the band, the radiation pattern becomes somewhat asymmetrical due to the asymmetry of the ground plane.
Sun, Cheung, and Yuk, 2014	Along the y-axis or in the vertical direction	• Lower resonance slightly shifts, which increases the lower cut-off frequency. At a higher frequency range, poor impedance matching occurs. • Stable radiation pattern.

TABLE 20.5
SAR analysis setup and results

Reference	Body part	Power input (mW)	Max. SAR averaged over 10 g of human tissue (W/Kg)	Min. gap between antenna and body (mm)
Nadh, Madhav, Kumar Rao, and Anilkumar, 2018	Three-layered phantom	100	1.55 at 10.5 GHz	5
	Arm		1.23 at 10.5 GHz	
	Head		1.29 at 10.5 GHz	
Simorangkir, Kiourti, and Esselle, 2018	Human muscle phantom	500	0.174 at 7 GHz	5
Sun, Cheung, and Yuk, 2014	Human body phantom	–	–	3
Poffelie, Soh, Yan, and Vandenbosch, 2015	Three-layer (Skin-Fat-Muscle) human body model	500	1.21 at 3 GHz	2

20.3.10 SAR Analysis

In this segment, the SAR analysis of the UWB wearable antenna is discussed. Table 20.5 shows the SAR analysis setup and results are shown in Table 20.6, on body inferences of wearable UWB antennas.

20.3.11 Change in Substrate Parameters

The effect of change in substrate parameters on antenna performance is presented in Table 20.7.

TABLE 20.6
On-body inferences of wearable/textile UWB antennas

Reference	On-body inferences
Samal, Soh, and Vandenbosch, 2014	• Ripples occur in both planes like YZ and XZ planes. • The human body plays the role of reflector at a higher frequency range, which distorts the radiation pattern. • A decrease in radiated efficiency. • Due to chest movement because of breathing, results are slightly degraded.
Sun, Cheung, and Yuk, 2014	• The gain of an on-body antenna is greater than the gain in free space since the human body makes antenna further direction. The efficiency of the on-body scenario is decreased because of the absorption of radiated power from the antenna to the human body. • Radiation power in E-plane is lesser as compared to free space situation. • Radiation patterns turn into directional in H-plane.
Poffelie, Soh, Yan, and Vandenbosch, 2015	• The entire UWB band is achieved. • Simulated radiation pattern does not change a lot in the upper half-space. • High on body fidelity values are obtained at 1-m distance.

TABLE 20.7
Effect of change in substrate parameters

Reference	Change in	Inferences
Zhang, Shandong, Yang, Qu, and Zong, 2020	Dielectric constant	Antenna retains wide bandwidth
Zhang, Shandong, Yang, Qu, and Zong, 2020	Thickness of substrate	Antenna maintains wide bandwidth

20.4 RESULTS

In this section, results of all the antennas discussed in the above sections are shown in Table 20.8 containing parameters like bandwidth, size, efficiency, gain, and radiation pattern.

20.5 SUBSTRATE AND CONDUCTIVE LAYER USED

The flexible substrates (in Table 20.9) used in wearable UWB antennas include polyimide, liquid crystal polymer (LCP), polyethylene terephthalate (PET), and so on. But, polyimide is more popular than others in flexible antennas due to various advantages such as extremely low thickness, high tensile strength, less dielectric loss, highly flexible, and a high temperature rating (Zhang, Shandong, Yang, Qu, and

TABLE 20.8
Results of wearable UWB antennas

Reference	Bandwidth (flat) (GHz)	Size (mm^2)	Gain (dBi)	Efficiency (%)	Radiation pattern
Sambandam, Kanagasabai, Ramadoss, Natarajan, Alsath, Shanmuganathan, Sindhadevi, and Palaniswamy, 2019	3.1–10.6	20 × 25	20.25	85–920.5	Directional
Zhang, Shandong, Yang, Qu, and Zong, 2020	3.06 to 13.58	38 × 30.4	>1.69	>59	Omnidirectional
Ali, Mansour, and Mohamed, 2016	3.5–15	41 × 41	3.11	–	Omnidirectional and dipole-like radiation pattern
Zahran, Abdalla, Gaafar, 2019	2–3.8; 5.6–20.2; 7–10.4	51 × 22	5.33	–	Omnidirectional
Hamouda, Wojkiewicz, Pud, Kone, Bergheul, and Lasri, 2017	1–8	48 × 34.9	3.1	–	E-plane omnidirectional, H-plane bidirectional
Chen, Ammann, Qing, Wu, See, and Cai, 2006	3–6; 7.8–8.2; 10–11.8	25 × 30	–	–	–
Abbasi, Rehman, Yang, Alomainy, Qaraqe, and Serpedin, 2013	3–12 with the notched band for WLAN	26 × 16	2–3	94	E-plane dipole, H-plane omnidirectional
Simorangkir, Kiourti, and Esselle, 2018	3.7–10.3	67 × 80	4	45	Directional
Sun, Cheung, Yuk, 2014	3–12	30 × 40	2–4	80	Omnidirectional
Ur-Rehman, Abbasi, Akram, and Parini, 2015	3.1–10.6 with the notched band for WLAN	26 × 16	2–3	94–99	E-plane donut shape H-plane omnidirectional
Poffelie, Soh, Yan, and Vandenbosch, 2015	3.1–10.6	80 × 61	1.1–7.2	–	Unidirectional
Nadh, Madhav, Kumar, Rao, and Anilkumar, 2018	3.1–11.5	30 × 25	5.15	–	Dipole like in both planes
Samal, Soh, and Vandenbosch, Jan. 2014	3.6–10.3	88 × 97	7.75	–	Boresight direction and omnidirectional
Sun, Cheung, and Yuk, 2014	3–12	30 × 40	2–4	80	Monopole-like radiation patterns
Poffelie, Soh, Yan, and Vandenbosch, 2015	3.18–11	80 × 61	7.2	–	Unidirectional

TABLE 20.9
Substrates used for UWB wearable antennas

Reference	Name	Dielectric constant	Loss tangent	Thickness (mm)
Zhang, Shandong, Yang, Qu, and Zong, 2020	Kapton polyimide	3.5	–	0.07
Ali, Mansour, and Mohamed, 2016	RO3003	3	0.025	1.524
Nadh, Madhav, Kumar, Rao, and Anilkumar, 2018	RT/Duroid5880	2.2	0.0009	0.8
Simorangkir, Kiourti, and Esselle, 2018	PDMS	2.7	0.02–0.07	3
Samal, Soh, and Vandenbosch, Jan. 2014	Felt	1.45	0.044	3
Sun, Cheung, Yuk, 2014	Cotton	1.75	0.08	0.75
Poffelie, Soh, Yan, and Vandenbosch, 2015	Felt	1.45	0.044	2

TABLE 20.10
Conductive layers used for UWB wearable antennas

Reference	Name	Conductivity (S/m)	Thickness (mm)
Simorangkir, Kiourti, and Esselle, 2018	Nickel-copper-silver-coated nylon ripstop	1.02×10^5	0.13
Samal, Soh, and Vandenbosch, 2014; Poffelie, Soh, Yan, and Vandenbosch, 2015	Shieldit Super	1.18×10^5	0.17
Simorangkir, Kiourti, and Esselle, 2018	Nickel-copper-coated ripstop	5.4×10^4	0.08

Zong, 2020). Also, different types of conductive layers are used, which are mentioned in Table 20.10.

20.6 CONCLUSION

UWB technology has unlocked the opportunities for WBAN applications, which provides a cheap and effective solution for many WBAN applications. There will be quick development of wearable UWB antennas with further practical applications and will arise as a significant field in wireless antennas in the upcoming years. Various UWB wearable antennas are presented in this chapter such as EBG-based antenna, full ground plane antenna, CPW fed antenna with different ground heights and tapered slots, and antenna with different radiator shapes such as circular, modified circular, and Ashoka Chakra. The main challenges for designing such antennas are working with a thick substrate, covering full UWB bandwidth, getting unidirectional radiation patterns, and so on. Some essential analyses are needed for better performance wearable antennas such as bending, crumpling, antenna misalignment, SAR, and variation with substrate parameters.

REFERENCES

Abbasi, Qammer H., Ur-Rehman, Masood, Yang, Xiaodong, Alomainy, Akram, Qaraqe, Khalid, and Serpedin, Erchin. (2013). "Ultrawideband band-notched flexible antenna for wearable applications." *IEEE Antennas and Wireless Propagation Letters*, 12, 1606–1609, 2013, doi: 10.1109/LAWP.2013.2294214.

Ali, Wael A., Mansour, Asmaa M., and Mohamed, Darwish A. (2016). "Compact UWB wearable planar antenna mounted on different phantoms and human body." *Microwave and Optical Technology Letters*, 58(10), 2531–2536.

Chen, Zhi Ning, Ammann, Max J., Qing, Xianming, Wu, Xuan Hui, See, Terence S. P., and Cai, Ailian. (2006). "Planar antennas." *IEEE Microwave Magazine*, 7(6), 63–73.

Hamouda, Zahir, Wojkiewicz, Jean-Luc, Pud, Alexander A., Kone, Lamine, Bergheul, Said, and Lasri, Tuami. (2017). "Flexible UWB organic antenna for wearable technologies application." *IET Microwaves, Antennas & Propagation*, 12(2), 160–166.

Nadh, B. Prudhvi, Madhav, B. T. P., Kumar, M. Siva, Rao, M. Venkateswara, and Anilkumar, T. (2018). "Circular ring structured ultra-wideband antenna for wearable applications." *International Journal of RF and Microwave Computer-Aided engineering*, vol. 33, 1–15.

Poffelie, Linda A. Yimdjo, Soh, Ping Jack, Yan, Sen, and Vandenbosch, Guy A. E. (2015). "A high fidelity all-textile UWB antenna with low back radiation for off-body WBAN applications." *IEEE Transactions on Antennas and Propagation*, 64(2), 757 –760.

Samal, P. B., Soh, P. J., and Vandenbosch, G. A. E. (Jan. 2014). "UWB all-textile antenna with full ground plane for off-body WBAN communications." *IEEE Transactions on Antennas and Propagation*, 62(1), 102–108.

Sambandam, Padmathilagam, Kanagasabai, Malathi, Ramadoss, Shini, Natarajan, Rajesh, Alsath, M. Gulam Nabi, Shanmuganathan, Shanmathi, Sindhadevi, M., and Palaniswamy, Sandeep Kumar. (2019). "Compact monopole antenna backed with fork-slotted EBG for wearable applications." *IEEE Antennas and Wireless Propagation Letters*, 19(2), 228 –232.

Simorangkir, Roy B. V. B., Kiourti, Asimina, and Esselle, Karu P. (2018). "UWB wearable antenna with a full ground plane based on PDMS-embedded conductive fabric." *IEEE Antennas and Wireless Propagation Letters*, 17(3), 493–496.

Sun, Yiye, Cheung, Sing Wai, and Yuk, Tung Ip. (2014). "Design of a textile ultra-wideband antenna with stable performance for body-centric wireless communications." *IET Microwaves, Antennas & Propagation*, 8(15), 1363–1375.

Ur-Rehman, Masood, Abbasi, Qammer H., Akram, Muhammad, and Parini, Clive. (2015). "Design of band-notched ultra-wideband antenna for indoor and wearable wireless communications." *IET Microwaves, Antennas & Propagation*, 9(3), 243 – 251.

Zahran, Sherif R., Abdalla, Mahmoud A., and Gaafar, Abdelhamid. (2019). "New thin wide-band bracelet-like antenna with low SAR for on-arm WBAN applications." *IET Microwaves, Antennas & Propagation*, 13(8), 1219–1225.

Zhang, Ying, Shandong, Li, Yang, Zhi-Qun, Qu, Xiao-Yun, and Zong, Wei-Hua. (2020). "A coplanar waveguide-fed flexible antenna for ultrawideband applications." *International Journal of RF and Microwave Computer-Aided Engineering*, 30, 1–11.

21 An Orthogonal Quad-Port Wearable MIMO Antenna for UWB Applications

G. Viswanadh Raviteja

21.1 INTRODUCTION

In this chapter, a modified deltoid-shaped antenna structure is proposed. Four elements in orthogonal configuration with individual port excitations are presented for obtaining the MIMO arrangement. The antenna is backed by a flexible FR4 epoxy dielectric substrate with relative permittivity of 4.4 and a loss tangent of 0.02. The impedance bandwidth of −10 dB is achieved from frequency ranges 3 GHz to 12 GHz covering the ultra-wideband applications. The proposed MIMO arrangement also showed good isolation of less than −20 dB between all the four elements. The important parameters, such as Total Active Reflection Coefficient, Mean Effective Gain, Envelope Correlation Coefficient, that are responsible for the efficient working of the MIMO are also investigated and are found to be well within the standards with MEG < 3 dB and ECC < 0.5. The antenna achieved a satisfactory gain over the entire UWB range with a peak gain close to 12 dB. Also, specific absorption rate analysis is carried out by creating the homogenous three-layer human body model consisting of skin, fat, and muscle layers. The results obtained showed that the MIMO antenna is well within the SAR limits of 1.6 W/Kg. The work is carried out using ANSYS High-Frequency Structure Simulator (HFSS) software. The requirement for high data rates and the need for uninterrupted data transmission led to the rapid rise in the developments of new age wireless communications. To satisfy all the requirements of a wireless standard, a single antenna is not sufficient for the present day scenario. This is where the multiple-input multiple-output (MIMO) antenna system finds its necessity. The design of these MIMO antennas is to make them perform their services for some intended operation of frequencies. The MIMO antenna

DOI: 10.1201/9781003093558-28

design for Wi-Max and WLAN applications is discussed in Ref. [1]. The effects of mutual coupling are reduced with the deployment of an electromagnetic bandgap (EBG) as an isolator. An inverted F-antenna working on dual-band frequency of operation for WLAN applications is shown in Ref. [2]. In this cited article, meander resonant band and T-shaped slot techniques are used to improve the MIMO working characteristics. A four- or eight-element MIMO antenna system for dual wideband operations focusing on WLAN applications is shown in Ref. [3]. On the other hand, intensive research is being conducted on a large scale for the fifth-generation communication systems. The use of MIMO systems is one of the primary aspects behind the evolution of this fifth-generation [4] technologies. In Ref. [5] an eight-element MIMO system with dual polarization is discussed for the use in 5G smartphones. To achieve this dual polarization, a pair of parasitic elements in the shape of circular rings is implemented.

In this chapter, a quad-port, four-element MIMO antenna is proposed. Thereby, the MIMO characteristics are studied by computing the parameters such as Envelope Correlation Coefficient (ECC), Mean Effective Gain (MEG), Total Active Reflection Coefficient (TARC), and radiation efficiency.

21.2 ANTENNA DESIGN STRUCTURE

An optimized modified deltoid-shaped microstrip monopole antenna structure is considered as shown in Figure 21.1. The structure is a stage-by-stage modification of the rectangular microstrip patch antenna with dimensions 43.48 × 41.54 mm^2 (width × length). The antenna is fed by a stripline of length 14.25 mm and width 2.4 mm. After the initial design, the antenna is extended for MIMO configuration with four deltoid-shaped antenna elements. These four antenna elements are designed to be symmetric and orthogonal in configuration with each other. The proposed structure is etched on a flexible FR4 epoxy dielectric material with a dielectric constant '4.4' and loss tangent of '0.02'. The height of the dielectric material is considered to be 1.6 mm. To address the UWB applications that cover the frequency range of 3.1 to

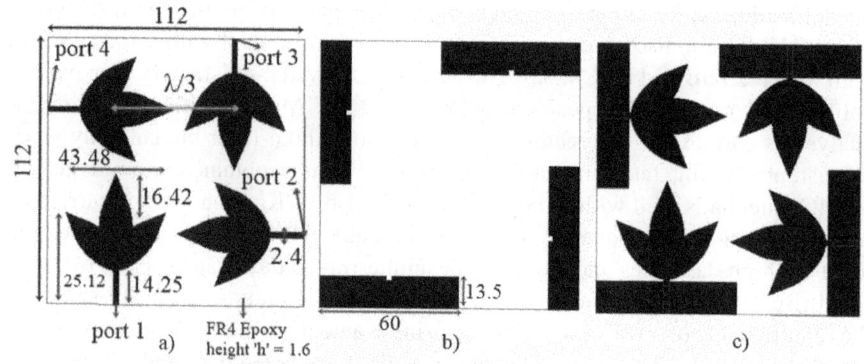

FIGURE 21.1 (a) Proposed MIMO structure (front view); (b) back view; (c) overall view.

10.6 GHz, the partial ground concept is used. The ground planc is taken to be 60 mm × 13.5 mm (width × length). All the parameters concerned with the radiating element and the ground plane are optimized to get the desired results. The modelling and the analysis of the proposed structure are carried out in ANSYS HFSS software.

To observe the isolation performance of the MIMO, the spacing between the symmetric orthogonal deltoid antenna elements is taken to be λ/ 3. The orthogonal configuration is considered to provide better isolation between the antenna elements.

21.3 ANALYSIS OF THE PROPOSED STRUCTURE

The S-parameters are analyzed for the MIMO antenna. Figure 21.2a shows the reflection coefficient for all the four MIMO antenna elements. It is evident from the figure that all the four antenna elements impedance bandwidth pertaining to −10 dB is well achieved from 3 GHz to 12 GHz.

The effect of mutual coupling which is a critical factor in the design of the MIMO antenna system is also examined. This factor is simulated when excitation is given at port 1 and all other ports are terminated with 50 ohms characteristic impedances. Figures 21.2b and 21.2c give necessary information regarding the mutual coupling from port 1 and all other ports as well. It is to be particularly noted from Figures 21.2b

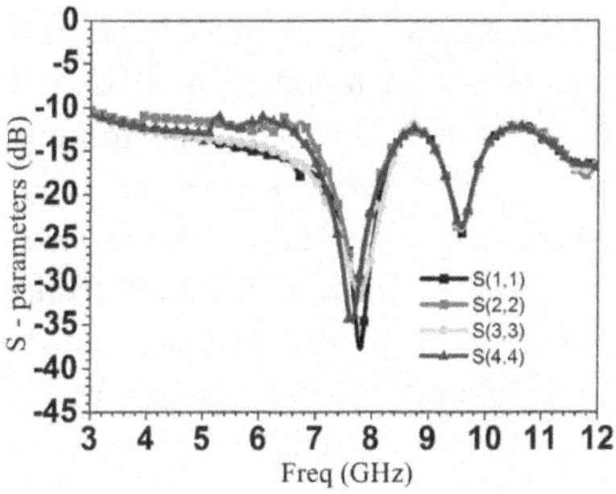

FIGURE 21.2a S-parameters for the quad-port MIMO antenna – reflection coefficients.

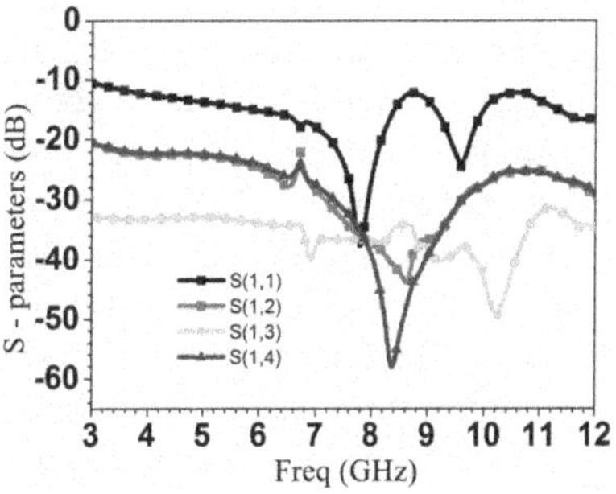

FIGURE 21.2b S-parameters for the quad-port MIMO antenna at port 1.

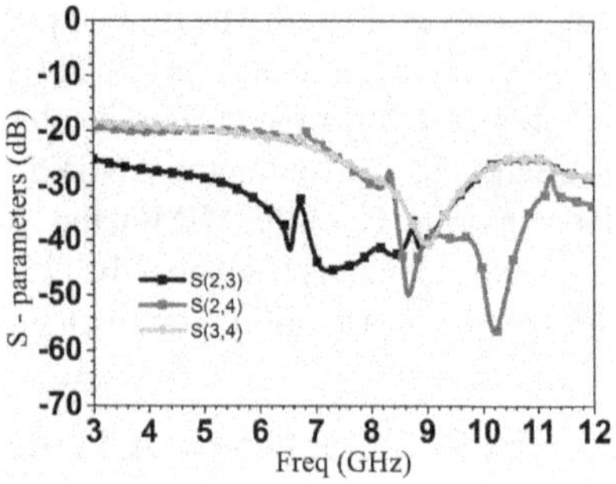

FIGURE 21.2c S-parameters for the quad-port MIMO antenna at other ports.

and 21.2c that the isolation or transmission coefficients are less than −20 dB for the entire working impedance bandwidth of 3 GHz to 12 GHz. S12 = S21 achieved peak isolation close to −45 dB within the range of 8 to 9 GHz. S13 = S31 showed an isolation improvement lower than −30 dB where the values are close to −50 dB for ranges between 10 and 11 GHz. Isolation close to −60 dB is achieved for S14 = S41 in the range of 8 to 9 GHz.

Wearable UWB antennas are being widely developed in recent times to address the on-body applications. These are usually mounted on the surface of the body targeting a specific application in the process of monitoring the health [6]. In particular, in the field of medical imaging, these UWB antennas prove to be very advantageous due to the characteristics that they possess such as planar, more compact, better radiation efficiency, and their ability (strengthwise) to work precisely in their allocated bands [7]. To make the antenna inline for the mentioned purpose, the conformal property should be used. The supremacy with the conformal arrangement is that it can be designed on the surface of any entity without changing the physical properties of the specimen considered over which the conformal antenna has to be installed [8–9]. Figure 21.3a shows the conformal arrangement of the proposed wearable MIMO antenna.

The curvature effect or the bend angle effect on the isolation properties of the MIMO antenna is particularly studied. Two bend angles are considered with an angle of curvature 20° and 25°. Figures 21.3b and 21.3c give the related simulation results for the conformal arrangement of the antenna. For both the situations (20° and 25° bend angles), the proposed MIMO antenna maintained good isolation properties with almost all the isolation and transmission coefficients being less than −20 dB for the entire UWB range. In particular, the conformal arrangement with the angle of curvature 20° showed better results with S14 = S41 less than −50 dB, S34 = S43 less than −30 dB, S24 = S42 and S12 = S21 less than −40 dB. With the angle of curvature increased to 25°, antenna showed minor variations at lower and upper frequencies concerning the isolation or transmission coefficients with S14 = S41 less than −25 dB, S24 = S42, and S12 = S21 less than −20 dB at the lower frequencies and less than −40 dB when moving to higher frequencies above 7 GHz. But, on a whole, the 25° curvature conformal arrangement also showed an overall good isolation characteristic

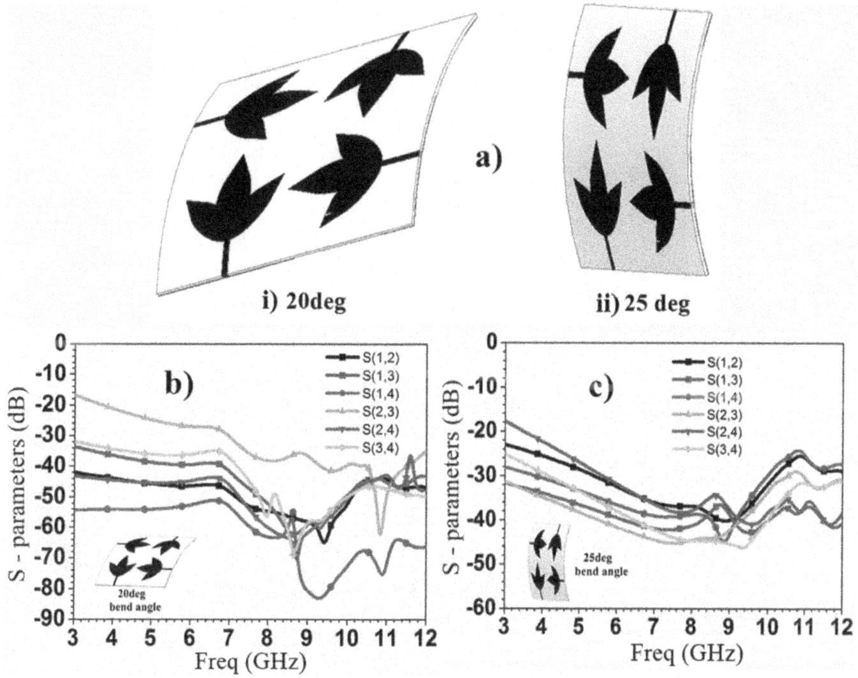

FIGURE 21.3 (a) Conformal arrangement of the proposed MIMO antenna (i) 20° and (ii) 25° bend angles; (b) S-parameters plot depicting the isolation or transmission coefficients information with 20° bend angle; (c) S-parameters plot depicting the isolation or transmission coefficients information with 25° bend angle.

with −20 dB over the entire UWB range. Therefore, the conformal MIMO antenna arrangement showed an overall satisfactory MIMO performance indicating good isolation or lower mutual coupling effect.

The surface current distribution is shown in Figure 21.4. The values are computed when each of the port is separately excited at the center frequency of 7.5 GHz. These stand at 146 A/m and 165 A/m when ports 1 and 2 are separately excited and 139 A/m and 138 A/m when the rest of the ports 3 and 4 are excited individually.

The diversity capabilities of the MIMO antenna are characterized by the computation of the Envelope Correlation Coefficient (ECC), Total Active Reflection Coefficient (TARC), and Mean Effective Gain (MEG). These parameters are calculated using the formulas given by

$$ECC = \frac{\left|S_{pp}^{*}S_{pq} + S_{qp}^{*}S_{qq}\right|^{2}}{\left|\left(1-\left|S_{pp}\right|^{2}-\left|S_{pq}\right|^{2}\right)\left(1-\left|S_{qp}\right|^{2}-\left|S_{qq}\right|^{2}\right)\right|^{*}} \tag{21.1}$$

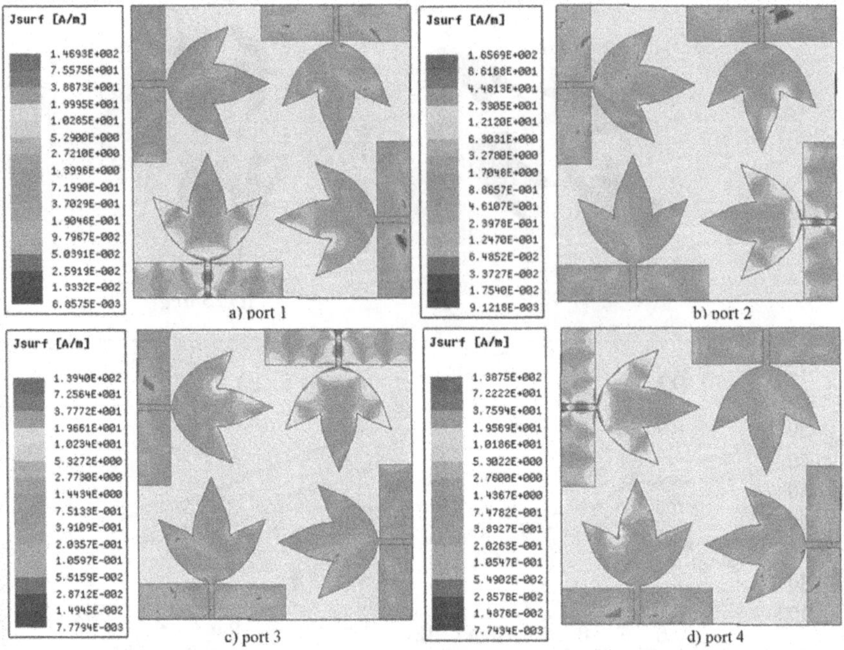

FIGURE 21.4 Surface current distribution for different port excitations.

$$TARC = \sqrt{\frac{\left(S_{pp} + S_{pq}\right)^2 + \left(S_{qp} + S_{qq}\right)^2}{2}} \tag{21.2}$$

Figure 21.5a deals with the ECC versus the frequency plot. From the figure, it is evident that the value of the ECC is limited to less than 0.045 for the entire operating frequency range of 3–12 GHz between any antenna elements. In general, if the ECC limit is 0.5 it is considered to be satisfactory and if the limit is less than 0.3, then it is considered to be good for the MIMO applications. The proposed modified deltoid-shaped MIMO antenna system is well within the standard limit and is as low as <0.045.

The Mean Effective Gain (MEG) as mentioned in Ref. [10] calculated between any two antenna elements is shown in Figure 21.5b. The MEG1, MEG2 confine the mean effective gain calculation between 1 and 2 antenna elements. Similarly, MEG3 and MEG4 for 1 and 3 elements and finally MEG5 and MEG6 for 1 and 4 antenna elements. The difference between these mean effective gains is to be limited to less than 3 dB as per the standards. In Figure 21.5b, it is seen that the MEG between two antenna elements is nearly equal to 1.0 for the entire impedance bandwidth which is way less than the standard value of <3 dB. The TARC versus frequency plot is shown in Figure 21.5c. The standard limit is < −10 dB and the plot shows that the entire operating range of the MIMO antenna is less than −10 dB. The TARC in specific is as low

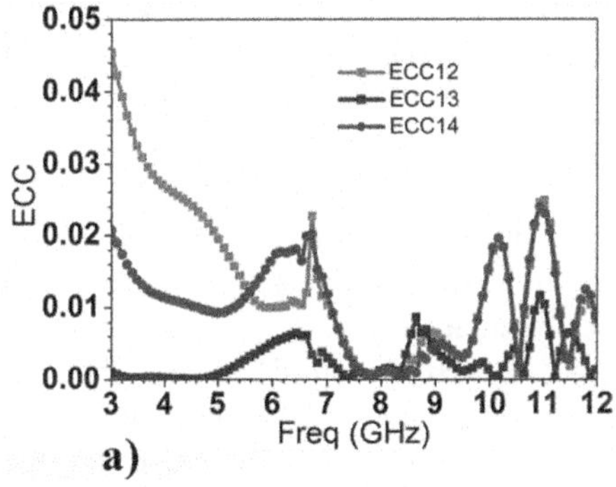

FIGURE 21.5a Envelope correlation coefficient vs frequency plot.

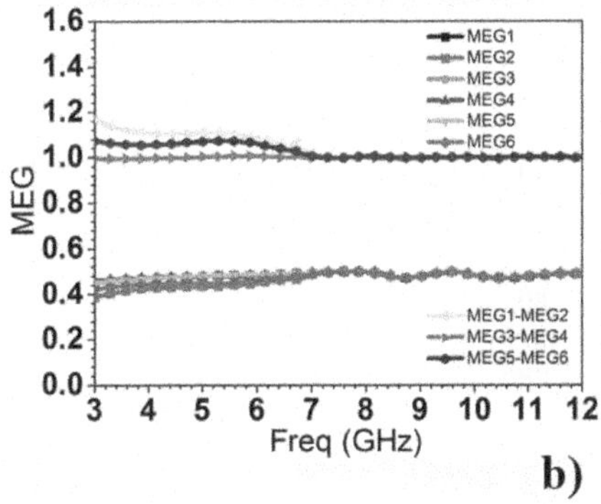

FIGURE 21.5b Mean effective gain vs frequency plot.

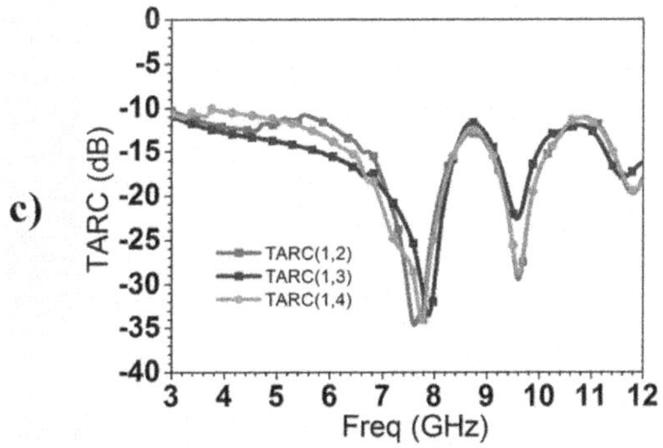

FIGURE 21.5c Total active reflection coefficient vs frequency plot.

as −35 dB in-between frequency ranges from 7 GHz to 8 GHz and up to −30 dB for the frequency ranges between 9 GHz and 10 GHz.

Figures 21.6a and 21.6b deal with the radiation efficiency, gain, and group delay of the MIMO antenna. The proposed MIMO antenna achieved a radiation efficiency of over 82% for the frequency range of 3–12 GHz. The efficiency is also slightly over 90% at the frequency ranges between 8 and 9 GHz. The gain on the other hand is computed and is found to be satisfactory with peak gain close to 12 dB in between the 8 and 9 GHz frequency range. The gain curve saw a decline at higher frequency ranges after 9 GHz up to 12 GHz. Figure 21.6b also shows the group delay curve where it is found that the group delay is almost constant for the entire proposed frequency range. Though two spikes are seen along the curve, they constitute less than 2 nanoseconds. Overall, good phase linearity is observed from the group delay that satisfies the demand for UWB operations.

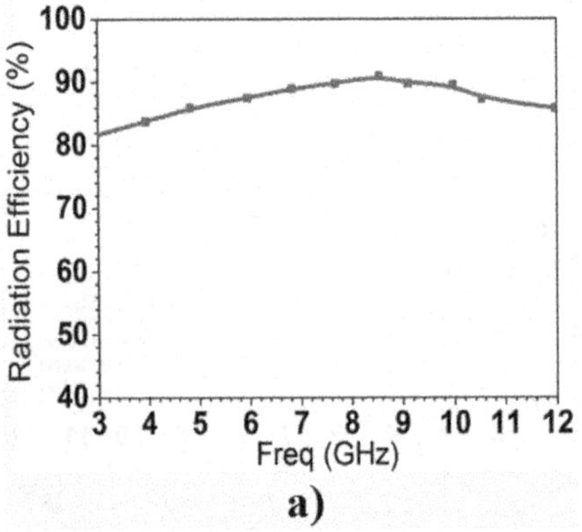

FIGURE 21.6a Radiation efficiency vs frequency plot.

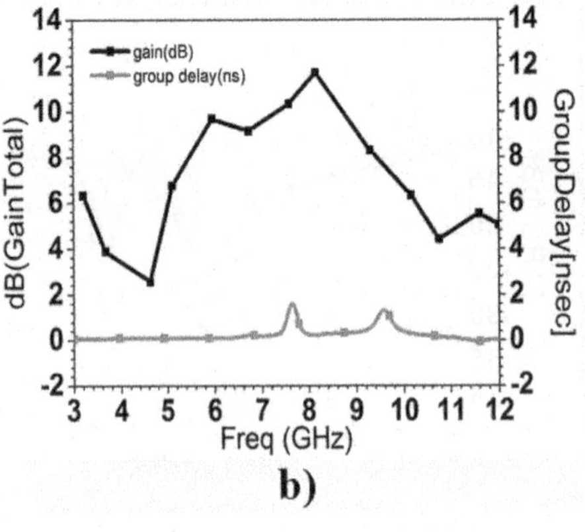

FIGURE 21.6b 2D gain vs frequency plot.

21.4 SPECIFIC ABSORPTION RATE ANALYSIS

The Specific Absorption Rate (SAR) effect on the human body is of primary focus when designing a wearable antenna. The mathematical formulation for calculating the SAR is given by [11]:

$$SAR = \sigma \frac{|E|^2}{\rho} \tag{21.3}$$

where 'σ' represents conductivity (s/m), E is the electric field (V/m), and 'ρ' is the biological tissue mass density often represented in (kg/m^3). In this work, the proposed MIMO antenna is placed at a distance of 20 mm from the target tissue. The three-layered homogeneous human body model as shown in Figure 21.7 is considered with skin, fat layer, and muscle. Each of these three layers has specific permittivity (ε_r), conductivity (s/m), loss tangent value, density (kg/m^3), and thickness (mm). These specifications are mentioned in table 2 of reference [11].

Figure 21.8 shows the computed SAR values for both 1 g and 10 g of tissue. Tables 21.1 and 21.2 represent the different SAR values for 1 g and 10 g of tissue. Also, the SAR values are computed using three different input power values of 1 mW, 25 mW, and 50 mW. It is seen from Figure 21.8 and Tables 21.1 and 21.2, the SAR values are within the Federal Communications Commission (FCC) recommended limit of 1.6 Watts/kg.

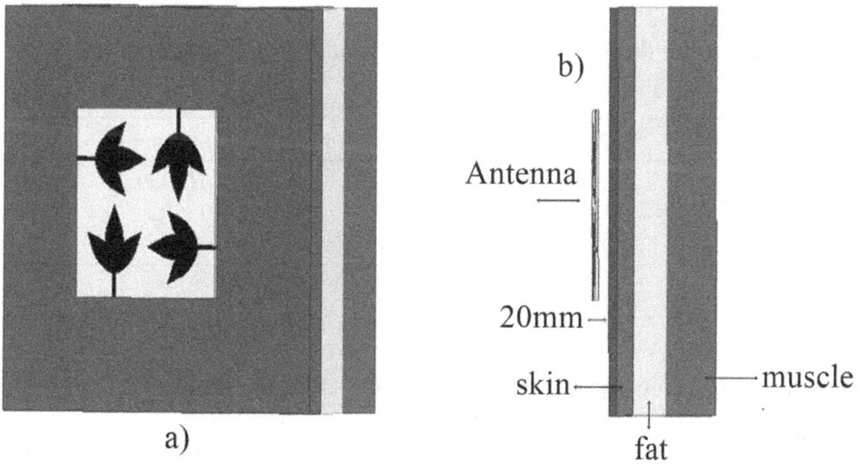

FIGURE 21.7 The MIMO antenna placed above three-layer human body model for SAR analysis.

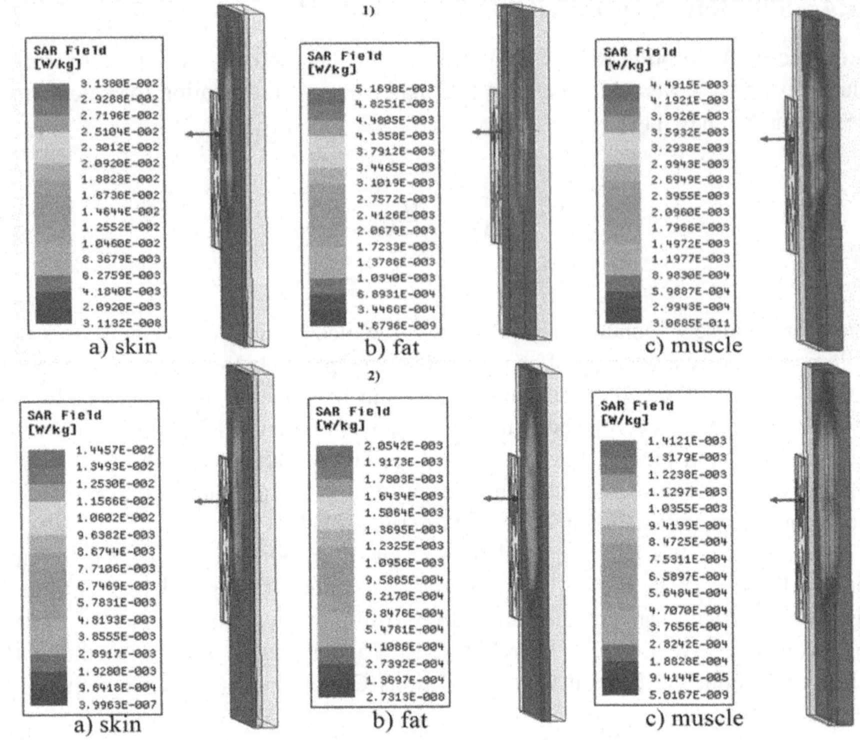

FIGURE 21.8 SAR values (1) 1 g of tissue and (2) 10 g of tissue.

TABLE 21.1
SAR values for the three-layer human body model (1 g of tissue)

Mass of tissue = 1 g	SAR Value (W/kg)		
Input power	**Skin**	**Fat**	**Muscle**
1 mW	0.0313	0.0051	0.00449
25 mW	0.784	0.129	0.1122
50 mW	1.56	0.258	0.224

TABLE 21.2
SAR values for three-layer human body model (10 g of tissue)

Mass of tissue = 10 g	SAR Value (W/kg)		
Input power	**Skin**	**Fat**	**Muscle**
1 mW	0.0144	0.002	0.00141
25 mW	0.361	0.0513	0.0353
50 mW	0.722	0.102	0.00706

21.5 CONCLUSION

A quad-port wearable MIMO antenna employing modified deltoid-shaped antenna elements in the orthogonal arrangement is proposed in this chapter. The MIMO antenna is designed to cover the ultra-wideband range of 3.1 to 10 GHz frequency range. The simulations showed the proposed MIMO antenna impedance bandwidth of less than −10 dB is achieved for frequency range covering 3 to 12 GHz which is way more than the FCC recommended UWB range. The proposed MIMO antenna also achieved good isolation of greater than −20 dB for the entire operating bandwidth. MIMO performance in terms of Envelope Correlation Coefficient (ECC), Mean Effective Gain (MEG), and Total Active Reflection Coefficient (TARC) is also computed which showed ECC < 0.045, MEG close to 1.0, and TARC < −10 dB for the mentioned frequency range. The radiation efficiency is over 82% from 3 to 12 GHz with efficiency slightly 90% at 8 to 9 GHz range. A peak gain close to 12 dB and group delay with good phase linearity is achieved. The Specific Absorption Rate (SAR) is also calculated using the three-layer homogenous human body model with varying inputs of 1 mW, 25 mW, and 50 mW for both 1 g and 10 g of tissue. The SAR values were also found to be well within the FCC recommended 1.6 W/Kg. Therefore, the proposed quad-port wearable MIMO antenna is an ideal one for UWB applications.

REFERENCES

[1] Roy, S., Ghosh, S., Pattanayak, S. S., and Chakarborty, U. (2020). "Dual-polarized textile-based two/four element MIMO antenna with improved isolation for dual wideband application." *International Journal of RF and Microwave Computer-Aided Engineering*, e22292.

[2] Deng, J., Li, J., Zhao, L., and Guo, L. (2017). "A dual-band inverted-F MIMO antenna with enhanced isolation for WLAN applications." *IEEE Antennas and Wireless Propagation Letters*, *16*, 2270–2273.

[3] Roy, S., Ghosh, S., and Chakarborty, U. (2019). "Compact dual wide-band four/eight elements MIMO antenna for WLAN applications." *International Journal of RF and Microwave Computer-Aided Engineering*, *29*(7), e21749.

[4] Yang, M. and Zhou, J. (2020). "A compact pattern diversity MIMO antenna with enhanced bandwidth and high-isolation characteristics for WLAN/5G/WiFi applications." *Microwave and Optical Technology Letters*, *62*(6), 2353–2364.

[5] Parchin, N. O., Al-Yasir, Y. I. A., Ali, A. H., Elfergani, I., Noras, J. M., Rodriguez, J., and Abd-Alhameed, R. A. (2019). "Eight-element dual-polarized MIMO slot antenna system for 5G smartphone applications." *IEEE Access*, *7*, 15612–15622.

[6] Wang, F., and Arslan, T. (2016, November). "A wearable ultra-wideband monopole antenna with flexible artificial magnetic conductor." In *2016 Loughborough Antennas & Propagation Conference (LAPC)* (pp. 1–5). IEEE.

[7] Alsharif, F., and Kurnaz, C. (2018, July). "Wearable microstrip patch ultra wide band antenna for breast cancer detection." In *2018 41st International Conference on Telecommunications and Signal Processing (TSP)* (pp. 1–5). IEEE.

[8] Wang, Q., Mu, N., Wang, L., Safavi-Naeini, S., and Liu, J. (2017). "5G MIMO conformal microstrip antenna design." *Wireless Communications and Mobile Computing*, 2017. vol. 2017, Article ID 7616825, p. 11. https://doi.org/10.1155/2017/7616825.

[9] Li, Z., Kang, X., Su, J., Guo, Q., Yang, Y. L. and Wang, J. (2004/2016). "Clutter limited detection performance of multi-channel conformal arrays." *International Journal of Antennas and Propagation*, *84*(9), 1481–1500.

[10] Nasir, J., Jamaluddin, M. H., Khalily, M., Kamarudin, M. R., Ullah, I., and Selvaraju, R. (2015). "A reduced size dual port MIMO DRA with high isolation for 4G applications." *International Journal of RF and Microwave Computer-Aided Engineering*, *25*(6), 495–501.

[11] Yadav, A., Kumar Singh, V., Kumar Bhoi, A., Marques, G., Garcia-Zapirain, B., and de la Torre Díez, I. (2020). "Wireless body area networks: UWB wearable textile antenna for telemedicine and mobile health systems." *Micromachines*, *11*(6), 558.

Part VIII

Microstrip Antenna Design for Miscellaneous Applications

22 Case Study

Design and Simulation of the 5G Microstrip Patch Antenna with a Phase Shifter and Dodecagonal Prism (12-Sided Prism) Shape Base Station

Vishal A. Ker, Vidyadhar S. Melkeri and Gauri Kalnoor

22.1 INTRODUCTION

In the world of mobile communication, fourth-generation (4G) technology has been a huge success and trend in the mobile communication field. However, for large amounts of data and wider bandwidth, there is still a need for a faster way of communication. Therefore, we propose a dual-band fifth-generation (5G) antenna with of frequencies 28 GHz and 38 GHz. The main antenna is designed for these frequencies using an H-shaped slot on the patch. The dielectric substrate used in the proposed design is Rogers RT 5880 (lossy) which has dielectric constant (εr) of 2.2 with loss tangent (tan δ) of 0.0009 and thickness (h) of the substrate is taken as 0.381 mm. The main patch or the single-element patch antenna has achieved the impedance matching

DOI: 10.1201/9781003093558-30

(S11-parameter) of −43.33 dB at 28 GHz and −28.695 dB at 38 GHz and at −10 dB bandwidth of 755 MHz at frequency 28 GHz and 645 MHz at frequency 38 GHz. The antenna array has been designed using four sub-array single-element patch antenna using quarter-wave transform power divider with left and right phase shift on either side of the sub-array. The three sub-arrays are arranged on each side on one face of dodecagonal prism to achieve dual polarization, i.e. vertical and horizontal polarization, that forms an I-shaped structure. The same I-shaped structure is arranged on each of the faces of dodecagonal prism. This has achieved the gain of 13.6 dBi at 28 GHz and 12.6 dBi at 38 GHz. 5G is the fifth generation of mobile communication technology as well as wireless communication. It is 40 times faster than 4G LTE due to its wider bandwidth. 5G technology uses millimeter wave (mm-wave) electromagnetic spectrum, which carry huge amounts of data but for a shorter range. 5G is divided into two types: (i) sub-5G that uses frequency range from 600 MHz to 6 GHz and (ii) 5G that uses millimeter wave band from 24 GHz to 100 GHz. The wider bandwidth of 5G helps in the quick transfer of large amounts of data and can be an added extra protocol layer for security and acknowledgment. The 5G communication technology increases the number of antenna installations in a geographical area. The 5G antennas can play a role in IoT devices for nearby communication and larger data transmission.

22.2 SINGLE-ELEMENT MSA

22.2.1 Design Process

We used Rogers RT 5880 (lossy) substrate, whose dielectric constant (ε_r) is 2.2 and loss tangent (tan δ) is 0.0009 [1]. We chose this substrate in order to achieve a smaller patch size, tighter fabrication tolerance and increased antenna efficiency [2, 3]. The substrate thickness (h) is taken as 0.381 mm. The width and length of the substrate and the ground are made of same parameters, i.e. W_g and L_g as mentioned in Table 22.1. The thickness (t) of ground is 0.01 mm. The width and length of patch are given by W and L, respectively, and values as mentioned in Table 22.1.

The main patch is designed for the frequency at 28 GHz and all the calculation of parameters are made considering this frequency. Slot on the patch helps in generating and radiating new frequency, i.e. 38 GHz. The width and length of slots are given by ww and ll. The gap-coupling of 0.05 mm is taken to match the impedance of insert feed and patch. The width of patch is taken as 1.1835 mm that matches the impedance of 50 Ω. Figure 22.1 shows the top view of single-element 5G microstrip patch antenna and Table 22.1 shows the parameter dimension and its values in millimeter (mm).

22.2.2 Simulation and Results

The design and simulation are done using CST Microwave Studio software and results are plotted in Origin Lab. The S11-parameter shows how good an impedance matching has occurred between the port and the feed. The proposed single-element 5G microstrip antenna (MSA) has achieved good impedance matching of −45.10 dB at 28 GHz and −30 dB at 38 GHz. It also shows that a bandwidth of 755 MHz is

TABLE 22.1
Parameter and dimension values

Parameters	Values (in mm)
W	4.18
L	3.6
W_g	6.466
L_g	5.886
W_f	1.1835
ww	3.2
ll	3
wy	0.05

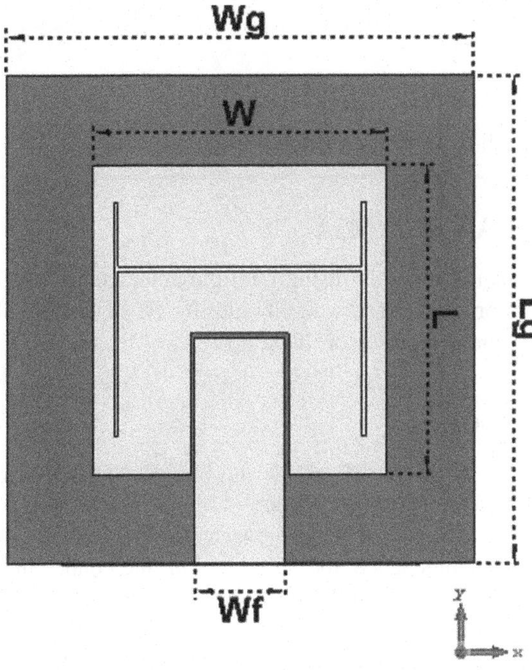

FIGURE 22.1 Single-element 5G microstrip patch antenna.

achieved at 28 GHz below −10 dB and 645 MHz at 38 GHz below −10 dB. Voltage Standing Wave Ratio (VSWR) defines the efficiency of power transmitted from the transmission line (feed) to a load (patch). A perfect VSWR has been achieved, i.e. the far-field gain represents the radiation pattern at the frequency defined and the maximum gain achieved. For the proposed antenna, we achieved a gain of 6.75 dBi at 28 GHz and 7.37 is seen at 38 GHz, which has a better result when compared to Ref. [4].

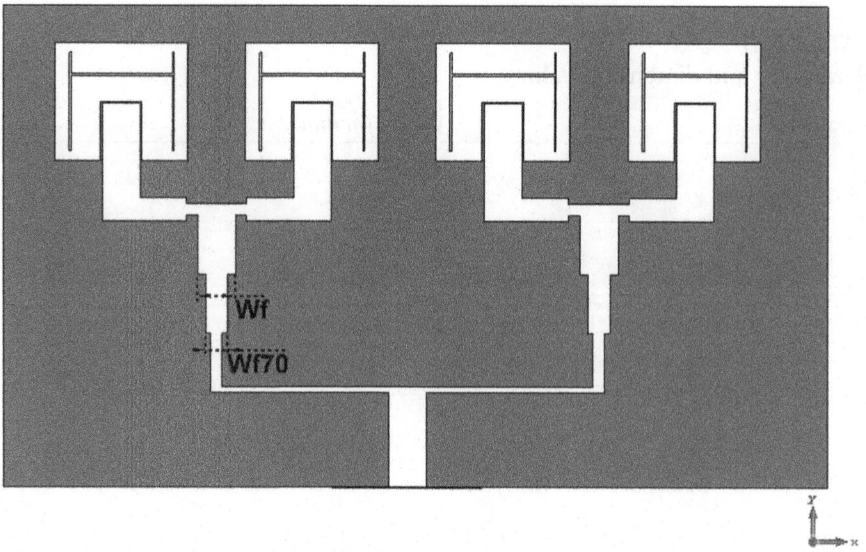

FIGURE 22.2 Four sub-array 5G microstrip patch antenna.

22.2.3 Four Sub-Array 5G MSA

The array of antennas is formed by using multiple antennae of identical parameters in order to achieve more gain, narrow bandwidth for far distances [5]. The array we propose is a single dimension array or linear array.

22.2.4 Design Process

The design of a four sub-array consists of four single-element patch antennas connected with the quarter wave transform power divider for an equal amount of power distributed among all four single-element antennas as revealed in Figure 22.2. The feed inserted into the patch has 50 Ω impedance, the transmission line connecting the insert feed is 70 Ω (W_f70) and followed by 100 Ω (W_f100) transmission lines. The width and length of the power divider transmission line is given by Equations (22.1) and (22.2). Table 22.2 shows the parameter dimensions and its values in millimeters (mm).

$$w = Z_o\sqrt{2} \ldots \tag{22.1}$$

$$l = \lambda_o/4 \ldots \tag{22.2}$$

22.3 SIMULATION AND RESULTS

Figure 22.3 shows the S11-parameter results of a four sub-array antenna, which has achieved similar results as a single-element patch antenna, i.e. –46 dB and VSWR less than 2, which is good when compared.

TABLE 22.2
Parameter and dimension values

Parameters	Values (in mm)
W_f	1.18
W_f70	0.8
W_f100	0.35

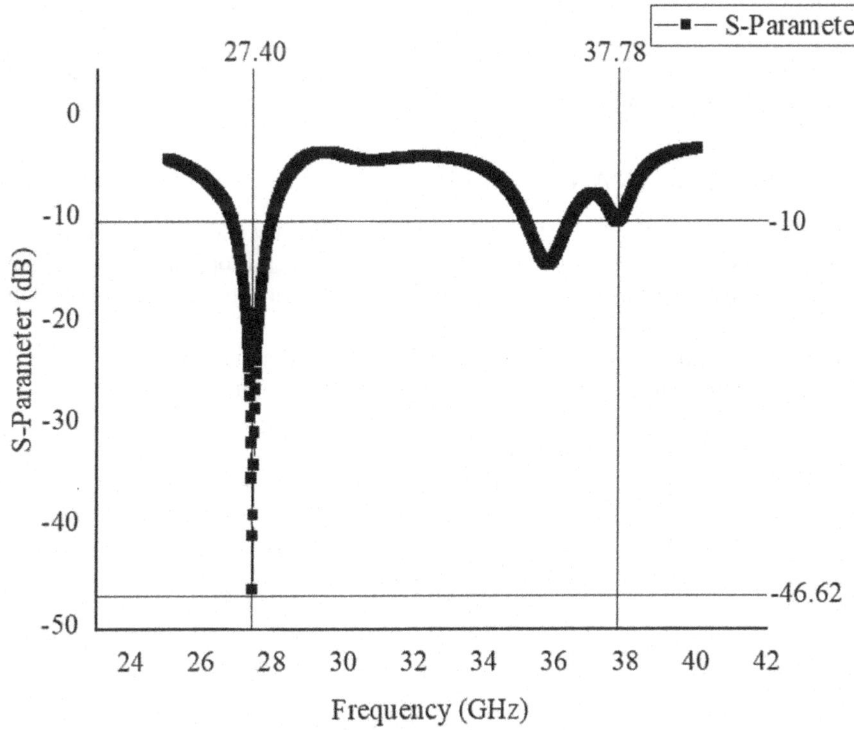

FIGURE 22.3 S11-parameter of four sub-array 5G MSAs.

In far-field radiation, gain of four sub-array 5G MSAs [6–10], we can see a narrower radiation pattern when compared to the single-element 5G MSA. Also, we can increase the maximum realized gain of 13.6 dBi at 28 GHz and 11.6 dBi at 38 GHz.

22.3.1 Phase Shift

In 5G systems, it is a prerequisite to have beam-steering ability since the antennas radiate narrow beams that surround the coverage area in addition to the constraints of wave propagation at mm-wave frequencies. The purpose of this addition is to offer the proposed 5G base station with the capability of fixed beam switching by placing different antenna configurations each with different beam direction. Fixed beam switching is less complex and cheaper alternative to the phased-array systems that require RF phase shifters that reduce the signal quality at mm-wave frequencies.

22.3.2 Design Process

We proposed a phase shift antenna embedded in the design to avoid physical beam-steering. To steer the main beam in a 10° left direction, a transmission line with a length of 5.393 mm is interleaved into the power divider transmission line and feed on the left side (Figure 22.4).

Similarly, for the right side, insert the transmission line on the right side of the feed and power divider, as in Figure 22.5.

22.4 SIMULATION AND RESULTS

The simulation of both phase shift arrays is performed and the radiation pattern is shown in Figure 22.6; similarly we have found for right shift operation. We can see the radiation pattern tilting toward the left compared to four sub-array antennas in left phase shift with gain of 12.8 dBi.

Right phase shift has been successfully achieved by radiation pattern tilting toward 10° right with the gain of 10.8 dBi, which is better when compared to Ref. [4].

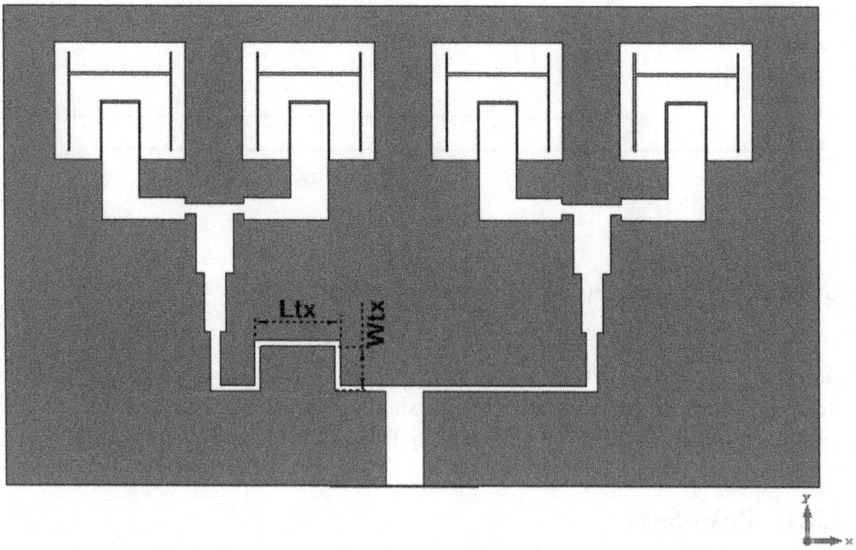

FIGURE 22.4 Left phase shift 5G microstrip patch antenna.

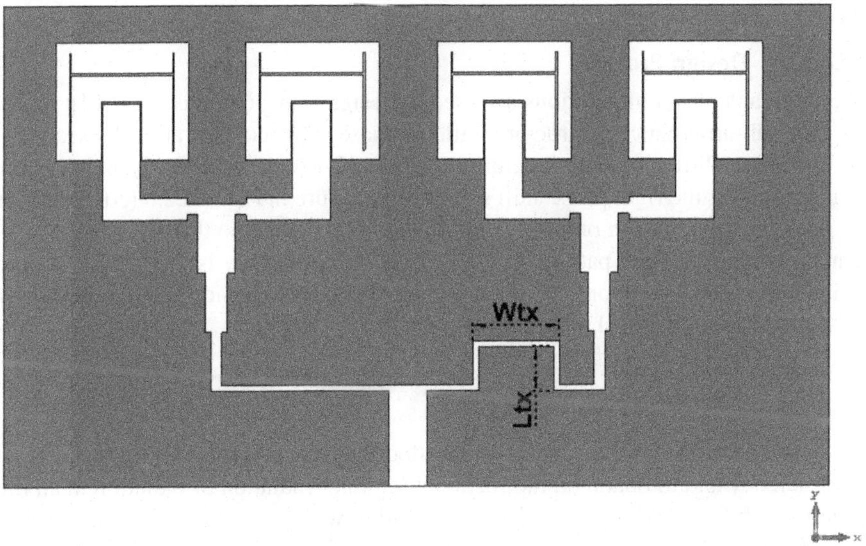

FIGURE 22.5 Right phase shift 5G microstrip patch antenna.

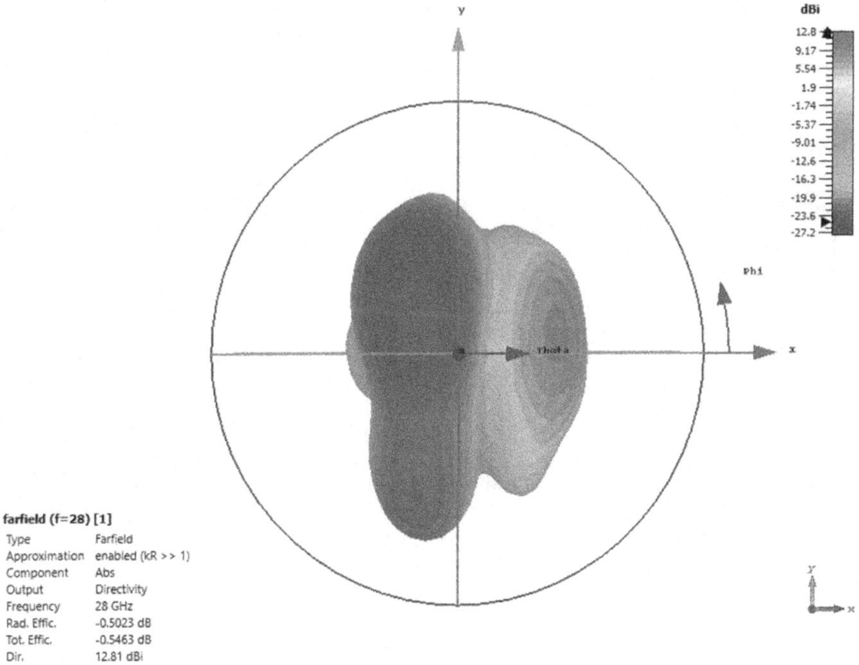

FIGURE 22.6 Far-field radiation pattern of left phase shift.

22.4.1 Base Station

22.4.1.1 Design Process

The three-antenna configurations are used to design I-shaped arrays [11–14]: first is the four sub-array antenna structure with no phase shift (configuration 1), second is the left phase shift sub-array structure in Figure 22.4 (configuration 2) and the third configuration is the right phase shift sub-array structure in Figure 22.5 (configuration 3). These are arranged in order of configurations 2, 1 and 3 so that we achieve ±10° than an actual radiation pattern. The new configuration array is arranged in a horizontal and vertical position to form an I-shape in order to achieve dual polarization, i.e. horizontal and vertical polarizations.

22.5 CONCLUSION

We have arranged I-shaped antennas in a dodecagonal prism (12 sides) to increase the efficiency and to obtain omnidirectional (circular radiation or radiation in all direction) as shown in Figure 22.7. The base station with a size of 3.13 cm consists of 12 I-shaped arrays or 144 sub-array 5G antennas. The functioning of the proposed base station antenna in terms of attained bandwidth, return loss and appreciated gain in addition to its features such as polarization switching and fixed beam switching marks it as an excellent candidate for future 5G communication systems.

The mutual couplings between pairs of antenna sub-arrays are simulated and the results are depicted in Figure 22.8. The results show that all the mutual couplings [15–16] are better than −10 dB; meanwhile, the sub-arrays are separated by acceptable distances. The maximum S11-parameter achieved by base station is less then −100 dB.

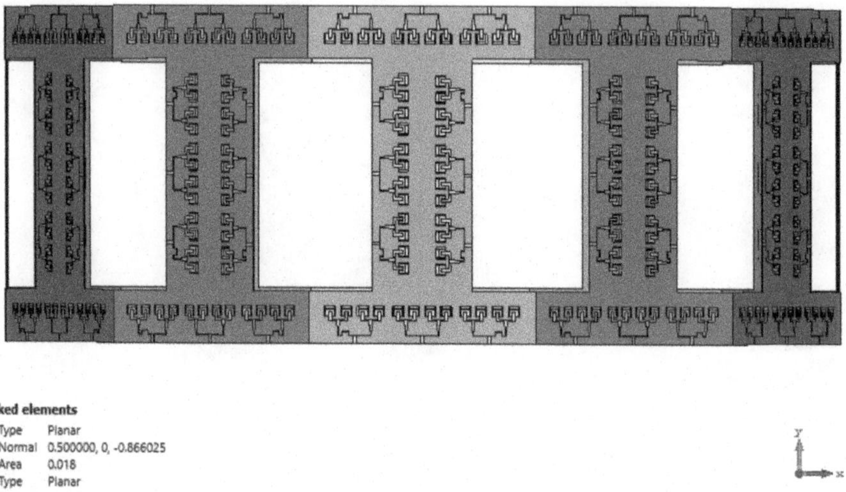

FIGURE 22.7 Dodecagonal prism 5G base station.

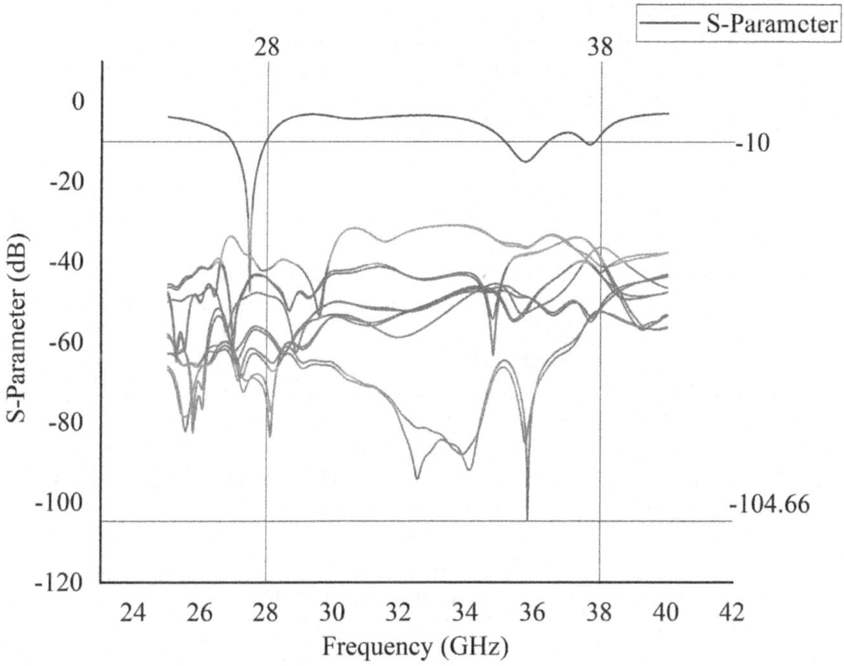

FIGURE 22.8 S11-parameter of 5G base station.

REFERENCES

[1] Tezpur University – *Computer Science and Information Technology*, Chapter 2, "Microstrip patch antenna and modelling techniques." 2008.

[2] J. Singh, "Insert feed microstrip patch antenna," *International Journal of Computer Science and Mobile Computing*, vol. 6, 452–456, 2018. 10.26438/ijcse/v6i11.452456.

[3] T. Elhabbash and T. Skaik, "Design of dual-band dual-polarized MIMO antenna for mm-wave 5G base stations with octagonal prism structure," *2019 IEEE 7th Palestinian International Conference on Electrical and Computer Engineering (PICECE),* 1–6, 2019. doi: 10.1109/PICECE.2019.8747180.

[4] A. Alieldin et al., "A triple-band dual-polarized indoor base station antenna for 2G, 3G, 4G and sub-6 GHz 5G applications," *IEEE Access*, vol. 6, 49209–49216, 2018.

[5] M. V. Komandla, G. Mishra, and S. K. Sharma, "Investigations on dual slant polarized cavity-backed massive MIMO antenna panel with beamforming," *IEEE Transactions on Antennas and Propagation*, vol. 65, no. 12, 6794–6799, December 2017.

[6] M. A. Saada, T. Skaik, and R. Alhalabi, "Design of efficient microstrip linear antenna array for 5G communications systems," *2017 International Conference on Promising Electronic Technologies (ICPET), Deir El-Balah*, 2017, pp. 43–47.

[7] B. Sadhu et al., "A 28-GHz 32-element TRX phased-array IC with concurrent dual-polarized operation and orthogonal phase and gain control for 5G communications," *IEEE Journal of Solid-State Circuits*, vol. 52, no. 12, 3373–3391, December 2017.

[8] S. I. Orakwue, R. Ngah, and T. A. Rahman, "A two dimensional beam scanning array antenna for 5G wireless communications," *2016 IEEE Wireless Communications and Networking Conference Workshops (WCNCW)*, Doha, 2016, pp. 433–436.

[9] S. I. Orakwue, R. Ngah, and T. A. Rahman, "A two dimensional beam scanning array antenna for 5G wireless communications," *2016 IEEE Wireless Communications and Networking Conference Workshops (WCNCW)*, Doha, 2016, pp. 433–436.

[10] S. F. Jilani and A. Alomainy, "Millimetre-wave T-shaped MIMO antenna with defected ground structures for 5G cellular networks," *IET Microwaves, Antennas & Propagation*, vol. 12, no. 5, 672–677, 2018.

[11] V. S. Melkeri, D. B. Ganure, S. N. Mulgi, R. M. Vani, and P. V. Hunagund. "Size reduction with multiband operating square microstrip patch antenna embedded with defected ground structure," *International Journal for Research in Applied Science and Engineering Technology (IJRASET)*, vol. 6, no 3, 1446–1454, March 2018.

[12] V. S. Melkeri, S. L. Mallikarjun, and P. V. Hunagund, "H-shaped-defected ground structure (DGS) embedded square patch antenna," *International Journal of Advance Research in Engineering and Technology (IJARET)*, vol. 6, no. 1, 73–79, January 2015.

[13] V. S. Melkeri, M. Lakshetty, and P. V. Hunagund, "Microstrip antenna with defected ground structure: A review," *International Journal of Electrical and Electronics and Telecommunication Engineering (IJEETE)*, vol. 46, no. 1, 1492–1496, February 2015.

[14] V. S. Melkeri, R. M. Vani, and P. V. Hunagund, "Chess board shape defected ground structure for multi band operation of microstrip patch antenna," *Journal of Computational Information Systems (JCIS)*, vol. 14, no. 5, 73–78, October 2018.

[15] V. S. Melkeri and P. V Hunagund, "Concentric ring defect on ground plane for enhancement of parameters and physical size reduction for 2.4 GHz antenna," *Journal of Computational Information Systems (JCIS)*, vol. 15, no. 1, 17–22, 2019.

[16] V. S. Melkeri and P. V. Hunagund, "Study and analysis of SQMSA with square shaped defected ground structure: Performance enhancement and virtual size reduction," *2017 IEEE International Conference on Antenna Innovations & Modern Technologies for Ground, Aircraft and Satellite Applications (iAIM), Bangalore,* 2017, pp. 1–5, doi: 10.1109/IAIM.2017.8402516.

Index

Note: Figures are indicated by *italics*. Tables are indicated by **bold**.

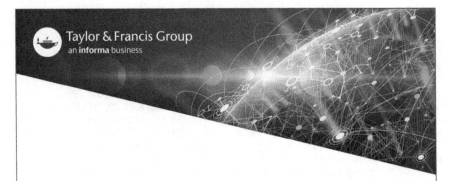